Laboratory Exercises in

Microbiology

Sixth Edition

John P. Harley
Eastern Kentucky University

Higher Education

Boston Burr Ridge, IL Dubuque, IA Madison, WI New York San Francisco St. Louis
Bangkok Bogotá Caracas Kuala Lumpur Lisbon London Madrid Mexico City
Milan Montreal New Delhi Santiago Seoul Singapore Sydney Taipei Toronto

Higher Education

LABORATORY EXERCISES IN MICROBIOLOGY, SIXTH EDITION

Published by McGraw-Hill, a business unit of The McGraw-Hill Companies, Inc., 1221 Avenue of the Americas, New York, NY 10020. Copyright © 2005, 2002, 1999, 1996 by The McGraw-Hill Companies, Inc. All rights reserved. No part of this publication may be reproduced or distributed in any form or by any means, or stored in a database or retrieval system, without the prior written consent of The McGraw-Hill Companies, Inc., including, but not limited to, in any network or other electronic storage or transmission, or broadcast for distance learning.

Some ancillaries, including electronic and print components, may not be available to customers outside the United States.

This book is printed on recycled, acid-free paper containing 10% postconsumer waste.

 3 4 5 6 7 8 9 0 QPD/QPD 0 9 8 7 6

ISBN 13: 978–0–07–255680–3
ISBN 10: 0–07–255680–3

Publisher: *Martin J. Lange*
Senior sponsoring editor: *Patrick E. Reidy*
Senior developmental editor: *Jean Sims Fornango*
Marketing manager: *Tami Petsche*
Senior project manager: *Jayne Klein*
Lead production supervisor: *Sandy Ludovissy*
Lead media project manager: *Judi David*
Media technology producer: *Renee Russian*
Designer: *Rick D. Noel*
Cover/interior designer: *Rokusek Design*
Cover image: *© Dennis Kunkel Microscopy, Inc., Penicillium Roqueforti (92386A)*
Lead photo research coordinator: *Carrie K. Burger*
Compositor: *Carlisle Communications, Ltd.*
Typeface: *11/13 Times Roman*
Printer: *Quebecor World Dubuque, IA*

www.mhhe.com

Contents

Take interest, I implore you, in those sacred dwellings which one designates by the expressive term: laboratories. Demand that they be multiplied, that they be adorned. These are the temples of the future— temples of well-being and of happiness. There it is that humanity grows greater, stronger, better.

Louis Pasteur
(French chemist, founder of microbiology, 1822–1895)

There are many excellent microbiology laboratory manuals on the market and many others that are called "in-house" productions because they are written for a microbiology course at a particular school. Why another microbiology manual? The answer is straightforward. Many instructors want a manual that is directly correlated with a specific textbook. As a result, this laboratory manual was designed and written to be used in conjunction with the textbook *Microbiology,* sixth edition, by Lansing M. Prescott, John P. Harley, and Donald A. Klein; however, it can be used with other textbooks with slight adaptation.

Since this manual correlates many of the microbiological concepts in the textbook with the various exercises, comprehensive introductory material is not given at the beginning of each exercise. Instead, just enough specific explanation is given to complement, augment, reinforce, and enhance what is in the textbook. I feel that time allocation is an important aspect of any microbiology course. Students should not be required to reread in the laboratory manual an in-depth presentation of material that has already been covered satisfactorily in the textbook.

Each exercise has been designed to be modular and short. This will allow the instructor to pick and choose only those exercises or parts of exercises that are applicable to a specific course. Several exercises usually can be completed in a two- or three-hour laboratory period. The exercises have also been designed to use commonly available equipment, with the least expense involved, and to be completed in the shortest possible time period.

Considering the above parameters, the purpose of this laboratory manual is to guide students through a process of development of microbiological technique, experimentation, interpretation of data, and discovery in a manner that will complement the textbook and make the study of microbiology both exciting and challenging. According to an old Chinese proverb:

Tell me and I will forget.
Show me and I might remember.
Involve me and I will understand.

And according to Albert Einstein, "Most of the fundamental ideas of science are essentially simple, and may, as a rule, be expressed in a language comprehensible to everyone."

These words convey my basic philosophy that it is experiences in the microbiology laboratory and the scientific method that help develop students' critical thinking and creativity and that increase their appreciation of the mechanisms by which microbiologists analyze information. The laboratory accomplishes this by having students become intensely and personally involved in the knowledge they acquire.

The array of exercises was chosen to illustrate the basic concepts of general microbiology as a whole and of the individual applied fields. The protocols vary in content and complexity, providing the instructor with flexibility to mold the laboratory syllabus to the particular needs of the students, available time and equipment, and confines and scope of the course. Furthermore, it provides a wide spectrum of individual exercises suitable for students in elementary and advanced general microbiology as well as those in various allied health programs.

In 1997, the American Society for Microbiology, through its Office of Education and Training, adopted a Laboratory Core Curriculum representing themes and topics considered essential to teach in every introductory microbiology laboratory, regardless of its emphasis. An instructor might add items appropriate to allied health, applied, environmental, or majors microbiology courses.

The Laboratory Core is not meant to be a syllabus or outline. The core themes and topics are meant to frame objectives to be met somewhere within the introductory microbiology laboratory. Depending on the

specific emphasis of the course, a single lab session could meet multiple core objectives, focus on one objective, or emphasize a topic that is not in the lab core but is important to that particular course.

Laboratory Skills

A student successfully completing basic microbiology will demonstrate the ability to

1. **Use a bright-field light microscope** to view and interpret slides, including
 a. correctly setting up and focusing the microscope
 b. proper handling, cleaning, and storage of the microscope
 c. correct use of all lenses
 d. recording microscopic observations

2. **Properly prepare slides** for microbiological examination, including
 a. cleaning and disposal of slides
 b. preparing smears from solid and liquid cultures
 c. performing wet-mount and/or hanging drop preparations
 d. performing Gram stains

3. **Properly use aseptic techniques** for the transfer and handling of microorganisms and instruments, including
 a. sterilizing and maintaining sterility of transfer instruments
 b. performing aseptic transfer
 c. obtaining microbial samples

4. **Use appropriate microbiological media and test systems,** including
 a. isolating colonies and/or plaques
 b. maintaining pure cultures
 c. using biochemical test media
 d. accurately recording macroscopic observations

5. **Estimate the number of microorganisms** in a sample using serial dilution techniques, including
 a. correctly choosing and using pipettes and pipetting devices
 b. correctly spreading diluted samples for counting
 c. estimating appropriate dilutions
 d. extrapolating plate counts to obtain correct CFU or PFU in the starting sample

6. **Use standard microbiology laboratory equipment correctly,** including
 a. using the standard metric system for weights, lengths, diameters, and volumes
 b. lighting and adjusting a laboratory burner
 c. using an incubator

Laboratory Thinking Skills

A student successfully completing basic microbiology will demonstrate an increased skill level in

1. **Cognitive processes,** including
 a. formulating a clear, answerable question
 b. developing a testable hypothesis
 c. predicting expected results
 d. following an experimental protocol

2. **Analysis skills,** including
 a. collecting and organizing data in a systematic fashion
 b. presenting data in an appropriate form (graphs, tables, figures, or descriptive paragraphs)
 c. assessing the validity of the data (including integrity and significance)
 d. drawing appropriate conclusions based on the results

3. **Communications skills,** including
 a. discussing and presenting laboratory results or findings in the laboratory

4. **Interpersonal and citizenry skills,** including
 a. working effectively in groups or teams so that the task, results, and analysis are shared
 b. effectively managing time and tasks to be done simultaneously, by individuals and within a group
 c. integrating knowledge and making informed judgments about microbiology in everyday life

Laboratories typically supplement and integrate closely with the lecture content in ways that are unique to each instructor. Consequently, the laboratory content that is considered essential for laboratory work by one instructor may be covered in the lecture portion of the course by another instructor, making it difficult to de-

fine specific topics that should be integral in all micro-biology laboratories. As a result, the ASM Laboratory Core Curriculum Committee developed themes, which are broadly based and will enable instructors to have the flexibility to use a wide variety of laboratories to meet the suggested core.

A student successfully completing basic microbiology will demonstrate mastery of the basic principles of the following themes and complete laboratory activities that focus on one or more of the topics under each theme.

Theme 1. Integrating themes—impact of microorganisms on the biosphere and humans; microbial diversity

Theme 2. Microbial cell biology, including cell structure and function, growth and division, and metabolism

Theme 3. Microbial genetics, including mutations

Theme 4. Interactions of microorganisms with hosts (humans, other animals, plants), including pathogenicity mechanisms and antimicrobial agents

In order to meet the above themes, topics, and skills (The American Society for Microbiology Laboratory Core Curriculum), this manual consists of 66 exercises arranged into 11 parts covering the following basic topics:

PART ONE, Microscopic Techniques, introduces the students to the proper use and care of the different types of microscopes used in the microbiology laboratory for the study of microorganisms.

PART TWO, Bacterial Morphology and Staining, presents the basic procedures for visualization and differentiation of microorganisms based on cell form and various structures.

PART THREE, Basic Laboratory and Culture Techniques, acquaints students with proper laboratory procedures in preparing microbiological media and in culture techniques that are used in isolating microorganisms.

PART FOUR, Biochemical Activities of Bacteria, introduces some of the biochemical activities that may be used in characterizing and identifying bacteria.

PART FIVE, Rapid Multitest Systems, acquaints students with some of the multitest systems that can be used to identify bacteria.

PART SIX, Unknown Identification, contains two exercises that guide students through the use of *Bergey's Manual of Systematic Bacteriology* in the identification of unknown bacteria.

PART SEVEN, Environmental Factors Affecting Growth of Microorganisms, acquaints students with some of the various physical and chemical agents that affect microbial growth.

PART EIGHT, Environmental and Food Microbiology, is concerned with the environmental aspects of water, milk, and food.

PART NINE, Medical Microbiology, presents an overview of some pathogenic microorganisms, and acquaints students with basic procedures used in isolation and identification of pathogens from infected hosts, including those from the student's own body.

PART TEN, Survey of Selected Eucaryotic Microorganisms, presents an overview that is intended to help students appreciate the morphology, taxonomy, and biology of the fungi.

PART ELEVEN, Microbial Genetics and Genomics, presents six experiments designed to illustrate the general principles of bacterial genetics and genomics.

The format of each exercise in this manual is intended to promote learning and mastery in the shortest possible time. To this end, each experiment is designed as follows:

Safety Considerations

This laboratory manual endeavors to include many of the safety precautionary measures established by the Centers for Disease Control and Prevention (CDC), Atlanta, Georgia; the Occupational Safety and Health Administration (OSHA); and the Environmental Protection Agency (EPA). Efforts are made to instruct the student on safety, and all exercises will contain precautionary procedures that these agencies are enforcing in hospitals, nursing homes, commercial laboratories, and industry. A **safety considerations box** is included for each exercise to help both the instructor and student prepare themselves for the possibility of accidents.

Both the instructor and student should keep in mind at all times that most technical programs, such as a microbiology laboratory, carry some measure of associated risk. The microbiology laboratory is a place where infectious microorganisms are handled, examined, and studied with safety and effectiveness. However, any of the microorganisms we work with may be pathogenic in an immunocompromised person. Therefore, rather than modifying the objectives in this laboratory manual to avoid any risk, the authors propose that instructors and students implement the Centers for Disease Control and

Prevention (CDC) principles of biosafety throughout. One way we propose is to simply modify the "Universal Precautions" (*see pp. xiii–xiv*) so the wording is appropriate for the classroom by simply changing "laboratory worker" to "student." In addition, a written safety policy consistent with CDC guidelines and adopted by your institution's governing body will protect you, your institution, and the students. As in any laboratory, safety should be a major part of the curriculum. Students should be required to demonstrate their knowledge of safety before they begin each laboratory exercise.

Materials per Student or Group of Students

To aid in the preparation of all exercises, each procedure contains a list of the required cultures with American Type Culture Collection catalog numbers (American Type Culture Collection, 12301 Parklawn Drive, Rockville, Maryland 29852-1776; *www.ATCC.org;* 703-365-2700), media, reagents, and other equipment necessary to complete the exercise in the allocated lab time either per student or group of students. Appendixes H and I provide recipes for reagents, stains, and culture media. Appendix J describes the maintenance of microorganisms and supply sources.

Learning Objectives

Each exercise has a set of learning objectives that define the specific goals of the laboratory session. It is to the student's advantage to read through this list before coming to class. In like manner, these objectives should be given special attention during the laboratory exercise. Upon conscientious completion of the exercise, the student should be able to meet all of the objectives for that exercise. Before leaving the class, students should check the objectives once again to see that they can master them. If problems arise, consult the instructor.

Suggested Reading in Textbook

These cross-references have been designed to save the student's time. By referring the student to sections, paragraphs, tables, charts, figures, and boxes within the textbook, unnecessary duplication is avoided.

Pronunciation Guide

This section contains the phonetic pronunciations for all organisms used in the exercise. If students take the time to sound out new and unfamiliar terms and say them aloud several times, they will learn to use the vocabulary of microbiologists.

Why Are the Above Bacteria, Slides, or Other Microorganisms Used in This Experiment?

The author has chosen specific viruses, bacteria, fungi, protozoa, algae, and various prepared slides for each exercise. This microbial material has been selected based on cost, ease of growth, availability, reliability, and most important, the ability to produce the desired experimental results. In order to communicate these guidelines to the student, this section explains why the author has chosen the microbial material being used and also gives additional biochemical, morphological, and taxonomic information about the microorganism(s) that the student should find helpful when performing the experiment.

Medical Application

Many students using this laboratory manual are either in one of the allied health disciplines, such as nursing, or in a preprofessional program such as premed, predent, or prevet and need to know the clinical relevance of each exercise performed. To satisfy this need, a *Medical Application* section is included for some of the medically oriented exercises. Medical applications are described for most clinical procedures as a specific application of the purpose of the exercise. For example, a procedure can be used for the identification of a particular microorganism or used in combination with other exercises in a diagnosis. For these exercises, some important pathogens with their diseases and their need for the test being performed in the exercise are listed.

Principles

This section contains a brief discussion of the microbiological principles, concepts, and techniques that underlie the experimental procedures being performed in the exercise.

Procedure

Explicit instructions are augmented by diagrams to aid students in executing the experiment as well as interpreting the results. Where applicable, actual results are shown so that the student can see what should be obtained.

Hints and Precautions

Additional information on what to watch out for, what can go wrong, and helpful tidbits to make the experiment work properly are presented in accompanying boxes.

Laboratory Report

Various pedagogical techniques are used for recording the obtained results. This part of the exercise can be turned in to the instructor for checking or grading.

Review Questions

Review questions are located at the end of each laboratory report. These were written so that students can test their understanding of the concepts and techniques presented in each exercise.

Dilution Ratios Used in This Manual

According to the *American Society for Microbiology Style Manual*, dilution ratios may be reported with either colons (:) or shills (/), but note there is a difference between them. A shill indicates the ratio of a part to a whole; e.g., 1/2 means 1 of 2 parts, with a total of 2 parts. A colon indicates the ratio of 1 part to 2 parts, with a total of 3 parts. Thus, 1/2 equals 1:1, but 1:2 equals 1/3.

Dilution Problems

Since dilution problems are such an integral part of any microbiology course, *Appendix A* gives an overview of the different types of dilution. This includes a variety of practice problems. Answers are provided.

Instructor's Guide

An instructor's guide has been prepared for the laboratory manual and is available on our lab central web site at **http://www.mhhe.com/biosci/ap/labcentral.** This guide also provides answers to the questions in this manual.

Finally, it is my hope that this manual will serve as a vehicle to (1) introduce the complexity and diversity of microorganisms and their relationships to one another; (2) provide a solid foundation for further study for those electing a career in science; and (3) convey something of the meaning, scope, and excitement of microbiology as a significant perspective from which to view the world.

I appreciate the many comments offered to me over the years by both faculty and students. In my desire to continue to improve this laboratory manual, I invite constructive comments from those using it. Please contact me through the Cell and Molecular Biology Editor, McGraw-Hill Publishers (**www.mhhe.com/prescott6**).

John P. Harley

Acknowledgments

Our special thanks go to the following reviewers, whose comments proved very helpful to us:

Ghayasuddin Ahmad
Seton Hall University
Alberta M. Albrecht
Manhattanville College
Mary A. Anderson
Gustavus Adolphus College
Susan T. Bagley
Michigan Tech University
Paul Blum
University of Nebraska–Lincoln
Geoffrey W. Gearner
Morehead State University
Robert J. Kearns
University of Dayton
Dana Kolibachuk
Rhode Island College
David Mardon
Eastern Kentucky University
Glendon Miller
Wichita State University
Rita Moyes
Texas A&M University

Raymond B. Otero
Eastern Kentucky University
Marcia Pierce
Eastern Kentucky University
Norbert A. Pilewski
Duquesne University School of Pharmacy
Ralph J. Rascati
Kennesaw State College
Jackie Reynolds
Richland College
Nancy Ricker
Capilano College
Ivan Roth
University of Georgia
Julie J. Shaffer
University of Nebraska at Kearney
Thomas Terry
University of Connecticut
Robert Twarog
University of North Carolina

A special thanks also goes to Kay Baitz, KEY Scientific Products, 1402 Chisholm Trail, Suite D, Round Rock, Texas 78681, for all of her help with the KEY products.

Orientation to the Laboratory: Rules of Conduct and General Safety

Many of the microorganisms used in this course may be pathogenic for humans and animals. As a result, certain rules are necessary to avoid the possibility of infecting yourself or other people. Anyone who chooses to disregard these rules or exhibits carelessness that endangers others may be subject to immediate dismissal from the laboratory. If doubt arises as to the procedure involved in handling infectious material, consult your instructor.

In 1997, the American Society for Microbiology, through its Office of Education and Training, adopted the following on laboratory safety. Each point is considered essential for every introductory microbiology laboratory, regardless of its emphasis.

A student successfully completing basic microbiology will demonstrate the ability to explain and practice safe

1. **Microbiological procedures,** including

 a. reporting all spills and broken glassware to the instructor and receiving instructions for cleanup
 b. methods for aseptic transfer
 c. minimizing or containing the production of aerosols and describing the hazards associated with aerosols
 d. washing hands prior to and following laboratories and at any time contamination is suspected
 e. never eating or drinking in the laboratory
 f. using universal precautions (see inside front and end covers of this laboratory manual)
 g. disinfecting lab benches prior to and at the conclusion of each lab session
 h. identification and proper disposal of different types of waste
 i. never applying cosmetics, including contact lenses, or placing objects (fingers, pencils) in the mouth or touching the face
 j. reading and signing a laboratory safety agreement indicating that the student has read and understands the safety rules of the laboratory
 k. good lab practice, including returning materials to proper locations, proper care and handling of equipment, and keeping the bench top clear of extraneous materials

2. **Protective procedures,** including

 a. tying long hair back, wearing personal protective equipment (eye protection, coats, closed shoes; glasses may be preferred to contact lenses), and using such equipment in appropriate situations
 b. always using appropriate pipetting devices and understanding that mouth pipetting is forbidden

3. **Emergency procedures,** including

 a. locating and properly using emergency equipment (eye-wash stations, first-aid kits, fire extinguishers, chemical safety showers, telephones, and emergency numbers)
 b. reporting all injuries immediately to the instructor
 c. following proper steps in the event of an emergency

In addition, institutions where microbiology laboratories are taught will

1. train faculty and staff in proper waste stream management
2. provide and maintain necessary safety equipment and information resources
3. train faculty, staff, and students in the use of safety equipment and procedures
4. train faculty and staff in the use of MSDS. The Workplace Hazardous Materials Information

System (WHMIS) requires that all hazardous substances, including microorganisms, be labeled in a specific manner. In addition, there must be a Material Safety Data Sheet (MSDS) available to accompany each hazardous substance. MSDS sheets are now supplied with every chemical sold by supply houses. The person in charge of the microbiology laboratory should ensure that adherence to this law is enforced.

All laboratory work can be done more effectively and efficiently if the subject matter is understood before coming to the laboratory. To accomplish this, read the experiment several times before the laboratory begins. Know how each exercise is to be done and what principle it is intended to convey. Also, read the appropriate sections in your textbook that pertain to the experiment being performed; this will save you much time and effort during the actual laboratory period.

All laboratory experiments will begin with a brief discussion by your instructor of what is to be done, the location of the materials, and other important information. Feel free to ask questions if you do not understand the instructor or the principle involved.

Much of the work in the laboratory is designed to be carried out in groups or with a partner. This is to aid in coverage of subject matter, to save time and expense, and to encourage discussion of data and results.

Many of the ASM's recommended precautions are represented by the specific safety guidelines given inside the cover of this laboratory manual.

I have read the above rules and understand their meaning

Signature

Date

Summary of Universal Precautions and Laboratory Safety Procedures

Universal Precautions

Since medical history and examination cannot reliably identify all patients infected with HIV or other blood-borne pathogens, blood and body-fluid precautions should be consistently used for all patients.

1. All health-care workers should routinely use appropriate barrier precautions to prevent skin and mucous-membrane exposure when contact with blood or other body fluids of any patient is anticipated. Gloves should be worn for touching blood and body fluids, mucous membranes, or nonintact skin of all patients, for handling items or surfaces soiled with blood or body fluids, and for performing venipuncture and other vascular access procedures. Gloves should be changed after contact with each patient. Masks and protective eyewear or face shields should be worn during procedures that are likely to generate droplets of blood or other body fluids to prevent exposure of mucous membranes of the mouth, nose, and eyes. Gowns or aprons should be worn during procedures that are likely to generate splashes of blood or other body fluids.
2. Hands and other skin surfaces should be washed immediately and thoroughly if contaminated with blood or other body fluids. Hands should be washed immediately after gloves are removed.
3. All health-care workers should take precautions to prevent injuries caused by needles, scalpels, and other sharp instruments or devices during procedures; when cleaning used instruments; during disposal of used needles; and when handling sharp instruments after procedures. To prevent needlestick injuries, needles should not be recapped, purposely bent or broken by hand, removed from disposable syringes, or otherwise manipulated by hand. After they are used, disposable syringes and needles, scalpel blades, and other sharp items should be placed in puncture-resistant containers for disposal.
4. Although saliva has not been implicated in HIV transmission, to minimize the need for emergency mouth-to-mouth resuscitation, mouthpieces, resuscitation bags, or other ventilation devices should be available for use in areas in which the need for resuscitation is predictable.
5. Health-care workers who have exudative lesions or weeping dermatitis should refrain from all direct patient care and from handling patient-care equipment.
6. The following procedure should be used to clean up spills of blood or blood-containing fluids: (1) Put on gloves and any other necessary barriers. (2) Wipe up excess material with disposable towels and place the towels in a container for sterilization. (3) Disinfect the area with either a commercial EPA-approved germicide or household bleach (sodium hypochlorite). The latter should be diluted from 1:100 (smooth surfaces) to 1:10 (porous or dirty surfaces); the dilution should be no more than 24 hours old. When dealing with large spills or those containing sharp objects such as broken glass, first cover the spill with disposable toweling. Then saturate the toweling with commercial germicide or a 1:10 bleach solution and allow it to stand for at least 10 minutes. Finally, clean as described above.

Precautions for Laboratories

Blood and other body fluids from all patients should be considered infective.

1. All specimens of blood and body fluids should be put in a well-constructed container with a secure lid to prevent leaking during transport. Care should be taken when collecting each specimen to avoid contaminating the outside of the container and of the laboratory form accompanying the specimen.

2. All persons processing blood and body-fluid specimens should wear gloves. Masks and protective eyewear should be worn if mucous-membrane contact with blood or body fluids is anticipated. Gloves should be changed and hands washed after completion of specimen processing.

3. For routine procedures, such as histologic and pathologic studies or microbiologic culturing, a biological safety cabinet is not necessary. However, biological safety cabinets should be used whenever procedures are conducted that have a high potential for generating droplets. These include activities such as blending, sonicating, and vigorous mixing.

4. Mechanical pipetting devices should be used for manipulating all liquids in the laboratory. Mouth pipetting must not be done.

5. Use of needles and syringes should be limited to situations in which there is no alternative, and the recommendations for preventing injuries with needles outlined under universal precautions should be followed.

6. Laboratory work surfaces should be decontaminated with an appropriate chemical germicide after a spill of blood or other body fluids and when work activities are completed.

7. Contaminated materials used in laboratory tests should be decontaminated before reprocessing or be placed in bags and disposed of in accordance with institutional policies for disposal of infective waste.

8. Scientific equipment that has been contaminated with blood or other body fluids should be decontaminated and cleaned before being repaired in the laboratory or transported to the manufacturer.

9. All persons should wash their hands after completing laboratory activities and should remove protective clothing before leaving the laboratory.

10. There should be no eating, drinking, or smoking in the work area.

Use of Pathogens

Recently, there has been much discussion about the advisability of using opportunistic or pathogenic microorganisms in an introductory microbiology laboratory. Ideally, it would be desirable never to use pathogens with first-semester students. However, there are certain biochemical tests (e.g., coagulase) that cannot be performed without the use of pathogens. For this reason, their use has been minimized in this manual, but they continue to be used when necessary.

It is the belief of the author that the most important lesson all introductory microbiology students can learn is that all microorganisms should be treated as potential pathogens. With careful emphasis on proper handling procedures and aseptic techniques, the use of microorganisms should not put the student at risk.

Microscopic Techniques

The most important discoveries of the laws, methods and progress of nature have nearly always sprung from the examination of the smallest objects which she contains.
Jean Baptiste Pierre Antoine Monet de Lamarck
(French naturalist, 1744–1829)

Microbiologists employ a variety of light microscopes in their work: bright-field, dark-field, phase-contrast, and fluorescence are most commonly used. In fact, the same microscope may be a combination of types: bright-field and phase-contrast, or phase-contrast and fluorescence. You will use these microscopes and the principles of microscopy extensively in this course as you study the form, structure, staining characteristics, and motility of different microorganisms. Therefore, proficiency in using the different microscopes is essential to all aspects of microbiology and must be mastered at the very beginning of a microbiology course. The next five exercises have been designed to accomplish this major objective.

After completing at least exercise 1, you will, at the minimum, be able to demonstrate the ability to use a bright-field light microscope. This will meet the American Society for Microbiology Core Curriculum skill number 1 (*see pp. vi–viii*): (a) correctly setting up and focusing the microscope; (b) proper handling, cleaning, and storage of the microscope; (c) correct use of all lenses; and (d) recording microscopic observations.

Antony van Leeuwenhoek (1632–1723)

Leeuwenhoek was a master at grinding lenses for his microscopes. Working in Delft, Holland, in the mid-1600s, he is considered the greatest early microscopist.

Leeuwenhoek was a manic observer, who tried to look at everything with his microscopes.

Those little animals were everywhere! He told the Royal Society of finding swarms of those subvisible things in his mouth—of all places: "Although I am now fifty years old," he wrote, "I have uncommonly well-preserved teeth, because it is my custom every morning to rub my teeth very hard with salt, and after cleaning my teeth with a quill, to rub them vigorously with a cloth. . . ."

From his teeth he scraped a bit of white stuff, mixed it with pure rainwater, stuck it in a little tube onto the needle of his microscope, closed the door of his study—

As he brought the tube into focus, there was an unbelievable tiny creature, leaping about in the water of the tube. . . . There was a second kind that swam forward a little way, then whirled about suddenly, then tumbled over itself in pretty somersaults. . . . There was a menagerie in his mouth! There were creatures shaped like flexible rods that went to and fro . . . there were spirals that whirled through the water like violently animated corkscrews. . . .

—Paul de Kruif
Microbe Hunters (1926)

Bright-Field Light Microscope and Microscopic Measurement of Organisms

MATERIALS PER STUDENT

> compound microscope
> lens paper and lens cleaner
> immersion oil
> prepared stained slides of several types of bacteria (rods, cocci, spirilla), fungi, algae, and protozoa
> glass slides
> coverslips
> dropper with bulb
> newspaper or cut-out letter *e*'s
> tweezers
> ocular micrometer
> stage micrometer

Learning Objectives

Each student should be able to

1. Identify all the parts of a compound microscope
2. Know how to correctly use the microscope—especially the oil immersion lens
3. Learn how to make and examine a wet-mount preparation
4. Understand how microorganisms can be measured under the light microscope
5. Calibrate an ocular micrometer
6. Perform some measurements on different microorganisms

SUGGESTED READING IN TEXTBOOK

1. The Bright-Field Microscope, section 2.2; see also figures 2.3–2.6.
2. See tables 2.1 and 34.1

Medical Application

In the clinical laboratory, natural cell size, arrangement, and motility are important characteristics in the identification and characterization of a bacterial pathogen.

Why Are Prepared Slides Used in This Exercise?

Because this is a microbiology course and most of the microorganisms studied are bacteria, this is an excellent place to introduce the student to the three basic bacterial shapes: cocci, rods, and spirilla. By gaining expertise in using the bright-field light microscope, the student should be able to observe these three bacterial shapes by the end of the lab period. In addition, the student will gain an appreciation for the small size and arrangement of procaryotic cell structure.

One major objective of this exercise is for the student to understand how microorganisms can be measured under the light microscope and to actually perform some measurements on different microorganisms. By making measurements on prepared slides of various bacteria, fungi, algae, and protozoa, the student will gain an appreciation for the size of different microorganisms discussed throughout both the lecture and laboratory portions of this course.

PRINCIPLES

The success of a student in a microbiology laboratory is directly related to the competent use of the compound bright-field light microscope. The **bright-field light microscope** is an instrument that magnifies images using two lens systems. Initial magnification occurs in the **objective lens.** Most microscopes have at least three objective lenses on a rotating base, and each lens may be rotated into alignment with the **eyepiece** or **ocular lens** in which the final magnification occurs. The objective lenses are identified as the **low-power, high-dry,** and **oil immersion objectives.** Each objective is also designated by other terms. These terms give either the **linear magnification** or the **focal length.** The latter is about equal to or greater than the **working distance** between the specimen when in focus and the tip of the objective lens. For example, the low-power objective is also called the **10×,** or **16 millimeter (mm), objective;** the high-dry is called the **40×,** or **4 mm, objective;** and the oil immersion is called the **90×, 100×,** or **1.8 mm objective.** As the magnification increases, the size of the lens at the tip of the objective becomes progressively smaller and admits less light. This is one of the reasons that changes in position of the **substage condenser** and **iris diaphragm** are required when using different objectives if the specimens viewed are to be seen distinctly. The condenser focuses the light on a small area above the stage, and the iris diaphragm controls the amount of light that enters the condenser. It is important to remember that, as a rule, one must increase the light as one increases the magnification. When the oil immersion lens is used, immersion oil fills the space between the objective and the specimen. Because immersion oil has the same **refractive index** as glass, the loss of light is minimized (figure 1.1). The **eyepiece,** or **ocular,** at the top of the tube magnifies the image formed by the objective lens. As a result, the total magnification seen by the observer is obtained by multiplying the magnification of the objective lens by the magnification of the ocular, or eyepiece. For example, when using the 10× ocular and the 43× objective, total magnification is 10 × 43 = 430 times.

Procedure for Basic Microscopy: Proper Use of the Microscope

1. Always carry the microscope with two hands. Place it on the desk with the open part away from you.
2. Clean all of the microscope's lenses only with lens paper and lens cleaner if necessary. Do not use paper towels or Kimwipes; they can scratch the lenses. Do not remove the oculars or any other parts from the body of the microscope.
3. Cut a lowercase *e* from a newspaper or other printed page. Prepare a wet-mount as illustrated in figure 1.2. Place the glass slide on the stage of the microscope and secure it firmly using stage clips. If

Figure 1.2 Preparation of a Wet-mount Slide. (a) Add a drop of water to a slide. (b) Place the specimen (letter *e*) in the water. (c) Place the edge of a coverslip on the slide so that it touches the edge of the water. (d) Slowly lower the coverslip to prevent forming and trapping air bubbles.

(a)　　　(b)

(c)　　　(d)

Figure 1.1 The Oil Immersion Objective. An oil immersion objective lens operating in air and with immersion oil. Light rays that must pass through air are bent (refracted), and many do not enter the objective lens. The immersion oil prevents the loss of light rays.

Slide　Air　Oil　Cover glass

your microscope has a mechanical stage device, place the slide securely in it. Move the slide until the letter *e* is over the opening in the stage.

4. With the low-power objective in position, lower the tube until the tip of the objective is within 5 mm of the slide. Be sure that you lower the tube while looking at the microscope from the side.

5. Look into the microscope and slowly raise the tube by turning the coarse adjustment knob counterclockwise until the object comes into view. Once the object is in view, use the fine adjustment knob to focus the desired image.

6. Open and close the diaphragm, and lower and raise the condenser, noting what effect these actions have on the appearance of the object being viewed. Usually the microscope is used with the substage condenser in its topmost position. The diaphragm should be open and then closed down until just a slight increase in contrast is observed (table 1.1).

7. Use the oil immersion lens to examine the stained bacteria that are provided (figure 1.3*a–d*). The directions for using this lens are as follows: First locate the stained area with the low-power objective and then turn the oil immersion lens into the oil and focus with the fine adjustment. An alternate procedure is to get the focus very sharp under high power, then move the revolving nosepiece until you are halfway between the high-power and oil immersion objectives. Place a

Figure 1.3 Examples of Bacterial Shapes as Seen with the Bright-field Light Microscope. (**a**) *Staphylococcus aureus* cocci; singular, coccus (×1,000). (**b**) *Bacillus subtilis* rods or bacilli; singular, bacillus (×1,000). (**c**) A single, large spirillum; plural, spiralla (*Spirillum volutans;* ×1,000). (**d**) Numerous, small spirilla (*Rhodospirillum rubrum;* ×1,000).

(a)

(b)

(c)

(d)

Table 1.1

Troubleshooting the Bright-Field Light Microscope

Common Problem	Possible Correction
No light passing through the ocular	Check to ensure that the microscope is completely plugged into a good receptacle
	Check to ensure that the power switch to the microscope is turned on
	Make sure the objective is locked or clicked in place
	Make sure the iris diaphragm is open
Insufficient light passing through the ocular	Raise the condenser as high as possible
	Open the iris diaphragm completely
	Make sure the objective is locked or clicked in place
Lint, dust, eyelashes interfering with view	Clean ocular with lens paper and cleaner
Particles seem to move in hazy visual field	Air bubbles in immersion oil; add more oil or make certain that oil immersion objective is in the oil
	Make sure that the high-dry objective is not being used with oil
	Make sure a temporary coverslip is not being used with oil. Oil causes the coverslip to float since the coverslip sticks to the oil and not the slide, making viewing very hazy or impossible

small drop of immersion oil in the center of the illuminated area on the slide. Continue revolving the nosepiece until the oil immersion objective clicks into place. The lens will now be immersed in oil. Sharpen the focus with the fine adjustment knob. Draw a few of the bacteria in the spaces provided.

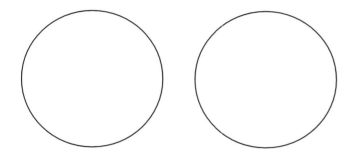

8. After you are finished with the microscope, place the low-power objective in line with the ocular, lower the tube to its lowest position, clean the oil from the oil immersion lens with lens paper and lens cleaner, cover, and return the microscope to its proper storage place.

PRINCIPLES OF MICROSCOPIC MEASUREMENT

It frequently is necessary to accurately measure the size of the microorganism one is viewing. For example, size determinations are often indispensable in the identification of a bacterial unknown. The size of microorganisms is generally expressed in metric units and is determined by the use of a microscope equipped with an ocular micrometer. An **ocular micrometer** is a small glass disk on which uniformly spaced lines of

unknown distance, ranging from 0 to 100, are etched. The ocular micrometer is inserted into the ocular of the microscope and then calibrated against a **stage micrometer,** which has uniformly spaced lines of known distance etched on it. The stage micrometer is usually divided into 0.01 millimeter and 0.1 millimeter graduations. The ocular micrometer is calibrated using the stage micrometer by aligning the images at the left edge of the scales.

The dimensions of microorganisms in dried, fixed, or stained smears tend to be reduced as much as 10 to 20% from the dimensions of the living microorganisms. Consequently, if the actual dimensions of a microorganism are required, measurements should be made in a wet-mount.

Procedure

Calibrating an Ocular Micrometer

1. If you were to observe the ocular micrometer without the stage micrometer in place, it would appear as shown in figure 1.4a. In like manner, the stage micrometer would appear as illustrated in figure 1.4b.
2. When in place, the two micrometers appear as shown in figure 1.4c. Turn the ocular in the body tube until the lines of the ocular micrometer are parallel with those of the stage micrometer (figure 1.4d). Match the lines at the left edges of the two micrometers by moving the stage micrometer.
3. Calculate the actual distance in millimeters between the lines of the ocular micrometer by observing how many spaces of the stage micrometer are included within a given number of spaces on

Figure 1.4 Calibrating an Ocular Micrometer.

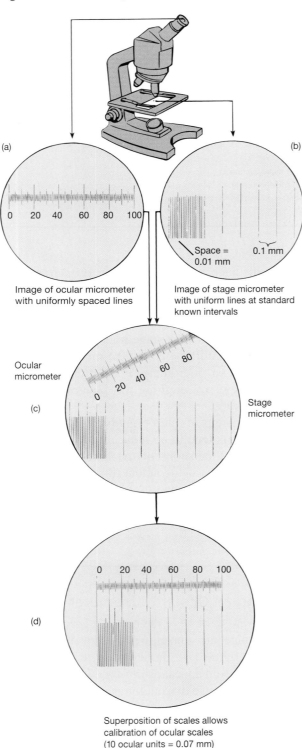

(a)

(b)

Image of ocular micrometer with uniformly spaced lines

Image of stage micrometer with uniform lines at standard known intervals

Space = 0.01 mm 0.1 mm

Ocular micrometer

(c)

Stage micrometer

(d)

Superposition of scales allows calibration of ocular scales (10 ocular units = 0.07 mm)

the ocular micrometer. You will get the greatest accuracy in calibration if you use more ocular micrometer spaces to match with stage micrometer lines.

Because the smallest space on the stage micrometer equals 0.01 millimeter or 10 μm (figure 1.4b), you can calibrate the ocular micrometer using the following:

10 spaces on the ocular micrometer = Y spaces on the stage micrometer.

Since the smallest space on a stage micrometer = 0.01 mm, then

10 spaces on the ocular micrometer = Y spaces on the stage micrometer \times 0.01 mm, and 1 space on the ocular micrometer = Y spaces on the stage micrometer $\dfrac{\times \ 0.01 \ mm}{10}$

For example, if 10 spaces on the ocular micrometer = 6 spaces on the stage micrometer, then

1 ocular space $\times \dfrac{6 \times 0.01 \ mm}{10}$,

1 ocular space = 0.006 mm or 6.0 μm.

This numerical value holds only for the specific objective-ocular lens combination used and may vary with different microscopes.

Calibrate for each of the objectives on your microscope and record below. Show all calculations in the space following the table; also show your calculations to your instructor.

Low power (10× objective)	1 ocular space = _____ mm
High-dry power (40× objective)	1 ocular space = _____ mm
Oil immersion (90× objective)	1 ocular space = _____ mm

HINTS AND PRECAUTIONS

(1) Forcing the fine or coarse adjustment knobs on the microscope beyond their gentle stopping points can render the microscope useless.

(2) A general rule for you to note is that the lower the magnification, the less light should be directed upon the object.

(3) The fine adjustment knob on the microscope should be centered prior to use to allow for maximum adjustment in either direction.

(4) If a slide is inadvertently placed upside down on the microscope stage, you will have no difficulty focusing the object under low and high power. However, when progressing to oil immersion, you will find it impossible to bring the object into focus.

(5) Slides should always be placed on and removed from the stage when the low-power ($4\times$ or $10\times$) objective is in place. Removing a slide when the higher objectives are in position may scratch the lenses.

(6) A note about wearing eyeglasses. A microscope can be focused; therefore, it is capable of correcting for near- or farsightedness. Individuals who wear eyeglasses that correct for near- or farsightedness do not have to wear their glasses. The microscope cannot correct for astigmatism; thus, these individuals must wear their glasses. If eyeglasses are worn, they should not touch the oculars for proper viewing. If you touch the oculars with your glasses, they may scratch either the glasses or the oculars.

(7) Because lens cleaner can be harmful to objectives, be sure not to use too much cleaner or leave it on too long.

The distance between the lines of an ocular micrometer is an arbitrary measurement that has meaning only if the ocular micrometer is calibrated for the specific objective being used. If it is necessary to insert an ocular micrometer in your eyepiece (ocular), ask your instructor whether it is to be inserted below the bottom lens or placed between the two lenses. Make sure that the etched graduations are on the upper surface of the glass disk that you are inserting. With stained preparations such as Gram-stained bacteria, the bacteria may measure smaller than they normally are if only the stained portion of the cell is the cytoplasm (gram-negative bacteria), whereas those whose walls are stained (gram-positive bacteria) will measure closer to their actual size.

Bright-Field Light Microscope (Basic Microscopy)

Parts of a Compound Microscope

1. Your microscope may have all or most of the features described below and illustrated in figure 2.3 in your textbook. By studying this figure and reading your textbook, label the compound microscope in figure LR1.1 on the next page. Locate the indicated parts of your microscope and answer the following questions.

 a. What is the magnification stamped on the housing of the oculars on your microscope? _____

 b. What are the magnifications of each of the objectives on your microscope? _____

 c. Calculate the total magnification for each ocular/objective combination on your microscope.

Ocular	×	Objective	=	Total Magnification
_____		_____		_____
_____		_____		_____
_____		_____		_____
_____		_____		_____

 d. List the magnification and numerical aperture for each objective on your microscope.

Magnification of Objective	Numerical Aperture (NA)
_____	_____
_____	_____
_____	_____
_____	_____

 e. With some compound microscopes, loosening a lock screw allows you to rotate the body tube 180°. What is the advantage of being able to rotate the body tube? _____

 f. Note the horizontal and vertical scales on the mechanical stage. What is the function of these scales?

 g. Where is the diaphragm on your microscope located? _____

 How can you regulate the diaphragm? _____

 h. Locate the substage condenser on your microscope. What is its function, and how can it be regulated?

 i. Can the light intensity of your microscope be regulated? Explain. _____

Microscopic Measurement of Microorganisms

2. After your ocular micrometer has been calibrated, determine the dimensions of the prepared slides of the following microorganisms.

Microorganism	Length	Width	Magnification
Bacterium name _____	_____	_____	_____
Fungus name _____	_____	_____	_____
Alga name _____	_____	_____	_____
Protozoan name _____	_____	_____	_____

3. Draw and label, as completely as possible, the microorganisms that you measured.

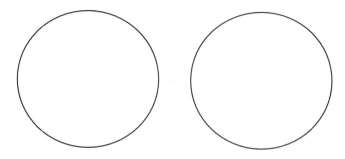

Genus and species: _____
Magnification: ✕ _____

Genus and species: _____
Magnification: ✕ _____

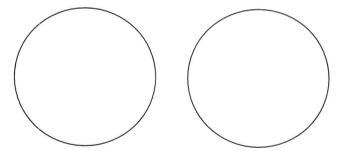

Genus and species: _____
Magnification: ✕ _____

Genus and species: _____
Magnification: ✕ _____

Review Questions

1. Differentiate between the resolving power and magnifying power of a lens. What is meant by the term "parfocal"?

2. Why is the low-power objective placed in position when the microscope is stored or carried?

3. Why is oil necessary when using the 90× to 100× objective?

4. What is the function of the iris diaphragm? The substage condenser?

5. What is meant by the limit of resolution?

6. How can you increase the bulb life of your microscope if the bulb's voltage is regulated by a rheostat?

7. In general, at what position should you keep your microscope's substage condenser lens?

8. What are three bacterial shapes you observed?

9. How can you increase the resolution on your microscope?

10. In microbiology, what is the most commonly used objective? Explain your answer.

11. In microbiology, what is the most commonly used ocular? Explain your answer.

12. If 5× instead of 10× oculars were used in your microscope with the same objectives, what magnifications would be achieved?

13. Why is it necessary to calibrate the ocular micrometer with each objective?

14. In the prepared slides, which organism was the largest?

15. When identifying microorganisms, why should a wet-mount be used when making measurements?

16. What is a stage micrometer?

17. Complete the following for the 10× objective:
 a. _____ ocular micrometer divisions = _____ stage micrometer divisions
 b. _____ ocular micrometer divisions = 1 stage micrometer division = _____ mm
 c. One ocular micrometer division = _____ stage micrometer divisions = _____ mm

18. Complete the following on units of measurement:

Unit	Abbreviation	Value
a. 1 centimeter	_____	10^{-2} meter
b. 1 millimeter	mm	_____
c. _____	μm	10^{-6} meter
d. 1 nanometer	_____	10^{-9} meter
e. 1 angstrom	_____	10^{-10} meter

19. In summary, what is the major purpose of this exercise?

The Hanging Drop Slide and Bacterial Motility

MATERIALS PER STUDENT

24- to 48-hour tryptic soy broth cultures of *Pseudomonas aeruginosa* (ATCC 10145, small, motile bacillus), *Bacillus cereus* (ATCC 21768, large, motile bacillus), and *Spirillum volutans* (ATCC 19554, spiral, motile bacterium)
microscope or phase-contrast microscope
lens paper and lens cleaner
immersion oil
clean depression slides and coverslips
petroleum jelly (Vaseline)
inoculating loop
toothpicks
Bunsen burner
methylene blue dye

Learning Objectives

Each student should be able to

1. Make a hanging drop slide in order to observe living bacteria
2. Differentiate between the three bacterial species used in this exercise on the basis of size, shape, arrangement, and motility

SUGGESTED READING IN TEXTBOOK

1. Flagella and Motility, section 3.6; see also figures 3.31–3.36.

Pronunciation Guide

Bacillus cereus (bah-SIL-lus SEE-ree-us)
Pseudomonas aeruginosa (soo-do-MO-nas a-ruh-jin-OH-sah)
Spirillum volutans (spy-RIL-lum VOL-u-tans)

Why Are the Above Bacteria Used in This Exercise?

The major objectives of this exercise are to allow students to gain expertise in making hanging drop slides and observing the motility of living bacteria. Motility is also an important characteristic used to identify microorganisms. To accomplish these objectives, the authors have chosen three bacteria that are easy to culture and vary in size, shape, arrangement of flagella, and types of motion. Specifically, *Pseudomonas aeruginosa* (L. *aeruginosa,* full of copper rust, hence green) is a straight or slightly curved rod (1.5 to 3.0 μm in length) that exhibits high motility by way of a polar flagellum; *Bacillus cereus* (L. *cereus,* waxen, wax colored) is a large (3.0 to 5.0 μm in length), rod-shaped, and straight bacillus that moves by peritrichous flagella; and *Spirillum volutans* (L. *voluto,* tumble about) is a rigid, helical cell (14 to 60 μm in length) that is highly motile since it contains large bipolar tufts of flagella having a long wavelength and about one helical turn. *P. aeruginosa* is widely distributed in nature and may be a saprophytic or opportunistic animal pathogen. *B. cereus* is found in a wide range of habitats and is a significant cause of food poisoning. *S. volutans* occurs in stagnant freshwater environments.

PRINCIPLES

Many bacteria show no motion and are termed **nonmotile.** However, in an aqueous environment, these same bacteria appear to be moving erratically. This

erratic movement is due to **Brownian movement.** Brownian movement results from the random motion of the water molecules bombarding the bacteria and causing them to move.

True **motility** (self-propulsion) has been recognized in other bacteria and involves several different mechanisms. Bacteria that possess flagella exhibit **flagellar motion.** Helical-shaped spirochetes have axial fibrils (modified flagella that wrap around the bacterium) that form axial filaments. These spirochetes move in a **corkscrew-** and **bending-type motion.** Other bacteria simply slide over moist surfaces in a form of **gliding motion.**

The above types of motility or nonmotility can be observed over a long period in a hanging drop slide. Hanging drop slides are also useful in observing the general shape of living bacteria and the arrangement of bacterial cells when they associate together (see figure 1.3). A ring of Vaseline around the edge of the coverslip keeps the slide from drying out.

Figure 2.1 Preparation of a Hanging Drop Slide.

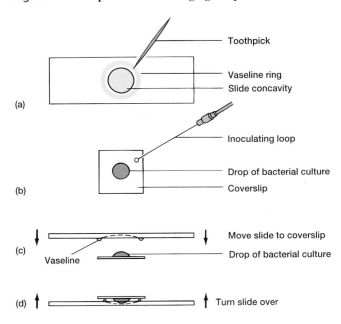

Procedure

1. With a toothpick, spread a small ring of Vaseline around the concavity of a depression slide (figure 2.1a). Do not use too much Vaseline.
2. After thoroughly mixing one of the cultures, use the inoculating loop to aseptically place a small drop of one of the bacterial suspensions in the center of a coverslip (figure 2.1b).
3. If desired, a loopful of methylene blue may be mixed with a loopful of bacteria to increase contrast in the hanging drop procedure.
4. Lower the depression slide, with the concavity facing down, onto the coverslip so that the drop protrudes into the center of the concavity of the slide (figure 2.1c). Press gently to form a seal.
5. Turn the hanging drop slide over (figure 2.1d) and place on the stage of the microscope so that the drop is over the light hole.
6. Examine the drop by first locating its edge under low power and focusing on the drop. Switch to the high-dry objective and then, using immersion oil, to

the 90 to 100× objective. In order to see the bacteria clearly, close the diaphragm as much as possible for increased contrast. Note bacterial shape, size, arrangement, and motility. Be careful to distinguish between motility and Brownian movement.
7. Discard your coverslips and any contaminated slides in a container with disinfectant solution.
8. Complete the report for exercise 2.

HINTS AND PRECAUTIONS

(1) Always make sure the specimen is on the top side of the slide.

(2) Particular care must be taken to avoid breaking the coverslip since it is more vulnerable when supported only around its edges.

(3) With depression slides, the added thickness of the slide and coverslip may preclude the use of the oil immersion objective with some microscopes.

(4) If your microscope is equipped with an automatic stop, it may be necessary to bring the image into focus by using the coarse adjustment knob.

Laboratory Report 2

Name: _____

Date: _____

Lab Section: _____

The Hanging Drop Slide and Bacterial Motility

1. Examine the hanging drop slide and complete the following table with respect to the size, shape, and motility of the different bacteria.

Bacterium	Size	Shape	Type of Motility	Cell Arrangement
B. cereus	_____	_____	_____	_____
P. aeruginosa	_____	_____	_____	_____
S. volutans	_____	_____	_____	_____

2. Draw a representative field for each bacterium.

B. cereus P. aeruginosa S. volutans

Magnification: ×_____ Magnification: ×_____ Magnification: ×_____

3. In summary, what is the major purpose of this exercise?

Review Questions

1. Why are unstained bacteria more difficult to observe than stained bacteria?

2. What are some reasons for making a hanging drop slide?

3. Describe the following types of bacterial movement:
 a. Brownian movement
 b. flagellar motion
 c. gliding motion

4. Why do you have to reduce the amount of light with the diaphragm in order to see bacteria in a hanging drop slide?

5. Can the hanging drop slide be used to examine other microorganisms? Explain which ones.

6. Which of the bacteria exhibited true motility on the slides?

7. How does true motility differ from Brownian movement?

Dark-Field Light Microscope

MATERIALS PER GROUP OF STUDENTS

dark-field light microscope
flat toothpicks
lens paper and lens cleaner
immersion oil
slides and coverslips
prepared slides of spirochetes (e.g., *Treponema denticola*), radiolarians, protozoa
tweezers

Learning Objectives

Each student should be able to

1. Understand the principles behind dark-field microscopy
2. Correctly use the dark-field microscope
3. Make a wet-mount and examine it for spirochetes with the dark-field microscope

SUGGESTED READING IN TEXTBOOK

1. The Dark-Field Microscope, section 2.2; see also figures 2.7 and 2.8.

Pronunciation Guide

Treponema denticola (trep-o-NE-mah dent-A-cola)

Why Is the Following Bacterium Used in This Exercise?

Treponema denticola (M.L. n, *denticola,* tooth dweller) often is a part of the normal microbiota of the oral mucosa; thus, this spirochete is readily available and does not have to be cultured. Most species stain poorly if at all with Gram's or Giemsa's methods and are best observed with dark-field or phase-contrast microscopy. Thus, *T. denticola* is an excellent specimen to observe when practicing the use of a dark-field microscope, and also allows the student to continue practicing the wet-mount preparation. *T. denticola* is a slender, helical cell, 6 to 16 μm in length. In a wet-mount, the bacteria show both rotational and translational movements due to two or three periplasmic flagella inserted at each end of the protoplasmic cylinder. Young cells rotate rapidly on their axes. Thus, by using *T. denticola,* the student is also able to observe bacterial motility.

PRINCIPLES

The compound microscope may be fitted with a dark-field condenser that has a numerical aperture (resolving power) greater than the objective. The condenser also contains a dark-field stop. The compound microscope now becomes a **dark-field microscope.** Light passing through the specimen is diffracted and enters the objective lens, whereas undiffracted light does not, resulting in a bright image against a dark background (figures 3.1–3.2). Since light objects against a dark background are seen more clearly by the eye than the reverse, dark-field microscopy is useful in observing unstained living microorganisms, microorganisms that are difficult to stain, and spirochetes (figure 3.2), which are poorly defined by bright-field microscopy.

Figure 3.1 Dark-field Microscopy. Dark-field microscopy can best visualize transparent, unstained specimens, which display only low contrast in bright-field. In this dark-field photomicrograph (×100), a mixture of radiolarian shells is shown. Notice their many unique and beautiful shapes.

Figure 3.2 Photomicrograph of *Treponema pallidum*, as Seen with Dark-field Microscopy (×500). (©*Arthur M. Siegelman/Visuals Unlimited*)

Procedure

1. Place a drop of immersion oil directly on the dark-field condenser lens.
2. Position one of the prepared slides so that the specimen is directly over the light opening.
3. Raise the dark-field condenser with the height control until the oil on the condenser lens just touches the slide.
4. Lock the 10× objective into position. Focus with the coarse and fine adjustment knobs until the spirochetes come into sharp focus. Do the same with the 40× objective.
5. Use the oil immersion objective lens to observe the spirochetes. Draw several in the space provided in the report for exercise 3.
6. Nonpathogenic spirochetes (*T. denticola*) may be part of the normal microbiota of the oral mucosa.

To make a wet-mount of these, gently scrape your gum line with a flat toothpick. Stir the scrapings into a drop of water on a slide. Gently lower a coverslip (*see figure 1.2*) to prevent trapping air bubbles. Examine with the dark-field microscope and draw several spirochetes in the space provided in the report for exercise 3.

HINTS AND PRECAUTIONS

(1) It is good practice to always clean the condenser lens before placing a drop of oil on it.

(2) Make sure the prepared slide is placed right side up (coverslip up) on the stage.

(3) If you have trouble focusing with the oil immersion lens, don't flounder—ask for help from your instructor.

(4) Always make sure that the substage condenser diaphragm is wide open for adequate illumination of the specimen.

Laboratory Report 3

Name: _____

Date: _____

Lab Section: _____

Dark-Field Light Microscope

1. Drawing of spirochetes from a prepared slide. Drawing of spirochetes from a wet-mount.

Magnification: ×_____ Magnification: ×_____

Genus and species: _____ Genus and species: _____

Shape: _____ Shape: _____

2. Label the following parts of a dark-field microscope. Use the following terms: dark-field stop, specimen, Abbé condenser, and objective.

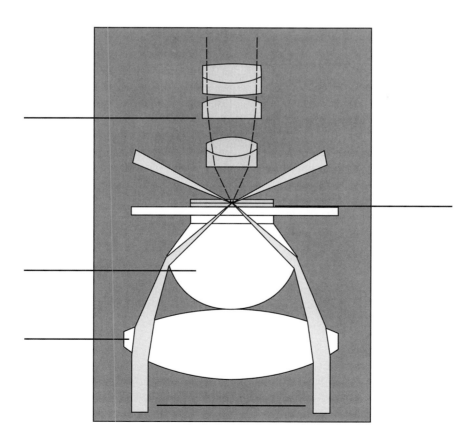

Review Questions

1. What is the principle behind dark-field microscopy?

2. When would you use the dark-field microscope?

3. Why is the field dark and the specimen bright when a dark-field microscope is used to examine a specimen?

4. Differentiate between bright-field and dark-field microscopy.

5. What is the function of the Abbé condenser in dark-field microscopy?

6. What is the function of the dark-field stop?

7. In dark-field microscopy, why is a drop of oil placed directly on the condenser lens?

Phase-Contrast Light Microscope

MATERIALS PER GROUP OF STUDENTS

pond water
phase-contrast light microscope
new microscope slides and coverslips
Pasteur pipette with pipettor
pictorial guides of common pond water
 microorganisms
methyl cellulose (Protoslo, Carolina Biological
 Supply)
tweezers
lens paper and lens cleaner
prepared slides of *Bacillus* or *Clostridium*
 showing endospores

Learning Objectives

Each student should be able to

1. Understand the basic principles behind phase-contrast microscopy
2. Correctly use the phase-contrast microscope
3. Make a wet-mount of pond water and observe some of the transparent, colorless microorganisms that are present

SUGGESTED READING IN TEXTBOOK

1. The Phase-Contrast Microscope, section 2.2; see also figures 2.8 and 2.9.

Pronunciation Guide

Bacillus (bah-SIL-lus)
Clostridium (klos-STRID-ee-um)

Why Are the Following Bacteria and Pond Water Used in This Exercise?

Most microorganisms and their organelles are colorless and often difficult to see by ordinary bright-field or dark-field microscopy. Phase-contrast microscopy permits the observation of otherwise indistinct, living, unstained bacteria and their associated structures (such as endospores). *Bacillus* species are rod shaped, often arranged in pairs or chains, with rounded or square ends. Endospores are oval or sometimes cylindrical. *Clostridium* species are often arranged in pairs or short chains, with rounded or sometimes pointed ends. The endospores often distend the cell. Thus, by using prepared slides of *Bacillus* and *Clostridium,* the student gains expertise in using the phase-contrast microscope and in observing specific bacterial structures, such as different endospores.

Pond water is usually teeming with bacteria and protists. By using the phase-contrast microscope and slowing down the many microorganisms with Protoslo, the student is able to observe the internal structure of protists such as *Paramecium.*

PRINCIPLES

Certain transparent, colorless living microorganisms and their internal organelles are often impossible to see by ordinary bright-field or dark-field microscopy because they do not absorb, reflect, refract, or diffract sufficient light to contrast with the surrounding environment or the rest of the microorganism. Microorganisms and their organelles are only visible when they absorb, reflect, refract, or diffract more light than their environment. The **phase-contrast microscope** permits the observation of otherwise invisible living, unstained microorganisms (figure 4.1a–d).

Figure 4.1 **Some Examples of Microorganisms Seen with Phase-contrast Microscopy** (**a**) A protozoan, *Paramecium caudatum*, from pond water stained to show internal structures (×200). (**b**) A bacterium, *Bacillus cereus*, stained to show spores (×1,000). (**c**) A yeast, *Saccharomyces cerevisiae*, stained to show budding (×1,000). (**d**) A filamentous green alga, *Spirogyra*, showing its helical chloroplasts (×200).

(a)

(b)

(c)

(d)

In the phase-contrast microscope, the condenser has an annular diaphragm, which produces a hollow cone of light; the objective has a glass disk (the phase plate) with a thin film of transparent material deposited on it, which accentuates phase changes produced in the specimen. This phase change is observed in the specimen as a difference in **light intensity.** Phase plates may either retard (positive phase plate) the diffracted light relative to the undiffracted light, producing **dark-phase-contrast microscopy,** or advance (negative phase plate) the undiffracted light relative to the directed light, producing **bright-phase-contrast microscopy.**

Procedure

1. Make a wet-mount of pond water. Add a drop of methyl cellulose (Protoslo) to slow the swimming of the microorganisms. Prepared slides of *Bacillus* or *Clostridium* may also be used.
2. Place the slide on the stage of the phase-contrast microscope so that the specimen is over the light hole.
3. Rotate the 10× objective into place.
4. Rotate into position the annular diaphragm that corresponds to the 10× objective. It is absolutely necessary that the cone of light produced by the an-

nular diaphragm below the condenser be centered exactly with the phase plate of the objective (*see figure 2.9 in textbook if you do not understand this procedure*). Consequently, there are three different annular diaphragms that match the phase plates of the three different phase objectives (10×, 40×, and 90× or 100×). The substage unit beneath the condenser contains a disk that can be rotated in order to position the correct annular diaphragm.

5. Focus with the 10× objective and observe the microorganisms.
6. Rotate the nosepiece and annular diaphragm into the proper position for observation with the 40× objective.
7. Do the same with the oil immersion lens.

8. In the report for exercise 4, sketch several of the microorganisms that you have observed.
9. If you examined pond water, use the pictorial guides provided by your instructor to assist you in identifying some of the microorganisms present.

HINTS AND PRECAUTIONS

(1) Make sure the specimen is directly over the light hole in the stage of the microscope.

(2) The phase elements must be properly aligned. Misalignment is the major pitfall that beginning students encounter in phase-contrast microscopy.

(3) If your microscope is not properly aligned, ask your instructor for help.

Laboratory Report 4

Name: _____

Date: _____

Lab Section: _____

Phase-Contrast Light Microscope

1. Some typical microorganisms in pond water as seen with the phase-contrast light microscope.

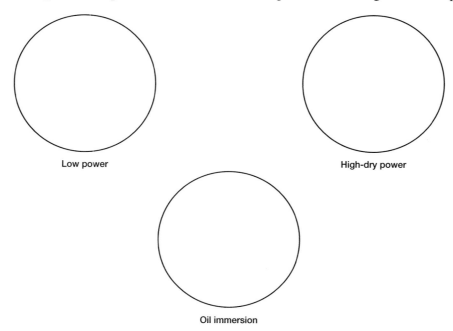

Low power High-dry power

Oil immersion

2. Drawings of *Bacillus, Clostridium,* or another bacterium showing endospores as seen with the phase-contrast microscope.

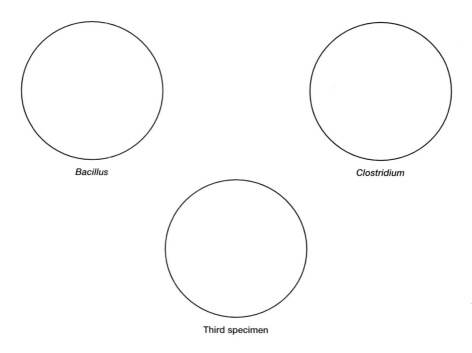

Bacillus *Clostridium*

Third specimen

Review Questions

1. In the phase-contrast microscope, what does the annular diaphragm do?

2. When would you use the phase-contrast microscope?

3. Explain how the phase plate works in a phase-contrast microscope that produces bright objects with respect to the background.

4. What happens to the phase of diffracted light in comparison to undiffracted light in a phase-contrast microscope?

5. What advantage does the phase-contrast microscope have over the ordinary bright-field microscope?

6. What is the difference between a bright-phase-contrast and a dark-phase-contrast microscope?

7. In microscopy, what does the term "phase" mean?

Exercise 5

Fluorescence Microscope

MATERIALS PER GROUP OF STUDENTS

fluorescence microscope
lens paper and lens cleaner
low-fluorescing immersion oil
protective glasses that filter UV light
prepared slides of known bacteria (*M. tuberculosis*) stained with fluorescent dye

Learning Objectives

Each student should be able to

1. Understand the principles behind the fluorescence microscope
2. Correctly use the fluorescence microscope by observing prepared slides of known bacteria stained with a fluorescent dye

SUGGESTED READING IN TEXTBOOK

1. The Fluorescence Microscope, section 2.2; see also figures 2.12, 2.13.

Pronunciation Guide

Mycobacterium tuberculosis (mi-ko-bak-TE-re-um too-ber-ku-LO-sis)

Why Is the Following Bacterium Used in This Exercise?

Mycobacterium tuberculosis (L. *tuberculum,* a small swelling + Gr. *-osis,* characterized by) is a human pathogen that causes tuberculosis. It is very slow growing and not readily stained by Gram's method. The cell is 1 to 4 μm in length, straight or slightly curved, occurring singly and in occasional threads. This bacterium can most readily be identified after staining with fluorochromes or specifically labelling it with fluorescent antibodies using complicated immunofluorescence procedures, which are both time consuming and expensive. By using commercially prepared slides, the student is able to immediately examine a pathogenic bacterium, such as *M. tuberculosis,* and gain expertise in using the fluorescence microscope. In this exercise, microscopic technique is more important than what is being observed.

Medical Applications

Fluorescence microscopy is commonly used in the clinical laboratory for the rapid detection and identification of bacterial antigens in tissue smears, sections, and fluids, as well as the rapid identification of many disease-causing microorganisms. For example, a sputum specimen can be quickly screened for *M. tuberculosis* by staining it with a fluorescent dye that binds specifically to *M. tuberculosis.* Only the stained bacterium of interest will be visible when the specimen is viewed under the fluorescence microscope.

PRINCIPLES

Fluorescence microscopy is based on the principle of removal of incident illumination by selective absorption, whereas light that has been absorbed by the specimen and re-emitted at an altered wavelength is

Figure 5.1 Fluorescence Microscopy. (**a**) *Escherichia coli* stained with DAPI and propidium iodine. These fluorochromes bind DNA (×1,000). The blue cells are viable and the red cells are dead. (**b**) *Giardia lamblia* stained with IFA (×1,000).

(a)

(b)

transmitted. The light source must produce a light beam of appropriate wavelength. An excitation filter removes wavelengths that are not effective in exciting the fluorochrome used. The light fluoresced by the specimen is transmitted through a filter that removes the incident wavelength from the beam of light. As a result, only light that has been produced by specimen fluorescence contributes to the intensity of the image being viewed (figure 5.1*a,b*).

Procedure

1. Turn on the UV light source at least 30 minutes before using the fluorescence microscope. NEVER LOOK AT THE UV LIGHT SOURCE WITHOUT PROTECTIVE GLASSES THAT FILTER UV LIGHT BECAUSE RETINAL BURNS AND BLINDNESS MIGHT RESULT.
2. Make sure that the proper excitation filter and barrier filter are matched for the type of fluorescence expected and are in place.
3. Place a drop of the low-fluorescing immersion oil on the condenser.
4. Place the prepared slide on the stage and position it so that the specimen is over the light opening. Raise the condenser so that the oil just touches the bottom of the slide.

5. After the mercury vapor arc lamp has been warmed up, turn on the regular tungsten filament light source and focus on the specimen.
6. Starting with the 10× objective, find and focus the specimen.
7. After finding the specimen, move to the 90× to 100× objective, switch to the mercury vapor arc, and view the specimen.
8. Compare what you see in the bright-field microscope with what you see in the fluorescence microscope by sketching the organisms in the report for exercise 5.

HINTS AND PRECAUTIONS

(**1**) The mercury vapor arc lamp requires about a 30-minute warm-up period. During a normal laboratory period, do not turn the microscope on and off.

(**2**) Make sure the proper filters are in place. If you are in doubt, ask your instructor.

(**3**) Note that there is no diaphragm control on the dark-field condenser.

(**4**) Never use ordinary immersion oil with a fluorescence microscope.

Laboratory Report 5

Name: _____

Date: _____

Lab Section: _____

Fluorescence Microscope

1. Bacterium as seen with the bright-field microscope. Bacterium as seen with the fluorescence microscope.

Genus and species: _____ Genus and species: _____

Magnification: ×_____ Magnification: ×_____

Shape: _____ Shape: _____

2. Label the following parts of a fluorescence microscope. Use the following terms: specimen and fluorochrome, heat filter, mercury vapor arc lamp, exciter filter, barrier filter, dark-field condenser.

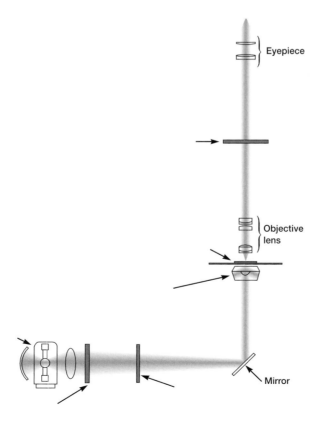

Review Questions

1. What kind of light is used to excite dyes and make microorganisms fluoresce?

2. List two fluorochromes that are used in staining bacteria.

3. What is a serious hazard one must guard against when working with mercury vapor arc lamps?

4. What is the function of each of the following?
 a. exciter filter
 b. barrier filter
 c. heat filter
 d. mercury vapor arc lamp

5. When is fluorescence microscopy used in a clinical laboratory?

6. Differentiate between phosphorescence and fluorescence.

7. What advantage is there to using fluorescence procedures in ecological studies? Give several examples.

Bacterial Morphology and Staining

The more I practice, the luckier I get.
Lee Trevino (Professional golfer, 1939–)

Living bacteria sometimes may be studied directly by bright-field or phase-contrast microscopy. This is useful when one needs to demonstrate, for example, bacterial motility. The first exercise in part two provides an opportunity for observing bacteria alive and unstained.

Since living bacteria are generally colorless and almost invisible because of their lack of contrast with the water in which they may reside, staining is necessary in order to make them readily visible for observation of intracellular structures as well as overall morphology. The exercises in this part of the manual have been designed to give the student exper-

tise in staining and slide preparation, an appreciation for bacterial morphology, and experience in how to specifically stain some specialized bacterial features such as endospores, capsules, and flagella.

After completing the exercises in part two, you will be able to demonstrate how to properly prepare slides for microbiological examination. This will meet the American Society for Microbiology Core Curriculum skill number 2 (*see pp. vi–viii*):

(a) cleaning and disposal of slides;

(b) preparing smears from solid and liquid cultures;

(c) performing wet-mounts and/or hanging drop preparations; and

(d) performing Gram stains.

Hans Christian Joachim Gram (1853–1938)

The most widely used staining procedure in microbiology is the Gram stain, discovered by the Danish scientist and physician Christian Gram. (*© Bettmann/Corbis*)

In 1884, Gram entitled his procedure "The Differential Staining of Schizomycetes in Sections and in Smear Preparations" (*Fortschr. Med.* 2:185–89, 1884), and described it as follows:

> After having been dehydrated in alcohol, the preparations are immersed in the aniline-gentian violet solution of Ehrlich for 1 to 3 minutes. . . . The preparations are then placed in an aqueous solution of iodine-potassium iodide directly or after a rapid rinsing in alcohol. They are allowed to remain there for 1 to 3 minutes, during which time the color of the preparations changes from dark blue-violet to deep purple-red. The preparations are then completely decolorized with absolute alcohol. Further clearing is achieved with clove oil. . . . Bacteria are stained intense blue while the background tissues are light yellow. . . .

The Gram stain, as devised by Gram in 1884, was his attempt to differentiate bacterial cells from infected tissue. Although Gram observed what is now called the "Gram reaction," he did not recognize the taxonomic value of his technique.

Negative Staining

MATERIALS PER STUDENT

24- to 48-hour tryptic soy broth cultures of *Bacillus subtilis* (ATCC 6051), *Micrococcus luteus* (ATCC 9341), and *Spirillum volutans* (ATCC 19554)

Dorner's nigrosin solution, India ink, or eosin blue

clean microscope slides

inoculating loop

immersion oil

microscope

lens paper and lens cleaner

wax pencil

Bunsen burner

Learning Objectives

Each student should be able to

1. Understand the reason for the negative staining procedure
2. Stain three different bacteria using the negative staining procedure

SUGGESTED READING IN TEXTBOOK

1. Dyes and Simple Staining, section 2.2.

Pronunciation Guide

Bacillus subtilis (bah-SIL-lus sub-til-us)
Micrococcus luteus (my-kro-KOK-us LOO-tee-us)
Spirillum volutans (spy-RIL-lum VOL-u-tans)

Why Are the Following Bacteria Used in This Exercise?

Because cultures of *Bacillus subtilis* (L. *subtilis,* slender) and *Spirillum volutans* were used in exercise 6, students are now familiar with the rod-shaped morphology of these bacteria. Thus, these same bacteria can be used to illustrate the negative staining technique. One new bacterium has been added to broaden the student's awareness of bacterial morphology by way of negative staining. *Micrococcus luteus* (L. *luteus,* golden yellow) is a spherical cell, 0.9 to 1.8 μm in diameter, occurring in pairs, tetrads, or irregular clusters but not chains. It is nonsporing and seldom motile. The bacterium is easy to culture since it grows on simple medium and forms yellow, yellowish green, or orange colonies. *M. luteus* occurs primarily on mammalian skin and in soil but can be easily isolated from food products and the air.

Medical Application

Treponema pallidum is the spirochete that causes the sexually transmitted disease syphilis. This bacterium is a very delicate cell that is easily distorted by heat-fixing; thus, negative staining is the procedure of choice in the clinical laboratory.

PRINCIPLES

Sometimes it is convenient to determine overall bacterial morphology without the use of harsh staining or heat-fixing techniques that change the shape of cells. This might be the case when the bacterium does not stain well (e.g., some of the spirochetes) or when it is desirable to confirm observations made on the shape

Figure 6.1 India Ink Stain of *Bacillus megaterium* (×1,000). Notice the dark background around the clear bacterial cells.

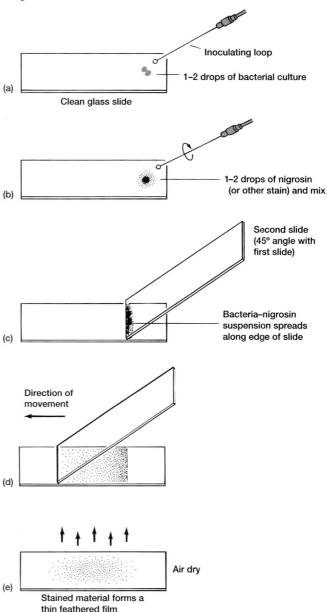

Figure 6.2 Negative Staining Procedure and Thin Smear Preparation.

(a) Inoculating loop — 1–2 drops of bacterial culture — Clean glass slide

(b) 1–2 drops of nigrosin (or other stain) and mix

(c) Second slide (45° angle with first slide) — Bacteria–nigrosin suspension spreads along edge of slide

(d) Direction of movement

(e) Air dry — Stained material forms a thin feathered film

and size of bacteria observed in either a wet-mount or hanging drop slide. Negative staining is also good for viewing capsules (*see exercise 11*).

Negative, indirect, or **background staining** is achieved by mixing bacteria with an acidic stain, such as nigrosin, India ink, or eosin, and then spreading out the mixture on a slide to form a film. The above stains will not penetrate and stain the bacterial cells due to repulsion between the negative charge of the stains and the negatively charged bacterial wall. Instead, these stains either produce a deposit around the bacteria or produce a dark background so that the bacteria appear as unstained cells with a clear area around them (figure 6.1).

Procedure

1. With a wax pencil, label the left-hand corner of three glass slides with the names of the respective bacteria.
2. Use an inoculating loop to apply a small amount of bacteria to one end of a clean microscope slide (figure 6.2*a*).
3. Add 1 to 2 loops of nigrosin, India ink, or eosin solution to the bacteria (figure 6.2*b*) and mix thoroughly.
4. Spread the mixture over the slide using a second slide. The second slide should be held at a 45° an-

gle so that the bacteria-nigrosin solution spreads across its edge (figure 6.2*c*). The slide is then pushed across the surface of the first slide in order to form a smear that is thick at one end and thin at the other (figure 6.2*d*). This is known as a **thin smear.**

5. Allow the smear to air dry (figure 6.2*e*). Do not heat-fix!
6. With the low-power objective, find an area of the smear that is of the optimal thickness for observation.
7. Use the oil immersion lens to observe, and draw the three bacterial species in the report for exercise 6.

HINTS AND PRECAUTIONS

(1) For a successful thin smear, the slides must be absolutely clean and free from oil and grease—including fingerprints.

(2) If an inconsistent smear is obtained, it is better to prepare a new slide than to search unsuccessfully for an appropriate area on a poorly stained slide.

(3) Do not use too much stain; use only a small drop of nigrosin.

(4) The mixture must be drawn over the slide, not pushed. Drawing the mixture over the slide will produce a more uniform film.

(5) Prepare a smear that consists of a thin layer of cells without clumps.

(6) View the thinner or clearer portions of the film.

Laboratory Report 6

Name: _____

Date: _____

Lab Section: _____

Negative Staining

1. Draw a representative field of your microscopic observation as seen with the oil immersion lens.

B. subtilis

M. luteus

S. volutans

Magnification: \times _____

Magnification: \times _____

Magnification: \times _____

2. Describe the microscopic appearance of the three bacteria.
 a. *B. subtilis*

 b. *M. luteus*

 c. *S. volutans*

3. In summary, what is the major purpose of this exercise?

Review Questions

1. When is negative staining used?

2. Name three stains that can be used for negative staining.

 a.

 b.

 c.

3. Why do the bacteria remain unstained in the negative staining procedure?

4. What is an advantage of negative staining?

5. Why didn't you heat-fix the bacterial suspension before staining?

6. Why is negative staining also called either indirect or background staining?

7. When streaking with the second slide, why must it be held at a 45° angle?

Smear Preparation and Simple Staining

MATERIALS PER STUDENT

24- to 48-hour tryptic soy broth or agar slants of *Bacillus subtilis* (ATCC 6051), *Corynebacterium pseudodiphtheriticum* (ATCC 7091), *Micrococcus luteus* (ATCC 9341), and *Spirillum volutans* (ATCC 19554)
microscope
clean microscope slides
bibulous paper
inoculating loop and needle
sterile distilled water
Bunsen burner
Loeffler's alkaline methylene blue
crystal violet (1% aqueous solution)
Ziehl's carbolfuchsin
wax pencil
immersion oil
lens paper and lens cleaner
slide holder or clothespin
slide warmer

Learning Objectives

Each student should be able to

1. Learn the proper procedure for preparing a bacterial smear
2. Do several simple staining procedures

SUGGESTED READING IN TEXTBOOK

1. Fixation, section 2.3.
2. Dyes and Simple Staining, section 2.3.
3. Size, Shape, and Arrangement, section 3.1; see also figures 3.1 and 3.2.

Pronunciation Guide

Bacillus subtilis (bah-SIL-lus sub-til-us)
Corynebacterium pseudodiphtheriticum (koh-rye-nee-back-TIR-ee-um soo-doh-dif-theh-RIT-ee-cum)
Micrococcus luteus (my-kro-KOK-us LOO-tee-us)
Spirillum volutans (spy-RIL-lum VOL-u-tans)

Why Are the Above Bacteria Used in This Exercise?

The same three cultures (*B. subtilis, M. luteus,* and *S. volutans*) that were used for the negative staining exercise will continue to be used in this exercise. The new bacterium is *Corynebacterium pseudodiphtheriticum. C. pseudodiphtheriticum* (M.L. n, *pseudodiphtheriticum,* relating to false diphtheria) is a straight or slightly curved slender rod 0.5 to 2.0 µm in length that has tapered or sometimes clubbed ends. Cells are arranged singly or in pairs, often in a "V" formation or in palisades of several parallel cells. *C. pseudodiphtheriticum* is primarily an obligate parasite of mucous membranes or the skin of mammals. By using Loeffler's alkaline methylene blue, crystal violet, and Ziehl's carbolfuchsin, the student gains expertise in using some simple stains to observe the morphology and characteristics of four different bacteria.

PRINCIPLES

While negative staining is satisfactory when making simple observations on bacterial morphology and size,

more specific stains are necessary if bacterial detail is to be observed. One way of achieving this detail involves smear preparation and simple staining. A **bacterial smear** is a dried preparation of bacterial cells on a glass slide. In a bacterial smear that has been properly processed, (1) the bacteria are evenly spread out on the slide in such a concentration that they are adequately separated from one another, (2) the bacteria are not washed off the slide during staining, and (3) bacterial form is not distorted.

In making a smear, bacteria from either a broth culture or an agar slant or plate may be used. If a slant or plate is used, a *small* amount of bacterial growth is transferred to a drop of water on a glass slide (figure 7.1*a*) and mixed. The mixture is then spread out evenly over a large area on the slide (figure 7.1*b*).

One of the most common errors in smear preparation from agar cultures is the use of too large an inoculum. This invariably results in the occurrence of large aggregates of bacteria piled on top of one another. If the medium is liquid, place one or two loops of the medium directly on the slide (figure 7.1*c*) and spread the bacteria over a large area (figure 7.1*d*). Allow the slide to air dry at room temperature (figure 7.1*e*). After the smear is dry, the next step is to attach the bacteria to the slide by **heat-fixing.** This is accomplished by gentle heating (figure 7.1*f*), passing the slide several times through the hot portion of the flame of a Bunsen burner. Most bacteria can be **fixed** to the slide and killed in this way without serious distortion of cell structure.

The use of a single stain or dye to create contrast between the bacteria and the background is referred to as **simple staining.** Its chief value lies in its simplicity and ease of use. Simple staining is often employed when information about cell shape, size, and arrangement is desired. In this procedure, one places the heat-fixed slide on a staining rack, covers the smear with a small amount of the desired stain for the proper amount of time, washes the stain off with water for a few seconds, and, finally, blots it dry. Basic dyes such as **crystal violet** (20 to 30 seconds staining time), **carbolfuchsin** (5 to 10 seconds staining time), or **methylene blue** (1 minute staining time) are often used. Once bacteria have been properly stained, it is usually an easy matter to discern their overall shape. Bacterial morphology is usually uncomplicated and limited to one of a few variations. For future reference, the most common shapes are presented in figure 7.2.

Figure 7.1 Bacterial Smear Preparation.

Procedure

Smear Preparation

1. With the wax pencil, mark the name of the bacterial culture in the far left corner on each of three slides.
2. For the broth culture, shake the culture tube and, with an inoculating loop, aseptically (*see figure 14.3*) transfer 1 to 2 loopfuls of bacteria to the center of the slide. Spread this out to about a ½-inch area. When preparing a smear from a slant or plate, place a loopful of water in the center of the slide. With the inoculating needle, aseptically pick up a *very small* amount of culture and mix into the drop

Figure 7.2 Common Bacterial Shapes.

Shape		Arrangement	
Spherical	coccus (pl., cocci)	diplococcus (pairs)	
		streptococcus (chains)	
		staphylococcus (random or grapelike clusters)	
		micrococcus (square groups of four cells)	
Rod-shaped	bacillus (pl., bacilli)	streptobacillus (chains)	
Spiral	spirillum (pl., spirilla)	sarcina (cubical packets of eight cells)	
Incomplete spiral	vibrio (pl., vibrios)		
Irregular or variable shape	pleomorphic		

Figure 7.4 Simple Staining Procedure.

(a) Staining loop — Sink or suitable receptacle — Stain — Staining bottle

(b) Water — Wash bottle

(c) Gentle blotting

Figure 7.3 A Typical Slide Warmer Used to Speed Up the Drying of Slides.

of water. Spread this out as above. (Three slides should be prepared; one each of *B. subtilis* or *C. pseudodiphtheriticum, M. luteus,* and *S. volutans.*)

3. Allow the slide to air dry, or place it on a slide warmer (figure 7.3).

4. Pass the slide through a Bunsen burner flame three times to heat-fix and kill the bacteria.

Simple Staining

1. Place the three fixed smears on a staining loop or rack over a sink or other suitable receptacle (figure 7.4*a*).

2. Stain one slide with alkaline methylene blue for 1 to 1½ minutes; one slide with carbolfuchsin for 5 to 10 seconds; and one slide with crystal violet for 20 to 30 seconds.

3. Wash stain off slide with water for a few seconds (figure 7.4*b*).

4. Blot slide dry with bibulous paper (figure 7.4*c*). Be careful not to rub the smear when drying the slide because this will remove the stained bacteria.

5. Examine under the oil immersion lens and complete the report for exercise 7.

6. You may want to treat smears of the same bacterium with all three stains in order to compare them more directly. It is also instructive to cover bacterial smears for varying lengths of time with a given stain in order to get a feel for how reactive they are and the results of overstaining or understaining a slide preparation. See figure 7.5 *a–c* for examples of bacteria stained with crystal violet.

Figure 7.5 **Bacteria Stained with Crystal Violet.** (**a**) *Bacillus subtilis* (×1,000). (**b**) *Spirillum volutans* (×1,000). (**c**) *Micrococcus luteus* (×1,000).

(a)

(b)

(c)

HINTS AND PRECAUTIONS

(**1**) When heat-fixing a smear, always make sure that the smear is on the top of the slide as you pass it through the flame.

(**2**) Bacteria growing on solid media tend to cling to each other and must be dispersed sufficiently by diluting with water. If this is not done, the smear will be too thick and uneven. Be careful not to use too much paste in making the smear. It is easy to ruin your results by using too many bacteria.

(**3**) Always wait until the slide is dry before heat-fixing.

(**4**) Fixing smears with an open flame may create artifacts.

(**5**) The inoculating loop must be relatively cool before inserting it into any broth. If the loop is too hot, it will spatter the broth and suspend bacteria into the air. Always flame the inoculating loop after using it and before setting it down.

(**6**) When rinsing with water, direct the stream of water so that it runs gently over the smear.

Name: _____

Date: _____

Lab Section: _____

Smear Preparation and Simple Staining

1. Complete the following drawings and table for the simple staining procedure.

Drawing of
representative
field

C. pseudodiphtheriticum

B. subtilis

M. luteus

S. volutans

Bacterium				
Magnification				
Stain				
Cell form (shape)				
Cell color				
Background color				
Cell grouping				

2. In summary, what is the major purpose of this exercise?

Review Questions

1. What are the two purposes of heat fixation?

 a.

 b.

2. What is the purpose of simple staining?

3. Why are basic dyes more successful in staining bacteria than acidic dyes?

4. Name three basic stains.

 a.

 b.

 c.

5. Why is time an important factor in simple staining?

6. How would you define a properly prepared bacterial smear?

7. Why should you use an inoculating needle when making smears from solid media? An inoculating loop from liquid media?

Gram Stain

MATERIALS PER STUDENT

18- to 24-hour tryptic soy broth cultures of formalinized (1 ml of concentrated formalin per 10 ml of culture) *Staphyloccus aureus* (ATCC 25923), *Escherichia coli* (ATCC 25922), and a mixture of *S. aureus* and *E. coli*

solutions of crystal violet, Gram's iodine (2 g potassium iodide in 300 ml distilled water plus 1 g iodine crystals), 95% ethanol and/or isopropanol-acetone mixture (3:1 v/v), and safranin

Bismark brown stain (for color-blind students)

clean glass slides

inoculating loop

Bunsen burner

bibulous paper

microscope

lens paper and lens cleaner

immersion oil

Hyphomonas (*Hyphomicrobium*) *neptunium* (ATCC 15444) grown in marine broth (Difco)

slide warmer

staining rack

Bacto Gram Stain Reagents from Difco for the three-step Gram stain

Learning Objectives

Each student should be able to

1. Understand the biochemistry underlying the Gram stain
2. Understand the theoretical basis for differential staining procedures
3. Perform a satisfactory Gram stain
4. Differentiate a mixture of bacteria into gram-positive and gram-negative cells

SUGGESTED READING IN TEXTBOOK

1. Differential Staining, section 2.3; see also figures 2.14 and 2.15.
2. Gram-Positive Cell Walls, section 3.5.
3. Gram-Negative Cell Walls, section 3.5.
4. The Mechanism of Gram Staining, section 3.5.
5. Budding and/or Appendaged Bacteria, section 22.1; see also figures 22.4 and 22.5.

Pronunciation Guide

Escherichia coli (esh-er-I-ke-a KOH-lee)
Hyphomonas (*Hyphomicrobium*) *neptunium* (hi-fo-MO-nas nep-TU-ne-um)
Staphylococcus aureus (staf-il-oh-KOK-us ORE-ee-us)

Why Are the Following Bacteria Used in This Exercise?

The major objective of this exercise is to enable the student to correctly use the Gram stain to differentiate a mixture of bacteria into gram-positive and gram-negative cells. The

classical standards for this differentiation are *Staphylococcus aureus* and *Escherichia coli*. *S. aureus* (L. *aureus*, golden) cells are spherical, 0.5 to 1.0 μm in diameter, occurring singly, in pairs, and in irregular clusters. This bacterium is gram-positive, nonmotile, and nonsporing. *S. aureus* is mainly associated with the skin and mucous membranes of warm-blooded vertebrates but is often isolated from food products, dust, and water. *E. coli* (Gr. *colon*, large intestine) cells are straight rods, 2.0 to 6.0 μm in length, occurring singly or in pairs. This bacterium is gram-negative. *E. coli* occurs as part of the normal flora in the lower part of the intestine of warm-blooded animals. *Hyphomonas* (*Hyphomicrobium*) *neptunium* is a rod-shaped, oval, or bean-shaped cell (1 to 3 μm in length) with a polar prostheca of varying length. This bacterium is gram-negative and provides the student the opportunity to Gram stain a large bacterium that differs in its morphology and reproduction. *H. neptunium* is widely distributed in freshwater, marine, and soil habitats.

Medical Application

Gram staining is the single most useful test in the clinical microbiology laboratory. It is the differential staining procedure most commonly used for the direct examination of specimens and bacterial colonies because it has a broad staining spectrum. The Gram stain is the first differential test run on a bacterial specimen brought into the laboratory for specific identification. The staining spectrum includes almost all bacteria, many fungi, and parasites such as *Trichomonas*, *Strongyloides*, and miscellaneous protozoan cysts. The significant exceptions include *Treponema*, *Mycoplasma*, *Chlamydia*, and *Rickettsia*, which are too small to visualize by light microscopy or lack a cell wall.

PRINCIPLES

Simple staining depends on the fact that bacteria differ chemically from their surroundings and thus can be stained to contrast with their environment. Bacteria also differ from one another chemically and physically and may react differently to a given staining procedure. This is the principle of **differential staining.** Differential staining can distinguish between types of bacteria.

The **Gram stain** (named after Christian Gram, Danish scientist and physician, 1853–1938) is the most useful and widely employed differential stain in bacteriology. It divides bacteria into two groups—**gram negative** and **gram positive.**

The first step in the procedure involves staining with the basic dye crystal violet. This is the **primary stain.** It is followed by treatment with an iodine solution, which functions as a **mordant;** that is, it increases the interaction between the bacterial cell and the dye so that the dye is more tightly bound or the cell is more strongly stained. The smear is then decolorized by washing with an agent such as 95% ethanol or isopropanol-acetone. Gram-positive bacteria retain the crystal violet-iodine complex when washed with the decolorizer, whereas gram-negative bacteria lose their crystal violet-iodine complex and become colorless. Finally, the smear is **counterstained** with a basic dye, different in color than crystal violet. This counterstain is usually safranin. The safranin will stain the colorless, gram-negative bacteria pink but does not alter the dark purple color of the gram-positive bacteria. The end result is that gram-positive bacteria are deep purple in color and gram-negative bacteria are pinkish to red in color (figure 8.1).

The Gram stain does not always yield clear results. Species will differ from one another in regard to the ease with which the crystal violet-iodine complex is removed by ethanol. Gram-positive cultures may often turn gram negative if they get too old. Thus, it is always best to Gram stain young, vigorous cultures rather than older ones. Furthermore, some bacterial species are **gram variable.** That is, some cells in the same culture will be gram positive and some, gram negative. Therefore, one should always be certain to run Gram stains on several cultures under carefully controlled conditions in order to make certain that a given bacterial "strain" is truly gram positive or gram negative.

Indistinct Gram-stain results can be confirmed by a simple test using KOH. Place a drop of 10% KOH on a clean glass slide and mix with a loopful of bacterial paste. Wait 30 seconds, then pull the loop slowly

Figure 8.1 Gram Stain. Light micrograph (×900) of a Gram-stained mixture of gram-positive *Staphylococcus aureus* (purple cocci) and gram-negative *Escherichia coli* (pink rods).

through the suspension and up and away from the slide. A gram-negative organism will produce a mucoid string; a gram-positive organism remains fluid.

In most introductory microbiology laboratories, the bacteria that are used in staining exercises are normally relatively small gram-negative or gram-positive cocci and rods. One usually does not have the opportunity to observe larger bacteria or those with differences in morphology and reproduction. Part of the Gram-staining exercise has been designed to help alleviate this deficiency by introducing you to a less typical bacterium, *Hyphomonas* (*Hyphomicrobium*) *neptunium.*

Hyphomicrobia are widely distributed in freshwater, marine, and soil habitats. Of particular concern in this Gram-stain exercise is the unique morphology and morphogenic cycle (figure 8.2) of these procaryotes.

A small, nonmotile swarmer cell about 0.5 μm in diameter matures into an ovoid cell, measuring 0.5 by 1.0 μm. This cell grows a stalk (hypha) about 0.3 μm wide and about 3.0 μm long. The stalk is just thick enough to be seen under the oil immersion lens, and success in viewing it provides a good test of one's ability to Gram stain correctly and focus the microscope. Through the tip of a growing hypha, a bud is formed, which grows a single flagellum. Completing the cycle, the bud separates from the parent and swims away (to

later differentiate into a stalked cell itself), while the mother cell continues to generate more buds. All morphological forms are gram negative.

Procedure for Traditional Gram-Stain Technique

1. Prepare heat-fixed smears of *E. coli, S. aureus,* and the mixture of *E. coli* and *S. aureus* (*see figure 7.1*).
2. Place the slides on the staining rack.
3. Flood the smears with crystal violet and let stand for 30 seconds (figure 8.3*a*).
4. Rinse with water for 5 seconds (figure 8.3*b*).

Figure 8.3 Gram-stain Procedure.

(a) Crystal violet; 30 seconds

(b) Rinse for 5 seconds

(c) Cover with Gram's iodine for 1 minute

(d) Rinse with water for 5 seconds

(e) Decolorize for 15–30 seconds

(f) Rinse with water for 5 seconds

(g) Counterstain with safranin for about 60–80 seconds

(h) Rinse for 5 seconds

(i) Blot dry with bibulous paper

Figure 8.2 *Hyphomonas (Hyphomicrobium) Neptunium.* Morphological forms of the life cycle: (1) nonmotile swarmer; (2) mature cell; (3) stalked cell with bud; (4) stalked cell with flagellated bud; (5) stalked cell; (6) motile swarmer.

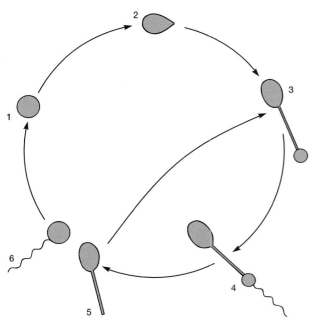

5. Cover with Gram's iodine mordant and let stand for 1 minute (figure 8.3c).

6. Rinse with water for 5 seconds (figure 8.3d).

7. Decolorize with 95% ethanol for 15 to 30 seconds. Do not decolorize too long. Add the decolorizer drop by drop until the crystal violet fails to wash from the slide (figure 8.3e). Alternatively, the smears may be decolorized for 30 to 60 seconds with a mixture of isopropanol-acetone (3:1 v/v).

8. Rinse with water for 5 seconds (figure 8.3f).

9. Counterstain with safranin for about 60 to 80 seconds (figure 8.3g). Safranin preparations vary considerably in strength, and different staining times may be required for each batch of stain. (If you are color-blind, use Bismark brown stain rather than safranin.)

10. Rinse with water for 5 seconds (figure 8.3h).

11. Blot dry with bibulous paper (figure 8.3i) and examine under oil immersion. Gram-positive organisms stain blue to purple; gram-negative organisms stain pink to red. There is no need to place a coverslip on the stained smear. See figure 8.1 for an example of gram-positive and gram-negative bacteria.

Control Procedure

1. Prepare two heat-fixed slides of the mixed culture of *E. coli* and *S. aureus*.

2. Stain one with crystal violet only (steps 3 to 6).

3. Carry the second slide through the decolorizing process (steps 3 to 8).

4. Examine these two slides and compare with the mixed culture slide that was carried all the way through the staining procedure (steps 1 to 10). Your observations should help you understand how the Gram stain works.

Hyphomonas (Hyphomicrobium) Neptunium

1. Gram stain this bacterium according to standard procedures (figure 8.3a–i).

Procedure for Three-Step Gram Stain

Difco Laboratories has introduced reagents for a three-step Gram stain. The advantages include less reagent usage versus conventional stains, reduced chance of overdecolorization, and saved time. The procedure recommended by the company is as follows:

1. Flood smear with Gram crystal violet primary stain and stain for 1 minute.

2. Wash off the crystal violet with cold water.

3. Flood the slide with Gram's iodine mordant and let sit for 1 minute.

4. Wash off the mordant with safranin decolorizer/counterstain solution. Then add more decolorizer/counterstain solution to the slide and stain for 20 to 50 seconds.

5. Wash off the decolorizer/counterstain with cold water.

6. Either blot or air dry.

If the three-step Gram-stain reagents are available, this new procedure may be used in place of the traditional approach.

Regardless of which procedure is used, run known cultures or controls. Smears of known cultures are available commercially (figure 8.4) or can be prepared in the laboratory. It is very important that controls be included in each staining run, preferably on the same

Figure 8.4 Gram Stain Control Slide. Notice the positive control at the top and negative control at the bottom. Each area contains a known gram-positive and gram-negative bacterium. *(Alpha Tec Systems, Inc.)*

slide using *Staphylococcus aureus* (ATTC 25923) and *Escherichia coli* (ATCC 25922). Both of these are also available from Difco as Bactrol™ Disks. When performing the Gram stain on a clinical specimen, particularly when the results will be used as a guide to the selection of a therapeutic agent, such a control system furnishes assurance that the iodine solution is providing proper mordant activity and that decolorization was performed properly.

Laboratory Report 8

Name: _____

Date: _____

Lab Section: _____

1. Draw the Gram-stained bacteria in the following circles.

 S. aureus *E. coli* Mixed culture
 (E. coli + S. aureus)

2. Control Gram-stain results.

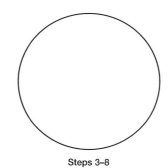

 Steps 3–6 Steps 3–8

Bacterial
color _____ _____

3. Gram stain of *H. neptunium* illustrating the different stages in its life cycle.

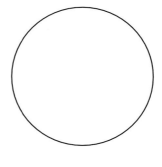

Stage _____ _____ _____

Review Questions

1. What is the difference between a simple and differential stain?

2. Name the reagent used and state the purpose of each of the following in the Gram stain:
 a. mordant
 b. primary stain
 c. decolorizer
 d. counterstain

3. Which step is the most crucial or most likely to cause poor results in the Gram stain? Why?

4. Why must young cultures be used when doing a Gram stain?

5. Why was *H. neptunium* Gram stained?

6. What is meant by gram variable?

7. What part of the bacterial cell is most involved with Gram staining, and why?

Acid-Fast Staining (Ziehl-Neelsen and Kinyoun) Procedure

MATERIALS PER STUDENT

tryptic soy broth culture of *Escherichia coli* (ATCC 11229) and nutrient agar slant culture of *Mycobacterium smegmatis* (ATCC 19420) or *Mycobacterium phlei* (ATCC 354)—5-day-old cultures

Ziehl's carbolfuchsin
carbolfuchsin prepared with either Tergitol No. 4 (a drop per 30 ml of carbolfuchsin) or Triton-X (2 drops per 100 ml of carbolfuchsin). Tergitol No.4 and Triton-X act as detergents, emulsifiers, and wetting agents.

alkaline methylene blue
acid-alcohol
clean glass slides
commercial slides showing acid-fast *Mycobacterium tuberculosis* (Carolina Biological Supply, Wards)
inoculating loop
hot plate
microscope
bibulous paper
paper toweling
lens paper and lens cleaner
immersion oil
staining racks
1-ml pipettes with pipettor

Learning Objectives

Each student should be able to

1. Understand the biochemical basis of the acid-fast stain
2. Perform an acid-fast stain
3. Differentiate bacteria into acid-fast and non-acid-fast groups

SUGGESTED READING IN TEXTBOOK

1. Differential Staining, section 2.3.
2. The Mycobacteria, section 24.5; see also figure 24.9.
3. Tuberculosis, section 39.1.
4. Leprosy, section 39.3.

Pronunciation Guide

Cryptosporidium (krip-toe-spoh-RED-jee-um)
Escherichia coli (esh-er-I-ke-a KOH-lee)
Mycobacterium phlei (mi-ko-bak-TE-re-um fee-ii)
M. smegmatis (M. smeg-MEH-tis)
M. tuberculosis (M. too-ber-ku-LO-sis)
Nocardia (no-KAR-dee-ah)

Why Are the Above Bacteria Used in This Exercise?

One of the major objectives of this exercise is to give the student expertise in acid-fast staining. To allow the student to differentiate between acid-fast and non-acid-fast bacteria,

the authors have chosen one of the cultures from exercise 8, *Escherichia coli. E. coli* is a good example of a non-acid-fast bacterium. *Mycobacterium smegmatis* and *M. phlei* are non-pathogenic members of the genus *Mycobacterium.* These bacteria are straight or slightly curved rods, 1 to 10 μm in length, acid-fast at some stage of growth, and not readily stained by Gram's method. They are also nonmotile, non-sporing, without capsules, and slow or very slow growers. The mycobacteria are widely distributed in soil and water; some species are obligate parasites and pathogens of vertebrates.

 ## Medical Application

In the clinical laboratory, the acid-fast stain is important in identifying bacteria in the genus *Mycobacterium;* specifically, *M. leprae* (leprosy) and *M. tuberculosis* (tuberculosis). This differential stain is also used to identify members of the aerobic actinomycete genus *Nocardia;* specifically, the opportunistic pathogens *N. brasiliensis* and *N. asteroides,* which cause the lung disease nocardiosis. The waterborne protozoan parasite *Cryptosporidium,* which causes diarrhea in humans (cryptosporidiosis), can also be identified by the acid-fast stain.

PRINCIPLES

A few species of bacteria in the genera *Mycobacterium* and *Nocardia,* and the parasite *Cryptosporidium* do not readily stain with simple stains. However, these microorganisms can be stained by heating them with carbolfuchsin. The heat drives the stain into the cells.

Once the microorganisms have taken up the carbol-fuchsin, they are not easily decolorized by acid-alcohol, and hence are termed **acid-fast.** Paul Erlich (*see part nine opener*) developed the acid-fast stain in 1882 while working with the tubercle bacillus *Mycobacterium tuberculosis.* This acid-fastness is due to the high lipid content (mycolic acid) in the cell wall of these microorganisms. The **Ziehl-Neelsen acid-fast staining procedure** (developed by Franz Ziehl, a German bacteriologist, and Friedrich Neelsen, a German pathologist, in the late 1800s) is a very useful differential staining technique that makes use of this difference in retention of carbolfuchsin. Acid-fast microorganisms will retain this dye and appear red (figure 9.1*a, b*). Microorganisms that are not acid-fast, termed **non-acid-fast,** will appear blue or brown due to the counterstaining with methylene blue after they have been decolorized by the acid-alcohol. A modification of this procedure that employs a wetting agent (Tergitol No. 7) rather than heat to ensure stain penetration is known as the **Kinyoun staining procedure** (developed by Joseph Kinyoun, German bacteriologist, in the early 1900s).

Procedure

Ziehl-Neelsen (Hot Stain) Procedure

1. Prepare a smear consisting of a mixture of *E. coli* and *M. smegmatis.*
2. Allow the smear to air dry and then heat-fix (*see figure 7.1*).
3. Place the slide on a hot plate that is within a chemical hood (with the exhaust fan on), and cover the

Figure 9.1 **Ziehl-Neelsen Stain of *Mycobacterium* Acid-fast Rods.** (**a**) *Mycobacterium leprae* stained red (×380). (**b**) In this photomicrograph, *Mycobacterium smegmatis* stains red and the background cells blue-brown.

(a)

(b)

Figure 9.2 Acid-fast Staining Procedure.

(a) Apply carbolfuchsin to saturate paper and heat for 5 minutes in an exhaust hood

(b) Cool and rinse with water for 30 seconds

(c) Decolorize with acid-alcohol until pink (10–30 seconds)

(d) Rinse with water for 5 seconds

(e) Counterstain with methylene blue for about 2 minutes

(f) Rinse with water for 30 seconds

(g) Blot dry with bibulous paper

smear with a piece of paper toweling that has been cut to the same size as the microscope slide. Saturate the paper with Ziehl's carbolfuchsin (figure 9.2a). Heat for 3 to 5 minutes. Do not allow the slide to dry out, and avoid excess flooding! Also, prevent boiling by adjusting the hot plate to a proper temperature. A boiling water bath with a staining rack or loop held 1 to 2 inches above the water surface also works well. (Instead of using a hot plate to heat-drive the carbolfuchsin into the bacteria, an alternate procedure is to cover the heat-fixed slide with a piece of paper towel. Soak the towel with the carbolfuchsin and heat, well above a Bunsen burner flame.)

4. Remove the slide, let it cool, and rinse with water for 30 seconds (figure 9.2b).

5. Decolorize by adding acid-alcohol drop by drop until the slide remains only slightly pink. This requires 10 to 30 seconds and must be done carefully (figure 9.2c).

6. Rinse with water for 5 seconds (figure 9.2d).

7. Counterstain with alkaline methylene blue for about 2 minutes (figure 9.2e).

8. Rinse with water for 30 seconds (figure 9.2f).

9. Blot dry with bibulous paper (figure 9.2g).

10. There is no need to place a coverslip on the stained smear. Examine the slide under oil immersion and record your results in the report for exercise 9. Acid-fast organisms stain red; the background and other organisms stain blue or brown. See figure 9.1 for an example of the Ziehl-Neelsen stain.

11. Examine the prepared slide of *Mycobacterium tuberculosis*.

Kinyoun (Cold Stain) Procedure

(This may be used instead of or in addition to the Ziehl-Neelsen procedure.)

1. Heat-fix the slide as previously directed.
2. Flood the slide for 5 minutes with carbolfuchsin prepared with Tergitol No. 4 (heat is not necessary).
3. Decolorize with acid-alcohol and wash with tap water. Repeat this step until no more color runs off the slide.
4. Counterstain with alkaline methylene blue for 2 minutes. Wash and blot dry.
5. Examine under oil. Acid-fast organisms stain red; the background and other organisms stain blue.

HINTS AND PRECAUTIONS

(1) Light (diaphragm and condenser adjustments) is critical in the ability to distinguish acid-fast-stained microorganisms in sputum or other viscous background materials.

(2) If the bacteria are not adhering to the slide, mix the bacteria with sheep serum or egg albumin during smear preparation. This will help the bacteria adhere to the slide.

(3) As in the Gram stain, not all acid-fast microorganisms are consistent in their reactions. Generally, young cultures of microorganisms may not be as acid-fast as older ones since they have not yet accumulated much lipid.

Name: _____

Date: _____

Lab Section: _____

Acid-Fast Staining (Ziehl-Neelsen and Kinyoun) Procedures

1. Complete the following table with respect to the acid-fast stain and draw representative specimens.

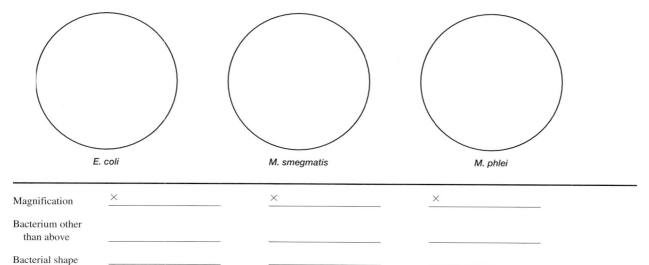

	E. coli	M. smegmatis	M. phlei
Magnification	× _____	× _____	× _____
Bacterium other than above	_____	_____	_____
Bacterial shape	_____	_____	_____
Cell color	_____	_____	_____
Acid-fast	_____	_____	_____

2. Are you satisfied with your results? _____ If not, what can you do to improve your technique the next time you prepare an acid-fast stain from a broth culture?

3. In summary, what is the major purpose of this exercise?

Review Questions

1. What is the purpose of the heat during the acid-fast staining procedure?

2. What is the function of the counterstain in the acid-fast staining procedure?

3. Are acid-fast bacteria gram positive or gram negative? Explain your answer.

4. For what diseases would you use an acid-fast stain?

5. What makes a microorganism non-acid-fast?

6. What chemical is responsible for the acid-fast property of mycobacteria?

7. Is a Gram stain an adequate substitute for an acid-fast stain? Why or why not?

Endospore Staining (Schaeffer-Fulton or Wirtz-Conklin)

MATERIALS PER STUDENT

24- to 48-hour nutrient agar slant cultures of *Bacillus megaterium* (ATCC 12872) and *Bacillus macerans* (ATCC 8244), and old (more than 48 hours) thioglycollate cultures of *Clostridium butyricum* (ATCC 19398) and *Bacillus circulans* (ATCC 4513)
clean glass slides
microscope
immersion oil
wax pencil
inoculating loop
hot plate or boiling water bath with staining rack or loop
5% malachite green solution
safranin
bibulous paper
paper toweling
lens paper and lens cleaner
slide warmer
forceps

Learning Objectives

Each student should be able to

1. Understand the biochemistry underlying endospore staining
2. Perform an endospore stain
3. Differentiate between bacterial endospore and vegetative cell forms

SUGGESTED READING IN TEXTBOOK

1. Staining Specific Structures, section 2.3.
2. The Bacterial Endospore, section 3.8; see also figures 3.40–3.44, 23.5, 23.6, 23.8.
3. Anthrax, section 39.3.
4. Tetanus, section 39.3.

Pronunciation Guide

Bacillus megaterium (bah-SIL-us meg-AH-ter-ee-um)
B. macerans (ma-ser-ANS)
B. circulans (sir-KOO-lanz)
Clostridium butyricum (klos-STRID-ee-um bu-TER-a-cum)

Why Are the Above Bacteria Used in This Exercise?

Because the major objective of this exercise is to provide experience in endospore staining, the authors have chosen several bacteria that vary in the size and shape of their endospores. *Bacillus megaterium* (M.L. n. *megaterium*, big beast) is a cylindrical to oval or pear-shaped cell about 1.2 to 1.5 μm in diameter and 2 to 5 μm long; it tends to occur in short, twisted chains. The spores are central and vary from short oval to elongate. Spores occur in the soil. *Bacillus macerans* (L. *macerans*, softening by steeping, rotting) is an elongated cell 0.5 to 0.7 μm wide and 2.5 to 5 μm in length with terminal spores. Spores are relatively scarce in the soil. *Bacillus circulans* (L. *circulans*, circling) is an elongate cell 2 to 5 μm in length and 0.5 to 0.7 μm wide. In most strains, the spore is terminal to subterminal; it is central in a spindle-shaped sporangium if the bacillus is short. In many strains, deeply stainable material persists on the surface of the free spores. The spores are found in the soil. *Clostridium butyricum* (Gr. *butyrum*, butter) is a straight or slightly curved rod, 2.4 to 7.6 μm in length and 0.5 to 1.7 μm wide, with

rounded ends. The cells occur singly, in pairs, in short chains, and occasionally as long filaments. They are motile with peritrichous flagella. Spores are oval and eccentric to subterminal and are found in the soil and animal feces.

Medical Application

Only a few bacteria produce endospores. Those of medical importance include *Bacillus anthracis* (anthrax), *Clostridium tetani* (tetanus), *C. botulinium* (botulism), and *C. perfringens* (gas gangrene). In the clinical laboratory, the location and size of endospores vary with the species; thus, they are often of value in identifying bacteria.

PRINCIPLES

Bacteria in genera such as *Bacillus* and *Clostridium* produce quite a resistant structure capable of surviving for long periods in an unfavorable environment and then giving rise to a new bacterial cell (figure 10.1). This structure is called an **endospore** since it develops within the bacterial cell. Endospores are spherical to elliptical in shape and may be either smaller or larger than the parent bacterial cell. Endospore position within the cell is characteristic and may be central, subterminal, or terminal.

Endospores do not stain easily, but, once stained, they strongly resist decolorization. This property is the basis of the **Schaeffer-Fulton** (Alice B. Schaeffer and

MacDonald Fulton were microbiologists at Middlebury College, Vermont, in the 1930s) or **Wirtz-Conklin method** (Robert Wirtz and Marie E. Conklin were bacteriologists in the early 1900s) of staining endospores. The endospores are stained with malachite green. Heat is used to provide stain penetration. The rest of the cell is then decolorized and counterstained a light red with safranin.

Procedure

1. With a wax pencil, place the names of the respective bacteria on the edge of four clean glass slides.
2. As shown in figure 14.3, aseptically transfer one species of bacterium with an inoculating loop to each of the respective slides, air dry (or use a slide warmer), and heat-fix.
3. Place the slide to be stained on a hot plate or boiling water bath equipped with a staining loop or rack. Cover the smear with paper toweling that has been cut the same size as the microscope slide.
4. Soak the paper with the malachite green staining solution. Gently heat on the hot plate (just until the stain steams) for 5 to 6 minutes after the malachite green solution begins to steam. Replace the malachite green solution as it evaporates so that the paper remains saturated during heating (figure 10.2*a*). Do not allow the slide to become dry.

Figure 10.1 **The Life Cycle of Endospore-forming Bacteria.**

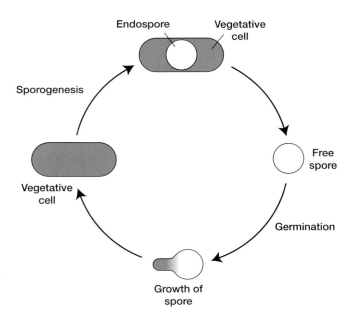

Figure 10.2 **Endospore Staining Procedure.**

(a) Apply malachite green to saturate paper and steam for 5 minutes

(b) Remove paper, cool, and rinse with water for 30 seconds

(c) Counterstain with safranin for 60–90 seconds

(d) Rinse with water for 30 seconds

(e) Blot dry with bibulous paper

5. Remove the paper using forceps, allow the slide to cool, and rinse the slide with water for 30 seconds (figure 10.2b).
6. Counterstain with safranin for 60 to 90 seconds (figure 10.2c).
7. Rinse the slide with water for 30 seconds (figure 10.2d).
8. Blot dry with bibulous paper (figure 10.2e) and examine under oil immersion. A coverslip is not necessary. The spores, both endospores and free spores, stain green; vegetative cells stain red. Draw the bacteria in the space provided in the report for exercise 10. See figure 10.3a–c for an example of endospore staining.

Figure 10.3 Examples of Endospores. (a) Central spores of *Bacillus* stained with malachite green and counterstained with safranin (\times1,000). Notice that the cells are rod shaped and straight, often arranged in pairs or chains, with rounded squared ends. The endospores are oval and not more than one spore per cell. **(b)** *Clostridium tetani* showing round, terminal spores that usually distend the cell (\times1,000). Notice that the cells are rod shaped and are often arranged in pairs or short chains with rounded or sometimes pointed ends. **(c)** *Bacillus megaterium* showing short oval to elongate spores. (b: *Arthur M. Siegelman/Visuals Unlimited*)

(a)

(b)

(c)

Endospore Staining (Schaeffer-Fulton or Wirtz-Conklin)

1. Make drawings and answer the questions for each of the bacterial endospore slides.

	B. megaterium	*B. macerans*	*B. circulans*	*C. butyricum*
Bacterium	_____	_____	_____	_____
Magnification	✕ _____	✕ _____	✕ _____	✕ _____
Bacterium other than above	_____	_____	_____	_____
Spore color	_____	_____	_____	_____
Color of vegetative cell	_____	_____	_____	_____
Location of endospore (central, terminal, subterminal)	_____	_____	_____	_____

2. Are you satisfied with the results of your endospore stain? _____ If not, how can you improve your results the next time you prepare an endospore stain?

3. In summary, what is the major purpose of this exercise?

Review Questions

1. Why is heat necessary in order to stain endospores?

2. Where are endospores located within vegetative cells?

3. In the Schaeffer-Fulton endospore stain, what is the primary stain? The counterstain?

4. Name two disease-causing bacteria that produce endospores.
 a.
 b.

5. What is the function of an endospore?

6. Why are endospores so difficult to stain?

7. What do endospore stains have in common with the acid-fast (Ziehl-Neelsen) stain?

Capsule Staining

MATERIALS PER STUDENT

18-hour skim milk cultures of *Klebsiella pneumoniae* (ATCC e13883) and *Alcaligenes denitrificans* (ATCC 15173)
Tyler's crystal violet (1% aqueous solution) or Gram's crystal violet (1% aqueous solution)
20% (w/v) solution of copper sulfate ($CuSO_4 \cdot 5H_2O$)
microscope
immersion oil
lens paper and lens cleaner
clean glass slides
wax pencil
bibulous paper
inoculating loop
Bon Ami
70% ethyl alcohol
India ink (Higgins no. 4465 black or Pelikan Drawing ink No. 17 black for technical pens) or SpotTest India ink ampules from Difco
safranin stain

Learning Objectives

Each student should be able to

1. Understand the biochemistry of the capsule stain
2. Perform a capsule stain
3. Distinguish capsular material from the bacterial cell

SUGGESTED READING IN TEXTBOOK

1. Capsules, Slime Layers, and S Layers, section 3.6; see also figure 3.27.

Pronunciation Guide

Alcaligenes denitrificans (al-kah-LIJ-e-neez de-ni-tri-fi-KANS)
Klebsiella pneumoniae (kleb-se-EL-lah nu-MO-ne-EYE)

 Why Are the Above Bacteria Used in This Exercise?

One of the major objectives of this exercise is to give the student experience in capsule staining. To help accomplish this objective, the authors have chosen one capsulated and one noncapsulated bacterium. *Klebsiella pneumoniae* (Gr. *pneumonia,* pneumonia) is a nonmotile, capsulated rod, 0.6 to 6 μm in length, and is arranged singly, in pairs, or in short chains. Cells contain a large polysaccharide capsule and give rise to large mucoid colonies. There are more than 80 capsular (K) antigens that can be used to serotype klebsiellae. *K. pneumoniae* occurs in human feces and clinical specimens, water, grain, fruits, and vegetables. *Alcaligenes denitrificans* (are able to reduce NO_3- to NO_2- and N_2) occurs as a rod, a coccal rod, or a coccus; is 0.5 to 2.6 μm in length; and usually occurs singly in water and soil. It is motile with 1 to 4 peritrichous flagella. No capsule is present.

 Medical Application

Many bacteria (e.g., *Bacillus anthracis* [anthrax], *Streptococcus mutans* [tooth decay], *Streptococcus pneumoniae* [pneumonia]) and the fungus *Cryptococcus neoformans* [cryptococcosis] contain a gelatinous covering called a capsule. In the clinical laboratory, demonstrating the presence of

a capsule is a means of diagnosis and determining the organism's virulence, the degree to which a pathogen can cause disease.

PRINCIPLES

Many bacteria have a slimy layer surrounding them, which is usually referred to as a **capsule** (figure 11.1*a*). The capsule's composition, as well as its thickness, varies with individual bacterial species. Polysaccharides, polypeptides, and glycoproteins have all been found in capsules. Often, a pathogenic bacterium with a thick capsule will be more virulent than a strain with little or no capsule since the capsule protects the bacterium against the phagocytic activity of the host's phagocytic cells. However, one cannot always determine if a capsule is present by simple staining procedures, such as using negative staining and India ink. An unstained area around a bacterial cell may be due to the separation of the cell from the surrounding stain upon drying. Two convenient procedures for determining the presence of a capsule are Anthony's (E. E. Anthony, Jr., a bacteriologist at the University of Texas, Austin, in the 1930s) capsule staining method (figure 11.1*b*) and the Graham and Evans (Florence L. Evans, a bacteriologist at the University of Illinois in the 1930s) procedure.

Anthony's procedure employs two reagents. The primary stain is crystal violet, which gives the bacterial cell and its capsular material a dark purple color. Un-like the cell, the capsule is nonionic and the primary stain cannot adhere. Copper sulfate is the decolorizing agent. It removes excess primary stain as well as color from the capsule. At the same time, the copper sulfate acts as a counterstain by being absorbed into the capsule and turning it a light blue or pink. In this procedure, smears should not be heat-fixed since shrinkage is likely to occur and create a clear zone around the bacterium, which can be mistaken for a capsule.

Procedure: Capsule Staining (Anthony's)

1. With a wax pencil, label the left-hand corner of a clean glass slide with the name of the bacterium that will be stained.
2. As shown in figure 14.3, aseptically transfer a loopful of culture with an inoculating loop to the slide. Allow the slide to air dry. Do not heat-fix! Heat-fixing can cause the bacterial cells to shrink and give a false appearance to the capsule.
3. Place the slide on a staining rack. Flood the slide with crystal violet and let stand for 4 to 7 minutes (figure 11.2*a*).
4. Rinse the slide thoroughly with 20% copper sulfate (figure 11.2*b*).
5. Blot dry with bibulous paper (figure 11.2*c*).
6. Examine under oil immersion (a coverslip is not necessary) and draw the respective bacteria in the report for exercise 11. Capsules appear as faint halos around dark cells.

Modified Capsule Stain (Graham and Evans)

1. Thoroughly clean the slide to be used with a household cleanser (such as Bon Ami) and alcohol.

Figure 11.1 Anthony's Capsule Staining Method. (**a**) Drawing of a single bacterium, capsule, and background material. (**b**) *Klebsiella pneumoniae* capsules; light micrograph (×1,000). Capsules appear as white halos around red backgrounds.

(a)

(b)

Figure 11.2 Capsule Staining Procedure.

(a) Flood the slide with crystal violet; let stand 4–7 minutes

(b) Rinse thoroughly with copper sulfate

(c) Blot dry with bibulous paper

2. Mix two loopfuls of culture with a small amount (1 to 2 drops) of India ink at one end of the slide.
3. Spread out the drop using a second slide in the same way one prepares a thin smear (*see figure 6.2*).
4. Dry the smear.
5. GENTLY rinse with distilled water so that the bacteria do not wash off the slide.
6. Stain for 1 minute with Gram's crystal violet.
7. Rinse again with water.
8. Stain for 1 1/2 minutes with safranin stain.
9. Rinse with water and blot dry.
10. If a capsule is present, the pink to red bacteria are surrounded by a clear zone. The background is dark.

HINTS AND PRECAUTIONS

(1) As with any materials stained with similar or identical colors, light adjustments under the microscope will be critical for optimal visualization of capsules.

(2) Be sure to use a very small amount of India ink or the capsules will not be clearly visible.

Laboratory Report *11*

Name: _____

Date: _____

Lab Section: _____

Capsule Staining

1. Fill in the table and make drawings of a representative field of each preparation as seen with the oil immersion lens.

K. pneumoniae *A. denitrificans*

Magnification	× _____	× _____
Bacterium other than above	_____	_____
Capsule size (in μm)	_____	_____
Capsule color	_____	_____
Vegetative cell color	_____	_____

2. Are you satisfied with the results of your capsule stain? _____ If not, how can you improve your results the next time you do a capsule stain?

3. In summary, what is the major purpose of this exercise?

Review Questions

1. What three chemical substances have been identified in bacterial capsules?

 a.

 b.

 c.

2. What is the relationship between the presence of capsules and bacterial pathogenicity?

3. What is the dual function of copper sulfate in capsule staining?

4. In staining bacterial capsules, why is heat-fixing omitted?

5. How is the capsule stain used in clinical microbiology?

6. Name several bacteria that have capsules.

7. Of what value is a capsule to a bacterium?

Flagella Staining: West and Difco's SpotTest Methods

SAFETY CONSIDERATIONS

Be careful with the boiling water bath and Bunsen burner flame. No mouth pipetting. West stain solutions A and B and Difco's SpotTest Flagella stain are irritants. Do not breathe vapors or get solutions on your skin. Always prepare solutions A and B while using an exhaust hood with the fan on. Also, always use the hoods when applying heat to the staining solutions. Discard slides in a container with disinfectant.

MATERIALS PER STUDENT

young, 18-hour tryptic soy agar slants of
 Alcaligenes faecalis (ATCC 8750,
 peritrichously flagellated) and *Pseudomonas
 fluorescens* (ATCC 13525, polarly flagellated)
wax pencil
inoculating loop
acid-cleaned glass slides with frosted ends
clean distilled water
microscope
immersion oil
lens paper and lens cleaner
boiling water bath (250-ml beaker with distilled
 water, ring stand, wire gauze pad, and Bunsen
 burner or hot plate)
Pasteur pipettes with pipettor
West stain
 solution A
 solution B
Difco's SpotTest Flagella stain

Learning Objectives

Each student should be able to

1. Understand the biochemical basis of flagella staining
2. Perform a flagella stain
3. Differentiate between two types of flagellar arrangement

SUGGESTED READING IN TEXTBOOK

1. Staining Specific Structures, section 2.3.
2. Flagella and Motility, section 3.6.
3. Flagella Ultrastructure, section 3.6.
4. Flagellar Synthesis, section 3.6.
5. The Mechanism of Flagellar Movement, section 3.6; see also figures 3.31–3.37.

Pronunciation Guide

Alcaligenes faecalis (al-kah-LIJ-en-eez fee-KAL-iss)
Pseudomonas fluorescens (soo-do-MO-nas floor-ES-sens)

Why Are the Above Bacteria Used in This Exercise?

After this exercise the student should be able to correctly stain bacteria to determine the presence of flagella and their arrangement. The authors have chosen two easily maintained bacteria that have different flagellar arrangements. *Alcaligenes faecalis* (M.L. adj. *faecalis,* fecal) cells exist as cocci or coccal rods 0.5 to 2.6 μm in length, usually occurring singly. Motility is with 1 to 8 (occasionally up to 12) peritrichous flagella. *A. faecalis* normally occurs in water and soil. *Pseudomonas fluorescens* (M.L. v., *fluoresco,* fluoresce) cells are rods 2.0 to 2.8 μm in length and 0.7 to 0.8 μm in width. They occur singly and in pairs. Motility is with one to several polar flagella. *P. fluorescens* is widely distributed in nature.

Medical Application

In the clinical laboratory, the presence, number, and arrangement of flagella are useful in identifying bacterial species. Important pathogens that are motile due to the presence of flagella include *Bordetella pertussis* (whooping

cough), *Listeria monocytogenes* (meningoencephalitis), *Proteus vulgaris* (urinary tract infections, bacteremia, pneumonia), *Pseudomonas aeruginosa* (skin and wound infections), *Salmonella typhi* (typhus or typhoid fever), and *Vibrio cholerae* (cholera).

PRINCIPLES

Bacterial **flagella** are fine, threadlike organelles of locomotion. They are slender (about 10 to 30 nm in diameter) and can be seen directly using only the electron microscope. In order to observe them with the light microscope, the thickness of the flagella is increased by coating them with mordants such as tannic acid and potassium alum, and staining them with basic fuchsin (Gray method), pararosaniline (Leifson method), silver nitrate (West method; named after Marcia West, a clinical microbiologist), or crystal violet (Difco's method). Although flagella staining procedures are difficult to carry out, they often provide information about the presence and location of flagella, which is of great value in bacterial identification (figure 12.1).

Difco's SpotTest Flagella stain employs an alcoholic solution of crystal violet as the primary stain, and

tannic acid and aluminum potassium sulfate as mordants. As the alcohol evaporates during the staining procedure, the crystal violet forms a precipitate around the flagella, thereby increasing their apparent size.

Procedure (West)

1. With a wax pencil, mark the left-hand corner of a clean glass slide with the name of the bacterium.
2. As shown in figure 14.3, aseptically transfer the bacterium with an inoculating loop from the turbid liquid at the bottom of the slant to 3 small drops of distilled water in the center of a clean slide that has been carefully wiped off with clean lens paper. Gently spread the diluted bacterial suspension over a 3-cm area, using the inoculating needle (figure 12.2*a*).
3. Let the slide air dry for 15 minutes (figure 12.2*b*).
4. Cover the dry smear with solution A (the mordant) for 4 minutes (figure 12.2*c*).
5. Rinse thoroughly with distilled water (figure 12.2*d*).
6. Place a piece of paper toweling on the smear and soak it with solution B (the stain). Heat the slide in a boiling water bath for 5 minutes in an exhaust hood with the fan on. Add more stain to keep the slide from drying out (figure 12.2*e*).
7. Remove the toweling and rinse off excess solution B with distilled water. Flood the slide with distilled water and allow it to sit for 1 minute while more silver nitrate residue floats to the surface (figure 12.2*f*).
8. Then, rinse gently with water once more and carefully shake excess water off the slide (figure 12.2*g*).
9. Allow the slide to air dry at room temperature (figure 12.2*h*).
10. Examine the slide with the oil immersion objective. The best specimens will probably be seen at the edge of the smear where bacteria are less dense (figure 12.3*a* and *b*). Record your results in the report for exercise 12.

Figure 12.1 Arrangement of Flagella on Bacterial Cells. (a) In monotrichous, polar, a single flagellum is located at one end of the cell; **(b)** in lophotrichous, polar, many flagella are grouped at one end of the cell; **(c)** in amphitrichous, polar, a single flagellum is located at both ends of the cell; and **(d)** in peritrichous, flagella are located all around the cell.

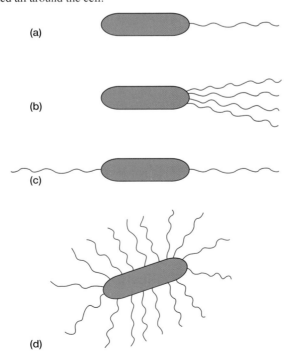

(a)

(b)

(c)

(d)

Procedure (Difco)

1. Draw a border around the clear portion of a frosted microscope slide with a wax pencil.
2. Place a drop of distilled water on the slide, approximately 1 cm from the frosted edge.
3. Gently touch a colony of the culture being tested with an inoculating loop and then lightly touch the drop of water without touching the slide. Do not mix.
4. Tilt the slide at a slight angle to allow the drop preparation to flow to the opposite end of the slide.

Figure 12.2 Flagella Staining Procedure (West).

(a) Place bacteria in 3 drops distilled water and spread out

(b) Air dry for 15 minutes

(c) Cover smear with mordant for 4 minutes

(d) Rinse thoroughly with distilled water

Boiling water bath

(e) Place paper towel over smear and soak with stain; heat for 5 minutes

(f) Flood slide with distilled water and allow to sit for 1 minute

(g) Shake excess water from slide

(h) Air dry at room temperature

Figure 12.3 West Silver-plating Staining. Examples of several patterns of flagellation as seen with the light microscope. (a) *Alcaligenes faecalis* with peritrichous flagella, West silver-plating staining. (b) *Pseudomonas fluorescens* with polar monotrichous flagella, West silver-plating staining (×1,000). *(a: Reprinted by permission of American Society for Microbiology; b: © E.S. Chan/Visuals Unlimited)*

(a)

(b)

5. Let the slide air dry at room temperature. Do not heat-fix.

6. Flood the slide with the contents of the Difco SpotTest Flagella stain ampule.

7. Allow the stain to remain on the slide for approximately 4 minutes. (Note: The staining time may need to be adjusted from 2 to 8 minutes depending on the age of the culture, the age of the stain, the temperature, and the depth of staining solution over the culture.)

8. Carefully rinse the stain by adding water from a faucet or wash bottle to the slide while it remains on the staining rack. Do not tip the slide before this is done.

9. After rinsing, gently tilt the slide to allow excess water to run off and let the slide air dry at room temperature or place on a slide warmer.

10. Examine the slide microscopically with the oil immersion objective. Begin examination at thinner areas of the preparation and move toward the center. Look for fields that contain several isolated bacteria, rather than fields that contain clumps of many bacteria. Bacteria and their flagella should stain purple.

HINTS AND PRECAUTIONS

(1) Do not vortex the cultures, and be gentle when making smears to avoid detaching the flagella.

(2) New tryptic soy agar slants must be used to prepare cultures so that there is still liquid at the bottom of the slant.

(3) Perform all steps as gently as possible; rough handling of bacteria could cause flagella to be broken off and lost.

Name: _____

Date: _____

Lab Section: _____

Flagella Staining: West and Difco's SpotTest Methods

1. Make a drawing of a representative microscope field of each preparation and fill in the table.

A. faecalis

P. fluorescens

	A. faecalis	P. fluorescens
Magnification	× _____	× _____
Bacterium other than above	_____	_____
Color of flagella	_____	_____
Color of vegetative cell	_____	_____
Arrangement of flagella (be specific)	_____	_____
Number of flagella	_____	_____

2. Are you satisfied with the results of your flagella stain? _____ If not, how can you improve your results the next time you do a flagella stain?

3. Make sketches that illustrate each of the following flagellar arrangements:

 Monotrichous Lophotrichous

 Amphitrichous Peritrichous

Review Questions

1. Why are flagella so difficult to stain?

2. Why did you use a young culture for the flagella stain?

3. Why must the glass slide be free of grease and oil before staining for flagella?

4. Name four types of flagella arrangements.

 a.

 b.

 c.

 d.

5. What is a mordant?

6. Compare the usefulness of a flagella stain with that of the hanging drop preparation.

7. What happens to the size of flagella when they are stained?

Basic Laboratory and Culture Techniques

What is education but a process by which a person begins to learn how to learn?

Peter Alexander Ustinov
(English actor and entertainer, 1921 –)

In this part of the manual you will be introduced to the basic techniques employed in microbiology laboratories to prepare and sterilize culture media, to isolate bacteria in pure culture from various types of specimens, and to subculture bacteria and fungi in the laboratory. You will also learn how to determine the number of bacteria in a given culture.

Along with parts one and two of this manual, part three continues your acquisition of basic microbiological techniques. A thorough understanding and applicability of microscopic, slide, and culture techniques are an important foundation on which the rest of this manual is built.

After completing the exercises in part three, you will be able to (1) demonstrate the proper use of aseptic techniques. This will meet the American Society for Microbiology Core Curriculum skill number 3 *(see pp. vi–viii):* **(a) sterilizing and maintaining sterility of transfer instruments; (b) performing aseptic transfers; and (c) obtaining microbial samples. (2) You will also be able to estimate the number of microorganisms in a sample using serial dilution techniques. This will meet the American Society for Microbiology Core Curriculum skill number 5 *(see pp. vi–viii):* (a) correctly choosing and using pipettes and pipetting devices; (b) correctly spreading diluted samples for counting; (c) estimating appropriate dilutions; and (d) extrapolating plate counts to obtain correct CFU or PFU in the starting sample.**

Alfred Theodore MacConkey (1861–1931).

Alfred MacConkey was an English bacteriologist who described the first and most famous solid differential medium—MacConkey's agar.

The first and most famous solid differential medium was described by MacConkey in 1900 in *The Lancet.* The medium is used for the detection of coliforms and enteric pathogens based on their ability to ferment lactose. Lactose-fermenting bacteria appear as red to pink colonies. Lactose nonfermenting bacteria appear as colorless or transparent colonies.

The story of the development of this medium is of interest. The first formula was for a bile salt litmus medium that contained glycocholate, lactose, and litmus, and that was to be incubated at 22°C. A year later MacConkey published a new formulation of the medium in the *Zentralblatt fur Bakteriologie.* The modified medium contained taurocholate in place of glycocholate and was to be incubated at 42°C. Finally, in 1905, MacConkey published a third formula for his medium in which neutral red was substituted for litmus as an indicator. It is this last formula that we use today for making MacConkey's medium, which is as follows:

MacConkey's Agar (pH 7.1)

Bacto peptone	17.0 g
Proteose peptone	3.0 g
Lactose	10.0 g
Bile salts mixture	1.5 g
Sodium Chloride	5.0 g
Agar	13.5 g
Neutral red	0.03 g
Crystal violet	0.001 g
Distilled water	1,000.0 ml

Microbiological Culture Media Preparation and Sterilization

MATERIALS PER GROUP OF STUDENTS

24- to 48-hour tryptic soy broth culture of *Escherichia coli* (ATCC 11229)
autoclave for the entire class
petri plates
culture tubes
test-tube rack or wire basket
test-tube caps
defined culture medium as in table 13.1
complex culture medium as in table 13.2
2-liter Erlenmeyer flask
10-ml pipettes with pipettor
weighing paper or boats
balance
agar
heat-proof Zetex fabric gloves
water bath set at 48° to 50°C
boiling water bath
aluminum foil
stirring rod or plate with stirring bar
Bunsen burner or hot plate
Difco Manual or *BBL Manual*

Learning Objectives

Each student should be able to

1. Describe the different types of culture media and their composition, and give several examples of what each is used for.
2. Describe the various ways culture tubes are capped.
3. Describe how to prepare and transfer culture media.
4. Prepare defined and undefined media, and prepare agar plates.
5. Describe the concept of sterility.
6. Describe how various media, supplies, and equipment can be sterilized.
7. Correctly and safely use the autoclave.

SUGGESTED READING IN TEXTBOOK

1. Definition of Frequently Used Terms, section 7.1.
2. The Use of Physical Methods in Control, section 7.4; see also figures 7.3–7.5.
3. Filtration, section 7.4.
4. Sterilizing Gases, section 7.5.
5. Culture Media, section 5.7.
6. Synthetic or Defined Media, section 5.7 and table 5.4.
7. Complex Media, section 5.7 and table 5.5.
8. Types of Media, section 5.7; see also table 36.1.

Pronunciation Guide

Escherichia coli (esh-er-I-ke-a KOH-lee)

Why Is the Above Bacterium Used in This Exercise?

One of the major objectives of this exercise is to prepare defined and complex media. After the media have been prepared, they can be inoculated with *Escherichia coli*. *E. coli* is facultatively anaerobic and chemoorganoheterotrophic, having both a respira-

tory and a fermentative type of metabolism. Glucose and other carbohydrates are catabolized with the formation of acid and gas.

PRINCIPLES

Microbiological Culture Media

The survival and growth of microorganisms depend on available nutrients and a favorable growth environment. In the laboratory, the nutrient preparations that are used for culturing microorganisms are called **media** (singular, **medium**). Three physical forms are used: **liquid, or broth, media; semisolid media;** and **solid media.** The major difference among these media is that solid and semisolid media contain a solidifying agent (usually **agar**), whereas a liquid medium does not. Liquid media, such as nutrient broth, tryptic soy broth, or brain-heart infusion broth (figure 13.1*a*), can be used to propagate large numbers of microorganisms in fermentation studies and for various biochemical tests. Semisolid media can also be used in fermentation studies, in determining bacterial motility, and in promoting anaerobic growth. Solid media, such as nutrient agar or blood agar, are used (1) for the surface growth of microorganisms in order to observe colony appearance, (2) for pure culture isolations, (3) for storage of cultures, and (4) to observe specific biochemical reactions.

While in the liquefied state, solid media can be poured into either a test tube or petri plate (dish). If the medium in the test tube is allowed to harden in a slanted position, the tube is designated an **agar slant** (figure 13.1*b, c*); if the tube is allowed to harden in an upright position, the tube is designated an **agar deep tube** (figure 13.1*d*); and if the agar is poured into a petri plate, the plate is designated an **agar plate** (figure 13.1*e*). **Agar pours** (the same as Agar deeps) containing about 15 to 16 ml of media are often used to prepare agar plates.

Table 13.1
A Chemically Defined Medium

Ingredient	Quantity
Dipotassium phosphate K_2HPO_4	7 g
Potassium phosphate, monobasic, KH_2PO_4	2 g
Hydrated magnesium sulfate, $MgSO_4 \cdot 7H_2O$	0.2 g
Ammonium sulfate, $(NH_4)_2SO_4$	1 g
Glucose	5 g
Distilled water	1 liter

Table 13.2
A Complex (Undefined) Medium—Tryptic Soy Broth

Ingredient	Quantity
Casein peptone	17 g
Soybean peptone	3 g
NaCl	5 g
Dipotassium phosphate K_2HPO_4	2.5 g
Glucose	2.5 g
Distilled water	1 liter

Microorganisms may be cultured using two different types of media. **Chemically defined, or synthetic, media** are composed of known amounts of pure chemicals (table 13.1). Such media are often used in culturing autotrophic microorganisms such as algae or nonfastidious heterotrophs. In routine bacteriology laboratory exercises, **complex, or nonsynthetic, media** are employed (table 13.2). These are composed of complex materials that are rich in vitamins and nutrients. Three of the most commonly used components are beef extract, yeast extract, and peptones. Commercial sources of media and the composition of culture media used in this manual are given in appendix J.

Figure 13.1 **Culture Media.** Different forms of culture media with the proper volume in each.

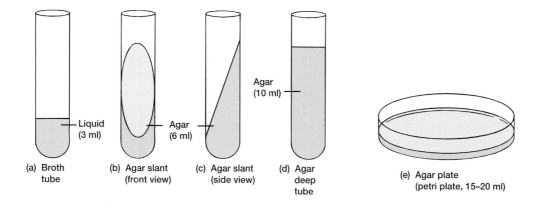

(a) Broth tube — Liquid (3 ml)
(b) Agar slant (front view) — Agar (6 ml)
(c) Agar slant (side view)
(d) Agar deep tube — Agar (10 ml)
(e) Agar plate (petri plate, 15–20 ml)

The preparation of media from commercial dehydrated products is simple and straightforward. Each bottle of dehydrated medium has instructions for preparation on its label. For example, to prepare a liter of tryptic soy broth, suspend 30 g of the dehydrated medium in 1,000 ml of distilled water. Mix thoroughly in a 2-liter Erlenmeyer flask (always use a flask that holds twice the volume of media you are preparing). Dispense and sterilize for 15 to 20 minutes at 121°C (15 lbs pressure). As noted, the amount of powder for 1,000 ml of water will be indicated.

If the medium lacks agar, the powder will usually dissolve without heating. If it contains agar, you must heat the medium until it starts to simmer or boil in order to completely dissolve the agar. Specific heating instructions are given for each type of medium. For example, to prepare a liter of Vogel-Johnson agar, suspend 61 g of the dehydrated medium in a liter of distilled water. Mix until a uniform suspension is obtained. Heat with constant agitation and simmer for 1 minute. Dispense in 100-ml amounts into 250-ml flasks and sterilize by autoclaving at 121°C for 20 minutes.

Most of the exercises you will be doing in this manual will involve the use of sterile media in culture tubes. Normally, 18×150 mm, 16×125 mm, or 13×100 mm bacteriologic culture tubes will be used. These tubes must be capped in order to maintain media sterility. This can be accomplished by using cotton plugs, plastic foam plugs, or plastic or metal caps (e.g., Morton closures or Bacti Capalls). All of these caps keep cultures free from contamination while allowing air into the culture tube and, at the same time, minimizing evaporation. It is sometimes desirable to use screw-cap culture tubes. This is especially true when the culture, such as in the case of slants, may be sealed and stored for long periods.

Culture broth can be dispensed with a pipetting machine, an automatic syringe, or a regular pipette. One can also pipette the proper volume (figure 13.1) of broth or agar into one culture tube and then pour approximately the same volume of media (using the initial tube as a guide) into a number of other tubes lined up in the same test-tube rack. This approach is fast, convenient, and relatively accurate.

After sterilization of slant tubes, the tubes are removed from the autoclave while the agar is still melted and carefully laid on a table with a piece of wood, vacuum tubing, or metal elevating the capped ends. Some test-tube racks are also specifically set up for this. The tubes are then allowed to remain undisturbed until the agar has cooled and hardened. Slants should be stored in a vertical position.

Agar deep tubes can be stored after sterilization for use in preparing petri plates when needed. Some agar

Figure 13.2 Pouring Agar Plates.

(a) Boiling water bath

(b) 48° to 50°C water bath

Thermometer

Test tube rack

(c) Wipe tube with paper towel

(d) Flame the top of the tube after removing cap

(e) Pour the agar into the petri dish bottom

(f) After cooling, store in an inverted position

Agar

deeps may be stored at room temperature for several days before use. If longer periods of storage are required, they should be placed in the refrigerator in order to prevent drying of the agar. When petri plates are needed, the agar deeps are melted either in a boiling water bath (figure 13.2a) or by bringing them to 121°C in an autoclave for 30 to 60 seconds and then releasing the steam under slow exhaust. After the agar has melted, the pours are transferred to a 48° to 50°C water bath and kept there for at least 5 to 10 minutes before use (figure 13.2b). The agar deeps should be cooled to about 50°C before they are used to minimize the amount of steam condensation on the petri plate lids after the agar has been poured. Agar does not solidify until its temperature drops to about 42°C. When the deeps have reached 50°C, one is taken from the bath and the outside is dried with a paper towel (figure 13.2c). Its cap is removed and the top is briefly flamed using a Bunsen burner (figure 13.2d). The agar is immediately

poured into a sterile, dry petri plate while holding the top carefully above the petri plate bottom in order to avoid contamination (figure 13.2e). Replace the top, allow the agar to cool and harden, and store the petri plates in an inverted position (figure 13.2f). When storing petri plates, do not stack them more than three high, or use a special petri plate storage holder.

Sterilization of Media and Equipment

Sterilization is the process of rendering a medium or material free of all forms of life. There are three basic ways in which sterilization of media and supplies can be achieved. The most useful approach is **autoclaving,** in which items are sterilized by exposure to steam at 121°C and 15 lbs of pressure for 15 minutes or longer, depending on the nature of the item. Under these conditions, microorganisms, even endospores, will not survive longer than about 12 to 13 minutes. This method is rapid and dependable. Modern autoclaves are designed to ensure that all of the air has been expelled and only steam is present in the autoclave chamber. They are carefully temperature controlled as well. Almost all media and anything else that will resist 121°C temperatures and steam can be sterilized in this way.

Often, dry glassware such as pipettes and petri plates must be sterilized. Steam tends to etch glassware and also leaves it damp. Therefore, such items are generally **dry-heat sterilized.** The glassware is placed in an electric oven set to operate between 160° and 170°C. Since dry heat is not as effective as wet heat, the glassware must be kept at this temperature for about 2 hours or longer. The oven temperature must not rise above 180°C or any cotton or paper present will char.

Sometimes media must be made from components that will not withstand heating at 121°C. Such a medium can be sterilized by passing it through a **bacteriological filter,** which physically removes bacteria and larger microorganisms from the solution and thereby sterilizes

them without heat. Scintered glass filters with ultrafine, fritted disks (0.9 to 1.4 μm pore size) and Seitz asbestos-pad filter funnels (3 mm thick with 0.1 μm pores) are both quite effective in sterilizing solutions. However, if pore sizes greater than 0.22 μm are used, there is an exceedingly high chance that the filtrate will not be sterile. By far, the most useful and popular approach is the use of specially prepared sterile, cellulose- or polycarbonate-based membranes of the appropriate pore size. Generally, membranes with 0.22 μm pores are employed in sterilization. A large number of different devices are commercially available for membrane sterilization of both large and small volumes. For example, one can use a filter flask with a vacuum or syringe with positive pressure to force liquid through a special membrane filter holder (figure 13.3).

Two other sterilization techniques use ultraviolet radiation and ethylene oxide. **Ultraviolet (UV) radiation** around 260 nm is quite lethal to many microorganisms but does not penetrate glass, dirt films, water, and other substances very effectively. Because of this disadvantage, UV is used as a sterilizing agent only in a few particular situations. For example, UV lamps are sometimes placed on the ceilings of rooms or in biological safety cabinets to sterilize the air and any exposed surfaces.

Many heat-sensitive items such as disposable plastic petri dishes and syringes, sutures, and catheters are now sterilized with **ethylene oxide** gas. Ethylene oxide is both microbicidal and sporicidal and kills by covalently attaching to cell proteins. It is a particularly effective sterilizing agent because it rapidly penetrates packing materials, even plastic wraps.

Procedure

Preparing a Chemically Defined Medium

1. Prepare 500 ml of glucose-mineral salts broth for the culture of *E. coli* using the recipe outlined in

Figure 13.3 Filter Apparatuses Used in Sterilizing Small Volumes of Media.

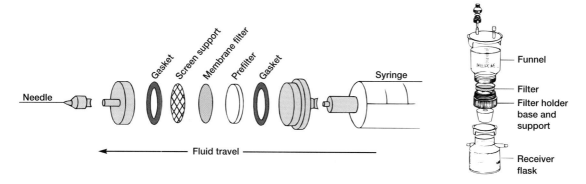

table 13.1. To a 1-liter Erlenmeyer flask add 375 ml of distilled water. Weigh out and add each ingredient to the water in the order listed and stir after each addition until the ingredient is completely dissolved. Remember to halve the quantity of each ingredient. Add the remaining 125 ml of water to wash the inside of the flask.

2. Adjust the pH to 7.2 to 7.4 by adding just enough HCl or NaOH dropwise (*see appendix E*).
3. Dispense 3 to 5 ml of the glucose-mineral salts broth into each of 10 test tubes using a 10-ml pipette and then loosely cap the tubes. Other students can use the remaining 450 ml of broth for their tubes. Place your tubes in a test-tube rack or basket. Place the basket or rack in the autoclave.

Preparing a Complex Medium

1. Prepare 500 ml of tryptic soy broth according to the recipe outlined in table 13.2.
2. Add 375 ml of distilled water to a 1-liter Erlenmeyer flask and add the ingredients individually (use half the amounts given); mix after each addition.
3. Add the remaining 125 ml of water to rinse the sides of the flask.
4. Adjust the pH to 7.3 by adding just enough HCl or NaOH dropwise (*see appendix E*).
5. Dispense 3 to 5 ml of the broth into each of 10 tubes and loosely cap them. Place the tubes in a test-tube rack or basket, and place in the autoclave.
6. To the remaining broth (450 ml), add 7.2 g of agar to give an agar concentration of approximately 1.6%. Heat the contents of the flask and gradually bring to a boil. Heat the agar until it is completely melted. Cover with aluminum foil and place in the autoclave. After autoclaving, cool the flask of sterile agar in a 48° to 50°C water bath. Line up the desired number of sterile petri plates on the bench top. Remove the aluminum foil cap from the flask and briefly flame the flask's neck. Lift the top of each plate, pour about 15 ml of agar, and quickly replace the top (the agar should be approximately 3 to 5 mm deep in the plate). Pour all plates without stopping.

Alternatively, after dissolving the agar medium, dispense 15-ml portions into 18 × 150 mm tubes; cap and autoclave the tubes. Cool them in a 48° to 50°C water bath and pour the agar as shown in figure 13.2.

Procedure for Autoclaving

1. Your instructor will demonstrate the use of the autoclave.
2. Load the autoclave with the freshly prepared culture media.
3. Close and lock the autoclave door.
4. Set the autoclave time for 15 minutes or longer and select a slow rate of exhaust.
5. Make certain that the autoclave temperature is set to 121°C.
6. Start the autoclave by pushing the start button or twisting the knob to the start position.
7. When the period of sterilization is completed and the pressure in the chamber reads 0, carefully open the door and remove the containers, using heat-proof gloves.

Preparing Agar Plates

1. As outlined previously and in figure 13.2*a–f,* use some of the sterilized tryptic soy agar to prepare agar plates.
2. When the plates are cool (agar solidified), invert them to prevent condensing moisture from accumulating on the agar surfaces.
3. All plates and tubes should be incubated for at least 24 hours to ensure sterility before you use them.

HINTS AND PRECAUTIONS

(1) Don't overload the autoclave chamber. Provide ample space between baskets of media to allow circulation of steam.

(2) You should bring media to a boil and then, using heat-proof gloves, quickly remove the media from the Bunsen burner or hot plate to prevent boiling over. Do not shake or swirl the flask as you remove it from the heat because such shaking may cause the media to boil over.

(3) Before opening the door to the autoclave, you should always wear heat-proof gloves, stand at arm's length, and slowly open the door. This will prevent two problems from occurring:

(a) the trapped steam will dissipate toward the ceiling in a controlled fashion without burning the skin, and

(b) the media will not boil out of the stoppered containers because of a too rapid change in internal pressure in the flask.

Name: _____

Date: _____

Lab Section: _____

Microbiological Culture Media Preparation and Sterilization

1. After at least 24 hours of incubation, do your prepared plates and broths appear to be sterile? Explain your answer.

2. List the steps you would go through to make tryptic soy agar slants.

 a. _____

 b. _____

 c. _____

 d. _____

 e. _____

 f. _____

3. List the steps you would go through to make Vogel–Johnson agar plates (refer to the *Difco Manual* or the *BBL Manual*).

 a. _____

 b. _____

 c. _____

 d. _____

 e. _____

 f. _____

4. Provide the requested information using a *Difco Manual* or *BBL Manual.*

 a. Quantity of starch in Mueller–Hinton agar _____

 b. Quantity of lactose in Levine eosin methylene blue agar _____

 c. Percentage of sodium chloride in mannitol salt agar _____

 d. Percentage of agar in MacConkey agar _____

 e. Quantity of beef extract in nutrient broth _____

5. Using a *Difco Manual* or *BBL Manual,* find the following:
 a. The purpose of Endo agar.

 b. The way in which bismuth sulfite agar selects for *Salmonella* and *Shigella* species.

 c. The use of Sabouraud dextrose agar and the role of pH in this selectivity.

 d. The roles of thioglycollate and methylene blue in thioglycollate medium.

 e. The function of autoclaving in the preparation of Salmonella-Shigella agar.

Review Questions

1. What are the three main types (in terms of their physical forms) of microbiological culture media?

 _____ _____ _____

2. Define culture medium, defined or synthetic medium, and complex or nonsynthetic medium.

3. Why are culture media sterilized before use?

4. Describe three ways for sterilizing culture media and supplies.
 a.
 b.
 c.

5. Why are petri plates inverted after they cool?

6. Why is culture medium cooled to about 48° to 50°C before it is poured into petri plates?

7. What is a buffer? What is the buffer system used in this exercise?

8. What is the source of carbon in the chemically defined medium in table 13.1? The source of nitrogen?

Exercise **14**

Culture Transfer Instruments, Techniques, and Isolation and Maintenance of Pure Cultures

SAFETY CONSIDERATIONS

To fill a pipette, always use a bulb or other mechanical device. **Never** use your mouth to pipette. **Never** lay a pipette on the bench top; use a pipette rack. If the pipette is contaminated, immediately place it in the proper container. Always carry the pipette can (or bag) to your work area before removing a pipette. Do not carry sterile pipettes through the lab. Be careful of the Bunsen burner flames and red-hot inoculating loops and needles.

MATERIALS PER STUDENT

vortex mixer (if available)
inoculating loop
inoculating needle
Bunsen burner
blow-out pipette with pipettor
to-deliver pipette
24-hour tryptic soy broth and tryptic soy agar
 slant cultures of either *Serratia marcescens,*
 pigmented (ATCC e13880) or *Micrococcus*
 roseus (ATCC 186) or streak-plate cultures
 from exercise 16
tryptic soy broth tubes
tryptic soy agar slants
tryptic soy agar deeps
wax pencil

Learning Objectives

Each student should be able to

1. Correctly use a blow-out pipette
2. Correctly use a to-deliver pipette
3. Correctly use an inoculating loop and needle
4. Use aseptic technique to remove and transfer bacteria for subculturing
5. Understand the reasoning behind pure culture preparations

6. Prepare a stock culture using the isolates from exercise 16
7. Describe how bacterial cultures can be maintained

SUGGESTED READING IN TEXTBOOK

1. Box 5.1.

Pronunciation Guide

Azotobacter (ah-ZO-to-bak-ter)
Bacillus (bah-SIL-lus)
Leuconostoc (loo-ko-NOS-tok)
Micrococcus (mi-kro-KOK-us)
Proteus (PRO-tee-us)
Pseudomonas aeruginosa (soo-do-MO-nas a-ruh-jin-
 OH-sah)
Serratia marcescens (se-RA-she-ah mar-SES-sens)
Streptococcus (strep-to-KOK-us)

Why Is the Above Bacterium Used in This Exercise?

Serratia marcescens (L. *marcescens,* fading away) is found in water, soil, and food, and is widely distributed in nature. It is easy to maintain in the laboratory, facultatively anaerobic, chemoorganotrophic, and has both a respiratory and a fermentative type of metabolism. It grows very quickly in tryptic soy broth or agar and often produces the red pigment prodigiosin when grown between 20° and 35°C.

PRINCIPLES FOR CULTURE TRANSFER INSTRUMENTS AND TECHNIQUES

A **pipette** is an instrument often used to transfer aliquots of culture, to prepare serial dilutions of microorganisms, and to dispense chemical reagents. Two types of measuring

85

Figure 14.1 Pipetting. (a) A blow-out (serological) pipette. **(b)** A to-deliver or Mohr pipette. **(c)** A plastic pump. The pump is attached to the pipette and the wheel turned to move fluid either into or out of the pipette. **(d)** A pipette bulb. While pressing the *A* valve, squeeze the bulb, *B*, and it will collapse. To draw fluid into the pipette, press the *S* valve; to release fluid, press the *E* valve. **(e)** Electronically powered pipetting device for dispensing 1 to 150 ml volumes. Just push buttons to fill and empty pipettes. **(f)** A safety bulb pipette filler. This rubber bulb with conical silicone attachment fits small pipettes through the largest volumetrics. Just squeeze the bulb and touch to pipette to form a seal. Let suction draw fluid, then lift bulb away and control the fluid level with index finger. **(g)** The volume in a pipette is read at the lowest level of the meniscus. (E. from Kleyn, Bickwell, Gilstrap Microbiology Experiments, 2nd ed. WCB McGraw-Hill. 1999. Fig. 6.2, page 54.)

pipettes are frequently used: the **blow-out pipette** (also called a **serological pipette**) and the **to-deliver pipette** (**Mohr measuring pipette,** named after Francis Mohr, American pharmaceutical chemist, nineteenth century) (figure 14.1*a,b*). With the blow-out pipette, the final few drops of liquid must be emptied in order to deliver the correct volume. With the to-deliver pipette, after the proper amount of liquid has been delivered, liquid will remain in the tip of the pipette and should not be eliminated.

To fill a pipette, use a bulb or other mechanical device (figure 14.1*c,d,e,f*). DO NOT USE YOUR MOUTH. Draw the desired amount of fluid into the pipette. The volume is read at the bottom of the meniscus (figure 14.1*g*).

Often the mouth end of a pipette is carefully plugged with a small piece of cotton before sterilization. This prevents cross-contamination of the bulb or mechanical device of the pipette.

Contamination problems are avoided by storing sterile pipettes in a pipette can. Clean, sterile pipettes are placed tip first in the pipette can. The bottom of the can should contain a wad of paper or cotton to protect

the pipette tips from breakage. The top of the can is then put on. If the top of the can sticks while it is being put on or taken off, a twisting motion will often unstick or free it. After the pipettes have been loaded into the cans, they can be autoclaved and dried by setting the autoclave on the "fast exhaust and dry" cycle. Pipettes can also be sterilized in a dry-heat oven.

To correctly use the pipette, hold the pipette can in a horizontal position and carefully remove the top with a twisting motion. The top should always be handled in such a way that its open end is pointing down. Remove the pipette. After removal, replace the top of the can while keeping the can horizontal at all times. Do not put the pipette down before it is used or it can no longer be considered sterile. After a pipette has been used, it should immediately be placed tip down in a container of a disinfectant such as 3 to 5% Lysol and completely immersed.

Inoculating needles and **loops** (figure 14.2*a,b*) are used to aseptically transfer microorganisms from broth, slant, or agar cultures to other media. Both may consist

Figure 14.2 **Microbiological Transfer Instruments.** (**a**) Inoculating needle, and (**b**) inoculating loop.

of handles, a shaft, and a turret, which holds a nickel-chromium or platinum wire. If the wire is straight, it is an inoculating needle; if a loop is present, it is an inoculating loop. Before using either, the end of the wire must be sterilized by passing it slowly through the tip of the flame from a Bunsen burner or into a Bacti—Cinerator (see figure 14.4). When done correctly, all parts of the wire will turn red with heat. The needle or loop should then be used before it becomes contaminated. After you have finished using an inoculating loop or needle, it should be thoroughly flame-sterilized.

Microorganisms are transferred from one culture medium to another by **subculturing,** using specific procedures and **aseptic technique. (Asepsis** means free from sepsis [a toxic condition resulting from the presence of microorganisms].) This aseptic technique is of such importance that it will be used in most of the exercises in this manual. Since microorganisms are always present in the laboratory, if aseptic technique is not followed, there is a good possibility that external contamination will result and will interfere with the results. Proper aseptic technique also protects the laboratory worker from contamination with the culture.

PRINCIPLES FOR ISOLATION OF PURE CULTURES AND THEIR MAINTENANCE

Once discrete, well-separated colonies develop on the surface of the streak plate, selected ones may be picked up with an inoculating needle and transferred to sepa-rate culture tubes, such as tryptic soy agar slants (the type of agar will depend on the microorganism). Where possible, bacteria from the center of a colony are transferred, because the center is less likely to be contaminated than the edges. Each slant now represents the growth of a single species of microorganism and is called a **pure** or **stock culture.**

One of the more important problems in a microbiology laboratory is the maintenance of pure stock cultures over a long period. Ideally, one should employ a technique that minimizes the need for **subculturing** the microorganism. This is achieved by storing the microorganism in a state of dormancy either by **refrigeration** or **desiccation.**

Short-term maintenance (generally one to three months) of aerobic bacteria can often be achieved by storing slant cultures in the refrigerator at 4° to 10°C. The use of screw-cap tubes for these slants will minimize desiccation during storage.

One way in which many cultures may be maintained for relatively long periods is by sealing them with sterile mineral oil in order to prevent moisture loss. The white mineral oil used can be sterilized by heating at 110°C for 1 hour in a drying oven. After an agar slant culture has grown, the slant surface is aseptically covered with the sterile oil. The mineral oil surface should be about 1/4 inch above the top of the slant. The oil-covered slant is then stored at the normal storage temperature. A number of genera may be stored under oil (e.g., *Bacillus,* most *Enterobacteriaceae,* some species of *Micrococcus, Proteus, Pseudomonas,* and *Streptococcus*). There are genera that may not be stored successfully under oil (e.g., *Azotobacter* and *Leuconostoc*). Table 14.1 summarizes maintenance conditions for a few representative bacteria.

In many cases, long-term maintenance of cultures does not even require mineral oil. *E. coli* and many other members of the family *Enterobacteriaceae, Pseudomonas aeruginosa,* staphylococci, and enterococci can often be successfully stored for years at room temperature with the following procedure. Stab inoculate screw-cap deeps containing either half-strength nutrient agar or 0.7% agar in distilled water. Incubate overnight at optimal temperature. Finally, screw down the caps tightly and seal the tubes with tape or paraffin. Store the cultures in a safe place at room temperature.

The best way to preserve many stock cultures for long periods is through **lyophilization (freeze-drying).** This eliminates the need for periodic transfers and reduces the chance of mutations occurring in the stock culture. In lyophilization, the bacterial culture is suspended in a sterile solution of some protective medium such as milk, serum, or 3% lactose. Small amounts of the thick

Table 14.1

Maintenance of Bacteria

Bacterium	Maintenance Media*	Storage Temperature (°C)	Storage Time (Months)
Aerobacter	1,2	4–10	2
Alcaligenes	1,2	4–10	3
Bacillus	1,2	4–10	12
Clostridium	3,4	25	12
Escherichia	1,2	4–10	3
Lactobacillus	4	25	1
Leuconostoc	4	25	1
Neisseria (saprophytic)	2,5	25	1
Proteus	1,2	4–10	3
Pseudomonas	1,2	4–10	3
Salmonella	1,2	4–10	3
Serratia	1,2	4–10	3
Staphylococcus	1,2,4,5	4–10	3
Streptococcus	3,4,5	25	1–3

*Maintenance media employed: (1) nutrient agar, (2) tryptic soy agar, (3) cooked meat medium, (4) thioglycollate medium with $CaCO_3$, and (5) CTA medium (BBL).

suspension are transferred to vials and then quickly frozen in a dry-ice/alcohol mixture. The frozen suspension is finally dried under vacuum while still frozen, and the vial sealed. These sealed, desiccated cultures may often be stored for years. Strict anaerobes and some facultative anaerobes will be injured by exposure to O_2. They can often be maintained as agar stab cultures. In this procedure, one allows a tube of the desired agar to solidify in an upright position and then inoculates it by thrusting an inoculation needle coated with bacteria into the center of the agar. The anaerobes will grow deep within the agar in the anaerobic environment it provides. After suitable growth, the stab may be refrigerated. Anaerobes can also be maintained in thioglycollate broth or cooked meat medium as described in exercise 20.

Commercial sources of cultures and more information on stock culture maintenance are given in appendix J.

Procedure for Culture Transfer Instruments and Techniques

Pipetting

1. Proper pipetting using both to-deliver and blow-out pipettes will be demonstrated in the laboratory by the instructor. After the demonstration, practice using both pipettes with some distilled water and the bulbs or mechanical devices provided.

Aseptic Technique

1. Using a wax pencil, label the tube or plate to be inoculated with the date, your name, and the name of the test microorganism (figure 14.3a).
2. Gently mix the primary culture tube in order to put the bacteria into a uniform suspension (figure 14.3b).

The tube can be tapped to create a vortex that will suspend the microorganisms, or if a vortex mixer is available, it can be used.

3. Place the stock culture tube and the tube to be inoculated in the palm of one hand and secure with the thumb. The tubes are then separated to form a V in the hand (figure 14.3c). They should be held at an angle so that the open ends are not vertical and directly exposed to airborne laboratory contaminants.
4. Using the other hand, flame the inoculating loop or needle over a Bunsen burner until the wire becomes red-hot (figure 14.3d) or in a Bacti–Cinerator (see figure 14.4).
5. Using the same hand that is holding the inoculating loop, remove the caps from the two tubes, hold them between your fingers, and briefly flame the necks of the tubes over a Bunsen burner (figure 14.3e) by passing them through the flame. However, DO NOT ALLOW THE TUBES TO BECOME RED-HOT.
6. Cool the hot loop in the broth culture until it stops "hissing." With the sterile inoculating loop, transfer 1 drop of culture from the stock culture tube into the new broth tube. At this point, one could also transfer to a glass slide, streak the surface of a slant, or streak the bacteria onto the surface of a petri plate (figure 14.3f). When picking up bacteria from a slant, cool the hot loop or needle by holding it against the top of the slant until it stops "hissing."
7. Reflame the neck of the tubes (figure 14.3g).
8. Recap the tubes (figure 14.3h).
9. Reflame or sterilize the loop or needle (figure 14.3i).

Figure 14.3 Aseptic Technique for Bacterial Removal and Subculturing.

(a) With a wax pencil, label the medium to be inoculated

(b) Shake the primary culture tube to suspend the bacteria

(c) Place both tubes in the palm of one hand to form a V

(d) Flame the inoculating loop or needle along full length

(e) Remove the caps from the tubes and flame the necks of the tubes. Do not place the caps on the lab bench

(f) Cool the loop or needle and pick up bacteria

or Streak the surface of a slant

or Place the bacteria on slide

or Streak the bacteria on petri plate

(g) Reflame the neck of the tubes

(h) Recap the tubes

(i) Reflame the loop or needle

Figure 14.4 A Bacti-Cinerator Sterilizer. This oven sterilizes needles, loops, and culture tube mouths in 5 to 7 seconds at optimum sterilizing temperature of 871°C (1600°F). This oven also eliminates microorganism spattering associated with flame sterilization. It consists of a ceramic funnel tube enclosed in a stainless-steel perforated guard and casting support stand.

Using aseptic technique, perform the following transfers:

a. With the inoculating loop, transfer the *S. marcescens* tryptic soy broth culture to a tryptic soy agar slant. Also, inoculate a tryptic soy broth tube with *S. marcescens,* using the inoculating loop.

b. With the inoculating needle, transfer the *S. marcescens* to a tryptic soy agar deep tube. This is done by plunging the inoculating needle of *S. marcescens* into the tube without touching the walls of the tube. Penetrate the medium to two-thirds of its depth. The inoculating needle is then withdrawn from the tube (figure 14.5*a–c*).

c. Using the inoculating loop, make a slant-to-slant transfer. This is done by gently streaking the surface of the slant in the form of a serpentine (wiggly or S-shaped) line (figure 14.5*d*). If there is liquid at the base of the slant, the tube may be tilted after inoculation so that the liquid runs over the slant surface. This will moisten the surface and spread out the bacteria.

Figure 14.5 Transferring Techniques. (a)–(c) Stab technique for transferring bacteria. Notice that the inoculating needle is moved into the tube without touching the walls of the tube, and the needle penetrates the medium to two-thirds of its depth. **(d)** Technique for streaking the surface of a slant with a loop.

Inoculating needle

Inoculating loop

(a) (b) (c) (d)

d. Place the tubes in a test-tube rack or a clean vegetable can and incubate at 35°C for 24 to 48 hours. Afterwards, examine all of the tubes for the presence of bacterial growth. Growth is indicated by turbidity (cloudiness) in the broth culture, and the appearance of an orange-to-red growth on the slant culture and along the line of inoculation in the agar deep tube. Also note if any contamination is present. This is indicated by growth that is not red to orange in color. Record your results in the report for exercise 14.

Procedure for Isolation of Pure Cultures and Their Maintenance

1. With a wax pencil, label the tryptic soy agar slants with the names of the respective bacteria. Do the same for the broth tubes. Add your name and date.

2. Using aseptic technique, select a well-isolated colony for each of the three bacteria and pick off as much of the center of the colony as possible with an inoculating loop. It may be necessary to obtain material from more than one colony. Prepare a slant culture and a tryptic soy broth tube for each of the bacteria. If screw-cap tubes are used, they must be loosened slightly before incubation to keep the slant aerobic.

3. After incubating 24 to 48 hours, you should have three pure slant and three pure broth stock cultures.

4. Observe the broth cultures (figure 14.6) while taking care not to agitate them. Record your observations in the report for exercise 14.

5. Place the pure cultures in the refrigerator for later use.

HINTS AND PRECAUTIONS

(1) Consider the material contained within the pipette contaminated if it is drawn up in the pipette until the liquid touches the cotton.

(2) Always check the loop size to see that it is approximately 3 mm in diameter, because a significantly larger or smaller loop often fails to hold liquids properly during transfer.

(3) When pipetting, always position your eyes so that they are horizontal with the top of the fluid column in the pipette. This avoids parallax (an apparent displacement of position of an object due to change in the observer's position) errors that can occur from misalignment of the meniscus with the graduated line on the pipette. Hold the pipette vertical and use your forefinger to control the flow. Remember to always use a pipetting aid to fill the pipette and do not pipette by mouth.

(4) Media containing fermentable carbohydrates should be avoided for the maintenance of cultures.

(5) Selective media should never be used.

(6) Cultures should not be allowed to dry out; tightly closed screw-cap tubes should be used for storage.

(7) Be sure to flame and cool needles between all inoculations to avoid incidental cross-contamination of cultures.

Figure 14.6 Some Typical Growth Patterns in Broth Media.

Growth turbid and diffuse throughout

Growth layered at surface only

Growth sedimented at bottom only

Growth layered below surface; none beneath center

Growth forms puff balls, layered below surface

Culture Transfer Instruments, Techniques, and Isolation and Maintenance of Pure Cultures

Type of Culture	Growth (+ or −)	Contamination (+ or −)
Tryptic soy agar deep	_____	_____
Tryptic soy agar slant	_____	_____
Tryptic soy broth	_____	_____

1. Examine the pure stock cultures for bacterial distribution and color of growth. Record your results by drawing exactly what you observed and completing the table.

B. subtilis

S. marcescens

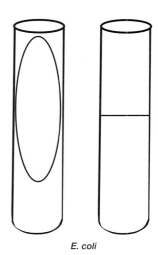

E. coli

Unique features _____ _____ _____

_____ _____ _____

_____ _____ _____

Review Questions

1. Describe how to use the two most common types of pipettes.

2. What is the purpose of flaming in the aseptic technique?

3. What is the purpose of subculturing?

4. In subculturing, when do you use the inoculating loop?

5. How is it possible to contaminate a subculture?

6. How would you determine whether culture media given to you by the laboratory instructor are sterile before you use them?

7. What are some signs of growth in a liquid medium?

8. Why did you use an inoculating loop instead of a needle to make the transfers from the culture plates to the culture tubes?

9. How do the pure broth cultures differ? The slant cultures?

10. What is the function of sterile mineral oil in the maintenance of stock cultures?

11. Describe how a culture can be lyophilized.

12. How can some anaerobes be maintained in pure cultures?

13. What are some signs of growth in a liquid medium?

14. In summary, what is the major purpose of this exercise?

Exercise 15

Spread-Plate Technique

MATERIALS PER STUDENT

 24- to 48-hour tryptic soy broth cultures of
 Bacillus subtilis (ATCC 6051, white or cream
 colonies), *Serratia marcescens* (ATCC 13880,
 red colonies) or *Micrococcus roseus* (ATCC
 186, red colonies), and a mixture of the two
 (*S. marcescens* [or *M. roseus*] and *B. subtilis*)
 Bunsen burner
 inoculating loop
 95% ethyl alcohol
 L-shaped glass rod
 wax pencil
 500-ml beaker
 pipettes with pipettor
 3 tryptic soy agar plates
 rulers

Learning Objectives

Each student should be able to

1. Understand the purpose of the spread-plate technique
2. Perform the spread-plate technique

SUGGESTED READING IN TEXTBOOK

1. Isolation of Pure Cultures, section 5.8.
2. The Spread Plate and Streak Plate, section 5.8; see also figures 5.7–5.9, 5.11.
3. Colony Morphology and Growth, section 5.8.

Medical Application

In the clinical laboratory, growth of a pure culture is absolutely necessary before any biochemical tests can be performed to identify a suspect microorganism.

Pronunciation Guide

Bacillus subtilis (bah-SIL-lus sub-til-lus)
Serratia marcescens (se-RA-she-ah mar-SES-sens)

Why Are the Above Bacteria Used in This Exercise?

After this exercise, the student should be able to use the spread-plate technique to separate a mixture of two or more bacteria into well-isolated colonies. The bacteria to be used are *Bacillus subtilis* and *Serratia marcescens* or *M. roseus*. *B. subtilis* is easy to culture since it grows on simple medium (e.g., tryptic soy agar) and produces dull white to cream colonies that are easy to see. *S. marcescens* was used in the last experiment and produces large red, pink, or magenta colonies. By using color and colony morphology, the student can see what a well-isolated colony of each of the above bacteria looks like. The isolated bacteria can then be picked up and streaked onto fresh medium to obtain a pure culture.

PRINCIPLES

In natural habitats, bacteria usually grow together in populations containing a number of species. In order to adequately study and characterize an individual bacterial species, one needs a **pure culture.** The **spread-plate technique** is an easy, direct way of

Figure 15.1 Bacterial Colony Characteristics on Agar Media as Seen with the Naked Eye. The characteristics of bacterial colonies are described using the following terms.

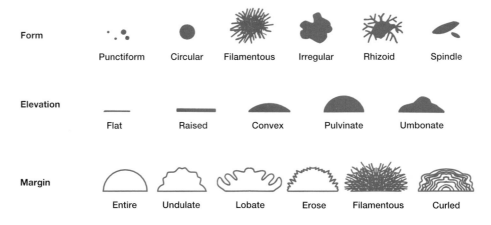

Appearance: Shiny or dull
Optical property: Opaque, translucent, transparent
Pigmentation: Pigmented (purple, red, yellow)
 Nonpigmented (cream, tan, white)
Texture: Rough or smooth

achieving this result. In this technique, a small volume of dilute bacterial mixture containing 100 to 200 cells or less is transferred to the center of an agar plate and is spread evenly over the surface with a sterile, L-shaped glass rod. The glass rod is normally sterilized by dipping it in alcohol and flaming it to burn off the alcohol. After incubation, some of the dispersed cells develop into isolated colonies. A **colony** is a large number of bacterial cells on solid medium, which is visible to the naked eye as a discrete entity. In this procedure, one assumes that a colony is derived from one cell and therefore represents a clone of a pure culture.

After incubation, the general form of the colony and the shape of the edge or margin can be determined by looking down at the top of the colony. The nature of the colony elevation is apparent when viewed from the side as the plate is held at eye level. These variations are illustrated in figure 15.1. After a well-isolated colony has been identified, it can then be picked up and streaked onto a fresh medium to obtain a pure culture.

Procedure

1. With a wax pencil, label the bottom of the agar medium plates with the name of the bacterium to be inoculated, your name, and date. Three plates are to be inoculated: (a) one with *B. subtilis,* (b) one with *S. marcescens,* and (c) one with the mixture.
2. Pipette 0.1 ml of the respective bacterial culture onto the center of a tryptic agar plate (figure 15.2*a*).
3. Dip the L-shaped glass rod into a beaker of ethanol (figure 15.2*b*) and then tap the rod on the side of the beaker to remove any excess ethanol.
4. Briefly pass the ethanol-soaked spreader through the flame to burn off the alcohol (figure 15.2*c*), and allow it to cool inside the lid of a sterile petri plate.
5. Spread the bacterial sample evenly over the agar surface with the sterilized spreader (figure 15.2*d*), making sure the entire surface of the plate has been covered. Also make sure you do not touch the edge of the plate.
6. Immerse the spreader in ethanol, tap on the side of the beaker to remove any excess ethanol, and re-flame.
7. Repeat the procedure to inoculate the remaining two plates.
8. Invert the plates and incubate for 24 to 48 hours at room temperature or 30°C.
9. After incubation, measure some representative colonies and carefully observe their morphology (figure 15.3). Record your results in the report for exercise 15.

Figure 15.2 Spread-Plate Technique.

(a)

(b)

(c)

(d)

Figure 15.3 Spread Plate. Macroscopic photomicrograph of a spread plate. Notice the many well-isolated colonies.

HINTS AND PRECAUTIONS

(1) When flaming the alcohol on the glass rod, touch it to the flame only long enough to ignite the alcohol, then remove it from the flame while the alcohol burns.

(2) Wait 5 to 10 seconds after flaming to allow the alcohol to burn off and to ensure that the glass is cool enough to spread the culture without sizzling. Hold the rod briefly on the surface of the agar to finish cooling. Do not return the flaming rod to the beaker. If you accidentally do this, remove the rod from the beaker and smother the flames with a book by quickly lowering the book on the beaker. Do not pour flaming alcohol into the sink. Do not pour water into the flaming alcohol.

(3) Avoid contamination of the petri plate cover and the culture by not placing the cover upon the table, desk, or other object while spreading. Hold the cover, bottom side down, above the agar surface as much as possible.

(4) Turning the plate while carefully spreading the culture (but not hitting the sides of the plate with the glass rod) will result in a more even separation of the bacteria.

(5) An inoculated plate is always incubated in an inverted position to prevent condensation from falling onto the surface of the plate and interfering with discrete colony formation.

(6) To prevent burns, avoid holding the glass rod so that alcohol runs onto your fingers.

(7) Keep all flammable objects, such as paper, out of reach of ignited alcohol.

Laboratory Report 15

Name: _____

Date: _____

Lab Section: _____

Spread-Plate Technique

1. Make drawings of several well-isolated colonies from each plate and fill in the table.

	B. subtilis	S. marcescens	Mixture
Form	_____	_____	_____
Elevation	_____	_____	_____
Margin	_____	_____	_____
Color of colony(ies)	_____	_____	_____

2. With your ruler, measure the diameter of the average colony appearing on each plate by placing the ruler on the bottom of the plate. Hold the plate and ruler against the light to make your readings and be sure to measure a well-separated colony.

 a. Size of *B. subtilis* colony _____

 b. Size of *S. marcescens* colony _____

3. In summary, what is the major purpose of this exercise?

Review Questions

1. What is a bacterial colony?

2. What is the purpose of the ethanol in the spread-plate technique?

3. Why is it necessary to use only diluted cultures that contain 100 to 200 cells for a successful spread plate?

4. Describe the form of some typical bacterial colonies.

5. What is the purpose of the spread-plate technique?

6. In all routine laboratory work, petri plates are labeled on the bottom. Why?

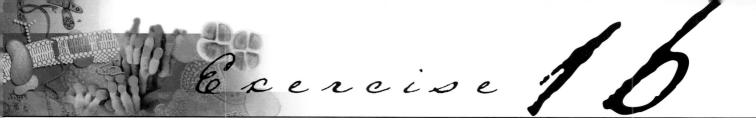
The Streak-Plate Technique and Differential Media

MATERIALS PER STUDENT

(Streak-Plate Technique)

24- to 48-hour tryptic soy broth cultures of *Escherichia coli* (ATCC 11229, white colonies), *Serratia marcescens* (ATCC 13880, red colonies; *Micrococcus roseus* ATCC 186 can also be used), and *Bacillus subtilis* (ATCC 6051, white or cream colonies)

3 tryptic soy agar pours
boiling water bath
48° to 50°C water bath
Bunsen burner
petri plates
inoculating loop
wax pencil

(Differential Media)

24- to 48-hour tryptic soy broth culture of a mixture of *Escherichia coli* (ATCC 11229), *Proteus vulgaris* (ATCC 13315), and *Staphylococcus aureus* (ATCC 25903)
mannitol salt agar pour
eosin methylene blue agar pour

Learning Objectives

Each student should be able to

1. Understand the purpose of the streak-plate technique and differential media
2. Perform a streak-plate technique and isolate discrete colonies for subculturing

SUGGESTED READING IN TEXTBOOK

1. Isolation of Pure Cultures, section 5.8.
2. The Spread Plate and Streak Plate, section 5.8.
3. Colony Morphology and Growth, section 5.8.
4. Differential Media, section 5.7.

Pronunciation Guide

Escherichia coli (esh-er-I-ke-a KOH-lee)
Bacillus subtilis (bah-SIL-lus sub-til-lus)
Serratia marcescens (se-RA-she-ah mar-SES-sens)

Why Are the Following Bacteria Used in This Exercise?

Another procedure that is used to obtain well-isolated, pure colonies is the streak-plate technique. Since *Serratia marcescens, Bacillus subtilis,* and *Escherichia coli* were used in the past few exercises, these same cultures are used in this exercise. Remember, *S. marcescens* produces red colonies; *B. subtilis,* white to cream colonies; and *E. coli,* white colonies. These same cultures will also be used in the next exercise (number 17).

Medical Application

In the clinical laboratory, growth of a pure culture is absolutely necessary before any biochemical tests can be performed to identify a suspect microorganism.

PRINCIPLES OF THE STREAK-PLATE TECHNIQUE

Isolated, pure colonies can also be obtained by the **streak-plate technique.** In this technique, the bacterial mixture is transferred to the edge of an agar plate with an inoculating loop and then streaked out over the surface in one of several patterns. At some point on the streaks, individual cells will be removed from the loop as it glides along the agar surface and will give rise to separate colonies (figure 16.1). Again, one assumes that one colony comes from one cell. The key principle of this method is that by streaking, a dilution gradient is established on the surface of the plate as cells are deposited on the agar surface. Because of this gradient, confluent growth occurs on part of the plate where the cells are not sufficiently separated, and individual, well-isolated colonies develop in other regions of the plate where few enough cells are deposited to form separate colonies that can be seen with the naked eye. Cells from the new colony can then be picked up with an inoculating needle and transferred to an agar slant or other suitable medium for maintenance of the pure culture.

Procedure

1. Melt three sterile, capped tubes of tryptic soy agar by heating them in a boiling water bath until melted (*see figure 13.2*a,b).
2. Cool the tubes in a 48° to 50°C water bath for 10–15 minutes.
3. Remove the cap, flame the lip of the tube, and pour the agar into a petri plate (*see figure 13.2*c–f). Be careful to keep the lid of the plate covering the bottom and the mouth of the tube while pouring the agar. Do the same for the other two plates.
4. After pouring the plates, allow them to cool for a few minutes on the bench top. With a wax pencil, mark on the bottom of the plate the name of the bacterium to be inoculated, your name, and date. Also draw four quadrants on the bottom of the plate, as illustrated in figure 16.2c, to aid you in keeping track of your streaks.
5. Aseptically remove a loopful of the bacterial mixture (*see figure 14.3*).
6. Streak out the loopful of bacteria on the agar plate that you have prepared as follows:

 a. Carefully lift the top of the petri plate just enough to insert your inoculating loop easily (figure 16.2a). The top should cover the agar surface as completely as possible at all times in order to avoid contamination. Insert the inoculating loopful of bacteria and spread it over a small area (area 1) at one edge of the plate as shown in figure 16.2b in order to make effective use of the agar surface. This is accomplished by letting the loop rest gently on the surface of the agar and then moving it across the surface each time without digging into the agar.

 b. Remove the inoculating loop and kill any remaining bacteria by flaming them. Then insert the loop under the lid and cool it at the edge of the agar near area 1.

 c. Rotate the plate while carefully keeping in mind where the initial streaks ended (use the marked quadrants as a guide) and cross over the streaks in area 1 (figure 16.2b). Streak out any bacteria picked up as shown in area 2.

 d. Remove the loop, flame it, cool in the agar as before, and repeat the streaking process (figure 16.2b, area 3).

 e. If necessary, you can repeat this sequence once more to make a fourth set of streaks (area 4). Use fewer cross-streaks here than in the previous quadrant.

 f. Repeat the above procedure (*a–e*) for the other two bacteria on two new petri plates.

7. Incubate the plates at 30° to 37°C for 24 to 48 hours in an inverted position. Afterwards, examine each of the agar plates to determine the distribution and amount of growth in the three or four streaked areas and record your results in the report for exercise 16.

Figure 16.1 Streak Plate. Notice the well-isolated colonies of *E.coli* (white) and *S. marcescens* (red).

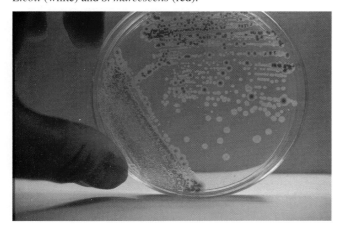

Figure 16.2 **Preparation of a Streak Plate.** Arrows indicate motion of the loop. In *b*, flame and cool the loop between 1 and 2, 2 and 3, and 3 and the end of the streak. The goal is to thin the numbers of bacteria growing in each successive area of the plate as it is rotated and streaked so that well-isolated colonies will appear in quadrant 3.

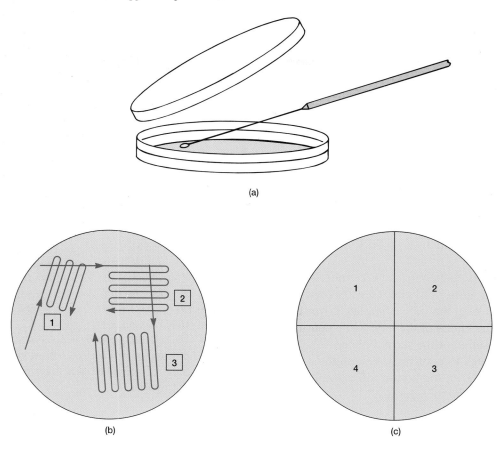

(a)

(b) (c)

PRINCIPLES FOR THE USE OF DIFFERENTIAL MEDIA

Many kinds of media can be used with streak plates. The first part of this exercise employed tryptic soy agar, a general purpose complex medium. Often it is most advantageous to prepare streak plates with selective and/or differential media. **Selective media** favor the growth of particular microorganisms. For example, bile salts or dyes like basic fuchsin and crystal violet favor the growth of gram-negative bacteria by inhibiting the growth of gram-positive bacteria without affecting gram-negative organisms. **Differential media** are media that distinguish between different groups of bacteria and even permit tentative identification of microorganisms based on their biological characteristics. Blood agar is both a differential medium and an enriched one. It distinguishes between hemolytic and nonhemolytic bacteria. Hemolytic bacteria (e.g., many streptococci and staphylococci isolated from throats) produce clear zones around their colonies because of red blood cell destruction.

Two very important differential and selective media that are used to isolate and partially identify bacteria are mannitol salt agar and eosin methylene blue agar. Mannitol salt agar is used to isolate staphylococci from clinical and nonclinical samples. It contains 7.5% sodium chloride, which inhibits the growth of most bacteria other than staphylococci. *Staphylococcus aureus* will ferment the mannitol and form yellow zones in the reddish agar because phenol red becomes yellow in the presence of fermentation acids (see *figure 54.5*). This differentiates it from *S. epidermidis,* which forms colonies with red zones or both zones (see *figure 54.6*). Eosin methylene blue (EMB) agar is widely used for the detection of *E. coli* and related bacteria in water supplies and elsewhere. It contains the dyes eosin Y and

methylene blue that partially suppress the growth of gram-positive bacteria. The dyes also help differentiate between gram-negative bacteria. Lactose fermenters such as *Escherichia coli* will take up the dyes and form blue-black colonies with a metallic sheen. Lactose nonfermenters such as *Salmonella, Proteus,* and *Pseudomonas* form colorless to amber colonies.

In this exercise, we will combine the streak-plate technique with differential and selective media to isolate and partly identify *Staphylococcus aureus* and *Escherichia coli.*

Procedure

1. Melt a sterile, capped tube of mannitol salt agar and a tube of EMB agar in a boiling water bath.
2. Cool the two tubes in a 48° to 50°C water bath for 10–15 minutes.
3. Remove the cap, flame the lip of the tube, and pour the agar into a sterile petri plate (*see figure 13.2c–f*). Be careful to keep the lid of the plate covering the bottom and the mouth of the tube while pouring the agar. Do the same for the second plate.
4. Allow the plates to cool for a few minutes on the bench top. Mark on the bottom of each plate your name, the date, and the agar used.
5. Aseptically remove a loopful of the bacterial mixture containing *E. coli, S. aureus,* and *Proteus vulgaris.* Prepare mannitol salt agar and EMB agar streak plates following the procedure described in step 6 on page 100.
6. Incubate the plates at 35–37°C for 24–48 hours in an inverted position. Examine them and evaluate the type of colony growth. Compare the colonies on the two plates and try to determine which bacteria are growing on each. Record your observations in the report for exercise 16.

HINTS AND PRECAUTIONS

(1) Each time the loop is flamed, allow it to cool on the agar at least 10 to 15 seconds before streaking the culture. Wait until the loop stops "hissing."

(2) Use a loopful of culture from the source tube only when applying the first streak in quadrant 1 of the petri plate. Do not return to the source tube for more culture when streaking quadrants 2 to 4.

(3) An inoculated plate is always incubated in an inverted position to prevent condensation from falling onto the surface of the plate and interfering with discrete colony formation.

The Streak-Plate Technique and Differential Media

1. Make a drawing of the distribution of the colonies on each petri plate.

 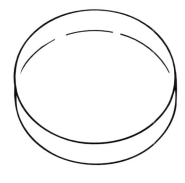

_____ _____ _____

2. Select one discrete colony, describe it (*see figure 15.1*), and identify the bacterium it contains.

Bacterium	_____	_____	_____
Colony form	_____	_____	_____
Colony elevation	_____	_____	_____
Colony margin	_____	_____	_____
Colony size	_____	_____	_____
Colony color	_____	_____	_____

3. Draw your streaking patterns. Did you obtain isolated colonies? If not, what went wrong? If you carried out the differential medium experiment, comment on the differences in growth on mannitol salt agar and EMB agar.

Review Questions

1. In the streak-plate technique, how are microorganisms diluted and spread out to form individual colonies?

2. Which area of a streak plate will contain the greatest amount of growth? The least amount of growth? Explain your answers.

3. Does each discrete colony represent the growth of one cell? Explain your answer. Why can a single colony on a plate be used to start a pure culture?

4. Why can mannitol salt agar and EMB agar be described as both selective and differential media?

5. How can a streak plate become contaminated?

Pour-Plate Technique

Exercise *17*

MATERIALS PER STUDENT

24- to 48-hour mixed tryptic soy broth culture of *Escherichia coli* (ATCC 11229), *Serratia marcescens* (ATCC 13880; *Micrococcus roseus* ATCC 186 also can be used), and *Bacillus subtilis* (ATCC 6051)
3 tryptic soy agar pour tubes
3 9-ml sterile 0.9% NaCl (saline) blanks
48° to 50°C water bath
boiling water bath
wax pencil
3 petri plates
inoculating loop
Bunsen burner
3 sterile 1-ml pipettes with pipettor

Learning Objectives

Each student should be able to

1. Understand the pour-plate technique
2. Perform a pour-plate technique to obtain isolated colonies

SUGGESTED READING IN TEXTBOOK

1. The Pour Plate, section 5.8.

Pronunciation Guide

Escherichia coli (esh-er-I-ke-a KOH-lee)
Bacillus subtilis (bah-SIL-lus sub-til-lus)
Serratia marcescens (se-RA-she-ah mar-SES-sens)

Why Are the Following Bacteria Used in This Exercise?

Another procedure that is used to obtain well-isolated, pure colonies is the pour-plate technique. Since *Serratia marcescens, Bacillus subtilis,* and *Escherichia coli* were used in the past three exercises, and the pure culture plates should have been saved, these same cultures are used in this exercise. Remember, *S. marcescens* produces red colonies; *B. subtilis,* white to cream colonies; and *E. coli,* white colonies.

PRINCIPLES

The **pour-plate technique** also will yield isolated colonies and has been extensively used with bacteria and fungi. The original sample is diluted several times to reduce the microbial population sufficiently to obtain separate colonies upon plating (figure 17.1). The small volumes of several diluted samples are added to sterile petri plates and mixed with liquid tryptic soy agar that has been cooled to about 48° to 50°C. Most bacteria and fungi will not be killed by the brief exposure to the warm agar. After the agar has hardened, each cell is fixed in place and will form an individual colony if the sample is dilute enough. Assuming no chaining or cell clusters, the total number of colonies are equivalent to the number of viable microorganisms in the diluted sample. To prepare **pure cultures,** colonies growing on the surface or subsurface can be inoculated into fresh medium.

Procedure

1. With a wax pencil, label three sterile saline blanks 1 to 3.

Figure 17.1 The Pour-Plate Technique. The original sample is diluted several times to decrease or dilute the population sufficiently. 1 ml of each dilution is then dispensed into the bottom of a petri plate. Agar pours are then added to each plate. Isolated cells grow into colonies and can be used to establish pure cultures. The surface colonies are circular and large, subsurface colonies are lenticular or lens-shaped and much smaller.

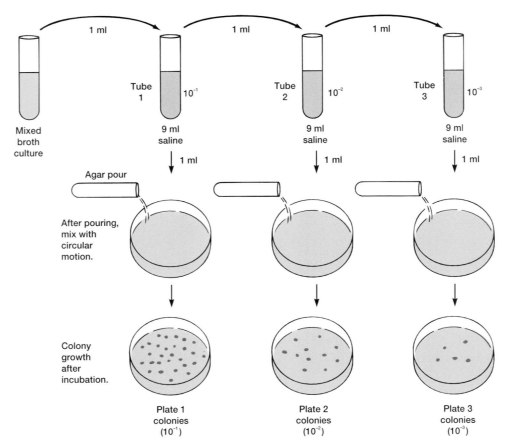

2. Melt the tryptic soy agar deeps in a boiling water bath and cool in a 48° to 50°C bath for at least 10 to 15 minutes (*see figure 13.2*).

3. With a wax pencil, label the bottoms of three petri plates 1 to 3, and add your name and date.

4. Inoculate saline tube 1 with 1 ml of the MIXED bacterial culture using aseptic technique (*see figure 14.3*) and MIX thoroughly. This represents a 10^{-1} dilution.

5. Using aseptic technique, immediately inoculate tube 2 with 1 ml from tube 1; a 10^{-2} dilution.

6. Using aseptic technique, mix the contents of tube 2 and use it to inoculate tube 3 with 1 ml; a 10^{-3} dilution.

7. After tube 3 has been inoculated, mix its contents, remove the cap, flame the top, and aseptically transfer 1 ml into petri plate 3. Then inoculate plates 1 and 2 in the same way, using 1 ml from tubes 1 and 2, respectively.

8. Add the contents of the melted tryptic soy agar pours to the petri plates. Gently mix each agar plate with a circular motion while keeping the plate flat on the bench top. Do not allow any agar to splash over the side of the plate! Set the plate aside to cool and harden.

9. Incubate the plates at 30° to 37°C for 24 to 48 hours in an inverted position.

10. Examine the pour plates and record your results in the report for exercise 17.

HINTS AND PRECAUTIONS

(1) Always allow sufficient time for the agar deeps to cool in the water bath after they have been boiled prior to the addition of bacteria.

(2) When the poured agar has solidified in the petri plates, it will become lighter in color and cloudy (opaque) in appearance. Wait until this occurs before attempting to move the plates.

(3) An inoculated plate is always incubated in an inverted position to prevent condensation from falling onto the surface of the plate and interfering with discrete colony formation.

Name: _____

Date: _____

Lab Section: _____

Pour-Plate Technique

1. Examine each of the agar plates for colony distribution and amount of growth. Look for discrete surface colonies and record your results. Do the same for the subsurface colonies. Color each species of bacterium a different color or label each. Fill in the table.

Plate 1 Plate 2 Plate 3

Surface Colonies			
Form	_____	_____	_____
Elevation	_____	_____	_____
Margin	_____	_____	_____
Number	_____	_____	_____
Subsurface Colonies			
Form	_____	_____	_____
Elevation	_____	_____	_____
Margin	_____	_____	_____
Number	_____	_____	_____

2. In summary, what is the major purpose of this exercise?

Review Questions

1. How do the results of the pour-plate method compare with those obtained using the streak-plate and spread-plate methods?

2. What is the main advantage of the pour-plate method over other methods of bacterial colony isolation? What are some problems?

3. Why are the surface colonies on a pour plate larger than those within the medium?

4. Why doesn't the 48° to 50°C temperature of the melted agar kill most of the bacteria?

5. Explain how the pour-plate method can be used to isolate fungi.

6. Why must tryptic soy agar be cooled below 50°C before pouring and inoculating?

7. Why is it important to invert the petri plates during incubation?

Cultivation of Anaerobic Bacteria

SAFETY CONSIDERATIONS

No mouth pipetting. Be careful with the Bunsen burner flame and water baths. Pyrogallol is poisonous; percutaneous absorption through the skin is possible—wear disposable gloves and use the spatula when transferring the crystals. Ten percent NaOH is caustic to the skin, and the vapors may cause damage to the respiratory tract. Use a pipette with pipettor when transferring this caustic soda. The various gas-generating envelopes that liberate hydrogen react rapidly with water to produce flammable gas. Observe appropriate precautions when handling these products. Further, the anaerobic jar used must not leak and the lid seal should be tight. Envelopes that have been activated with water but not used in anaerobic jars should be kept in a ventilating hood away from open flames or sparks for approximately 30 minutes until the reaction has subsided.

MATERIALS PER GROUP OF STUDENTS

24- to 48-hour Eugon broth cultures of *Pseudomonas aeruginosa* (ATCC e10145), and *Escherichia coli* (ATCC 11229); thioglycollate broth cultures of *Clostridium sporogenes* (ATCC 3584)
3 Eugon agar deeps
boiling water bath
48° to 50°C water bath
4 thioglycollate broth tubes
inoculating loop
2 tryptic soy agar plates
4 petri plates containing Brewer's anaerobic agar
sterilized Brewer's anaerobic covers
GasPak Jar, GasPak Disposable Anaerobic System (BBL Microbiology Systems), GasPak Pouch (BBL), Difco Gas Generating

envelopes, BioBag (Marion Scientific), Anaero-Pack System (KEY ScientificProducts)
Oxyrase For Broth, Oxyrase For Agar, OxyDish, OxyPlates (Oxyrase, Inc., P.O. Box 1345, Mansfield, OH 44901; 419-589-8800; www.oxyrase.com)
3 trypticase soy broth tubes containing 0.1-ml Oxyrase For Broth
3 OxyDishes containing Oxyrase For Agar
4 tryptic soy agar slants
scissors
cotton plugs
pyrogallol crystals (poisonous)
10% NaOH (caustic)
test tubes
rubber stoppers
test-tube rack
wax pencil
disposable gloves
spatula for handling pyrogallol crystals and soil
1-ml pipette with pipettor
garden soil

Learning Objectives

Each student should be able to

1. Appreciate why some bacteria need an anaerobic environment to grow
2. Understand some of the different methods that are used to cultivate anaerobic bacteria
3. Successfully cultivate several anaerobic bacteria

SUGGESTED READING IN TEXTBOOK

1. Oxygen Concentration, section 6.4.
2. Anaerobic Respiration, section 9.6.
3. Fermentations, section 9.3.

Pronunciation Guide

Clostridium sporogenes (klos-STRID-ee-um spo-ROJ-ah-nees)

Escherichia coli (esh-er-I-ke-a KOH-lee)

Pseudomonas aeruginosa (soo-do-MO-nas a-ruh-jin-OH-sah)

Why Are the Following Bacteria Used in This Exercise?

This exercise is designed to give the student expertise in the cultivation of anaerobic bacteria. Thus, the authors have chosen an obligate anaerobe (*C. sporogenes*), a facultative anaerobe (*E. coli*), and for comparison, a strict aerobe (*P. aeruginosa*). *Pseudomonas aeruginosa* (L. *aeruginosa*, full of copper rust, hence green) is a straight or slightly curved motile rod (1.5 to 3.0 μm in length) that has a polar flagellum. This bacterium is aerobic, having a strictly respiratory type of metabolism with oxygen as the usual terminal electron acceptor. It is widely distributed in nature. *Clostridium sporogenes* (M.L. n. *spora*, spore + Gr. v. *gennaio*, to produce) is a straight rod 1.3 to 16.0 μm in length, motile with peritrichous flagella. Endospores are oval and subterminal and distend the cell. *C. sporogenes* is obligatorily anaerobic; if growth occurs in air, it is scanty and sporulation is inhibited. It is widespread in the environment. *Escherichia coli* grows readily on nutrient agar. The colonies may be smooth, low convex, and moist. *E. coli* is facultatively anaerobic having both a respiratory and a fermentative type of metabolism.

Medical Application

From a clinical laboratory perspective, the most common pathogenic anaerobic bacteria vary depending on body site. Examples are as follows: blood—*Bacteroides fragilis;* intestine—*Clostridium, Bacteroides;* genital area—*Actinomycetes, Bacteroides, Fusobacterium, Clostridium, Mobiluncus;* and skin and soft tissue—*Clostridium perfringens, Bacteroides fragilis, Peptostreptococcus.*

PRINCIPLES

One of the environmental factors to which bacteria and other microorganisms are quite sensitive is the presence of O_2. For example, some microorganisms will grow only in the presence of O_2 and are called **obligate aerobes. Facultative anaerobes** will grow either aerobically or in the absence of O_2, but better in its presence. Strict **obligate anaerobes** will grow only in the absence of O_2 and are actually harmed by its presence. **Aerotolerant anaerobes** are microorganisms that cannot use O_2 but are not harmed by it either. Finally, microorganisms that require a small amount of O_2 for normal growth but are inhibited by O_2 at normal atmospheric tension are called **microaerophiles.** These variations in O_2 requirements can be easily seen by inoculating a tube of molten agar with the bacterium in question, mixing the agar thoroughly without aerating it, and allowing it to solidify. The bacteria will grow in the part of the **agar deep culture** that contains the proper O_2 concentration (figure 18.1).

The damaging effects of O_2 on anaerobic bacteria create difficult culturing problems. Ideally, one should provide not only an O_2-free environment, but also one that has an adequate amount of moisture for bacterial growth. It is also necessary to have CO_2 present for the growth of many anaerobic bacteria. There are a number of ways in which anaerobic bacteria may be cultured. Four of the most useful will be described.

One of the most convenient approaches is to employ a specially designed commercial anaerobic broth. Two of the most useful are cooked meat medium and thioglycollate broth. Thioglycollate medium can be purchased with methylene blue or resazurin as an oxidation-reduction indicator. When this medium begins to turn bluish or reddish, it is becoming too aerobic for the culture of anaerobic bacteria.

For those bacteria that are not really fastidious anaerobes, growth can occur on nutrient agar slants if anaerobic conditions are created. This setup is called a **Wright's tube** (named after James H. Wright, American physician, 1901–1978) (figure 18.2). The anaerobic condition is created using pyrogallol and NaOH. In the presence of NaOH, pyrogallol is oxidized and removes O_2 very effectively in the process. After the anaerobic bacterium has been streaked out on the surface of the agar slant, the cotton plug is trimmed and then pushed into the culture tube until it rests just above the top of the slant. The space between the top of the cotton plug and the open end of the culture tube is then filled with pyrogallol crystals, and 1 ml of 10% NaOH is added. The tube is closed with a rubber stopper and is immediately inverted. It is incubated upside down.

Anaerobic bacteria may also be grown in special petri plates without the use of complex and expensive incubators. One of the most convenient plate methods uses the **Brewer's** (named after John H. Brewer, an industrial bacteriologist, in 1942) **anaerobic petri**

Figure 18.1 The Appearance of Various Agar Deep Cultures. Each dot represents an individual bacterial colony within the agar or on its surface. The surface, which is directly exposed to atmospheric oxygen, will be aerobic. The oxygen content of the medium decreases with depth until the medium becomes anaerobic toward the bottom of the tube.

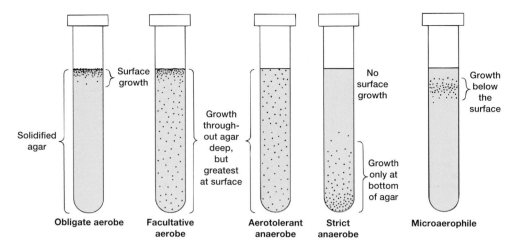

Figure 18.2 Preparation of an Anaerobic Wright's Tube. Pyrogallol is a reducing agent that is activated by NaOH to remove oxygen from the tube and create anaerobic conditions.

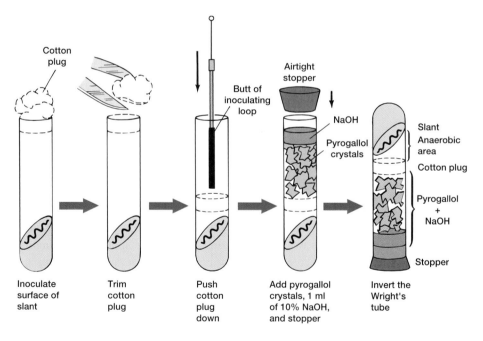

plate, together with special anaerobic agar (figure 18.3). Brewer's special cover fits on a normal petri plate bottom in such a way that its circular ridge rests on the agar, thereby protecting most of the surface from the exposure to O₂. Brewer's anaerobic agar contains a high concentration of thioglycollic acid. The free

Figure 18.3 Brewer's Petri Plate. An anaerobic space is created between the Brewer cover and the thioglycollate agar.

Figure 18.4 The GasPak System.

Lid Lockscrew Clamp

Catalyst chamber
Contains palladium pellets

$2H_2 + O_2 \longrightarrow 2H_2O$

Rubber gasket seal

H_2 CO_2

Oxygen is removed from chamber by combining with hydrogen to form water. This reaction is catalyzed by the palladium pellets.

Gas generator envelope
Water is added to chemicals in envelope to generate H_2 and CO_2. Carbon dioxide promotes more rapid growth of organisms.

Anaerobic indicator strip
Methylene blue becomes colorless in absence of O_2.

Figure 18.5 Procedure for GasPak Pouch. (a) Dispense GasPak liquid activating reagent into channel of pouch. (b) Place plates inside the pouch. (c) Lock in anaerobic environment with sealing bar and incubate.

(a)

(b)

(c)

sulfhydryl groups of thioglycollate reduce any O_2 present and create an anaerobic environment under the Brewer cover.

Another way of culturing bacteria anaerobically on plates is the **GasPak Anaerobic System.** In the GasPak System (figure 18.4), hydrogen and CO_2 are generated by a GasPak envelope after the addition of water. A palladium catalyst (pellets) in the chamber lid catalyzes the formation of water from hydrogen and O_2, thereby removing O_2 from the sealed chamber.

For greater convenience and visibility, **GasPak pouches** (figure 18.5) can be used instead of the regular GasPak incubation chamber. In this procedure, a special activating reagent is dispensed into the reagent channel (figure 18.5a). Inoculated plates are then put into the pouch (figure 18.5b). The anaerobic environment is locked in with a sealing bar (figure 18.5c) and the pouch incubated. Growth can be observed at one's convenience.

One of the simplest methods for growing anaerobes has recently been introduced by Oxyrase, Inc. One

simply adds 0.1 mL of **Oxyrase For Broth** to 5.0 ml of broth medium (Mueller-Hinton, Eugon, Trypticase Soy, Nutrient, Schaedler, Columbia, or Brain Heart Infusion). Most anaerobes can be inoculated immediately after this addition. Also from Oxyrase, Inc. are the **OxyDish** and **Oxyrase For Agar** that make growing and recovering anaerobes on agar plates easy. Simply mix Oxyrase For Agar with your agar medium and pour the still liquid agar into an OxyDish. The OxyDish contains an inner ring that forms a tight seal with the agar surface, which can be easily broken and reformed. Within minutes, the enzyme system and substrates in the Oxyrase For Agar reduce oxygen in the agar medium and the trapped headspace in the dish. The dish can be opened and closed several times while still maintaining an anaerobic environment. This system eliminates the complications and expense of bags, jars, anaerobic incubators, and chambers.

A newer methodology for growing anaerobes from Oxyrase is the **OxyPlate.** OxyPlates are prepoured plates with formulations identical to the traditional

anaerobic media with the addition of Oxyrase enzymatic preparation. The enzyme continues to reduce the oxygen within the media and in the space between the agar surface and the lid. Because the burden of achieving anaerobic conditions is placed directly into the media, no barrier apparatuses such as jars, pouches, and chambers are required. Simply streak the OxyPlate and incubate aerobically. The plates can be stacked and stored in an ambient air incubator alongside aerobic plates. Side-by-side comparison of the aerobic and anaerobic plates allows for early recognition of anaerobic colonies.

Procedure

The Relationship of O_2 to Bacterial Growth

1. Melt three Eugon agar deeps and heat them in a boiling water bath for a few minutes in order to drive off any O_2.
2. Cool the deeps in a water bath 48° to 50°C.
3. With a wax pencil, label each tube with the name of the bacterium to be inoculated, your name, and date.
4. Using aseptic technique (*see figure 14.3*), inoculate each cooled deep with 1 or 2 loopfuls of one of each of the three different bacteria (*P. aeruginosa, C. sporogenes,* and *E. coli*).
5. Mix the bacteria throughout the agar without aerating it by rolling each tube between the palms of your hands.
6. Allow the agar to harden and incubate the three tubes for 24 to 48 hours at 35°C.
7. Observe each tube for growth and record your results in the report for exercise 18.

Broth Culture of Anaerobic Bacteria

1. With a wax pencil, label three freshly steamed thioglycollate broth tubes with *P. aeruginosa, C. sporogenes,* and *E. coli,* as well as your name and date.
2. Using aseptic procedures, inoculate the three broth tubes. Do not shake these tubes to avoid oxidizing the medium! Methylene blue or resazurin is present in the medium as an oxidation-reduction indicator. If more than one-third of the broth is bluish or reddish in color, the tube should be reheated in a water bath in order to drive off the O_2 before use.
3. Incubate the tubes at 35°C for 24 to 48 hours.
4. Observe each tube for growth and record your results in the report for exercise 18.

Plate Culture of Anaerobic Bacteria

1. With a wax pencil, divide the bottom of each of the culture plates (one regular tryptic soy agar petri plate, one petri plate containing Brewer's anaerobic agar, and one Brewer's plate with cover) in half and label one of the halves *P. aeruginosa* and the other half *C. sporogenes.* Do the same with three more plates and label *P. aeruginosa* and *E. coli* (these plates will show facultative growth).
2. With an inoculating loop, streak the center of each half with the proper bacterium.
3. Incubate the tryptic soy agar plate inverted at 35°C in a regular aerobic incubator.
4. Carefully cover the Brewer's plate with a sterile Brewer anaerobic cover. The circular rim of the cover should press against the agar surface, but not sink into it. Incubate inverted at 35°C.
5. Place the regular anaerobic agar plate in a GasPak Anaerobic Jar or a GasPak Disposable Anaerobic Pouch.
6. Follow the instructions supplied with these products.
7. Incubate at 35°C for 24 to 48 hours.
8. Record your results in the report for exercise 18.

Slant Culture of Anaerobic Bacteria in a Wright's Tube

1. With a wax pencil, label four tryptic soy agar slants—two with *P. aeruginosa* and two with *C. sporogenes,* and your name and date.
2. Using aseptic technique, inoculate each slant with the respective bacterium.
3. Incubate two slants (one of each bacterium) aerobically at 35°C for 24 to 48 hours.
4. With the other two slants, cut off the cotton plug with a pair of scissors (figure 18.2), and with the butt of the inoculating loop, push it into the tube until it almost touches the top of the slant. While wearing gloves, fill the space above the cotton with pyrogallol crystals and add 1 ml of 10% NaOH. Quickly stopper the tube with a rubber stopper and invert the tube.
5. Place in a test-tube rack and incubate inverted at 35°C for 24 to 48 hours.
6. Record your growth results in the report for exercise 18.

Isolation of Anaerobes from the Soil

1. Place a spatula of soil in a thioglycollate broth tube.
2. Incubate at either 30° or 37°C for one day (teams can alternate temperatures).

3. Observe culture tubes for growth in the form of turbidity and the production of fermentation gases.
4. If time permits, streak the culture on a petri plate containing Brewer's anaerobic agar and incubate in a GasPak incubator. Observe colony morphology.
5. Record your results in the report for exercise 18.

Simplified Method for Growing Anaerobes

1. Using aseptic technique, inoculate each trypticase soy broth tube containing Oxyrase For Broth with *P. aeruginosa, C. sporogenes,* and *E. coli.* Repeat the inoculation on the OxyDish containing tryptic soy agar and Oxyrase For Agar.

2. Incubate for 24 to 48 hours at 37°C.
3. Record your growth results in the report for exercise 18.

HINTS AND PRECAUTIONS

(1) If screw-cap tubes are used with any tubed media (such as thioglycollate broth), the cap should be tightly closed during incubation outside of an anaerobic incubator to avoid unnecessary penetration of oxygen from the atmosphere.

(2) In an anaerobic GasPak jar, the caps should be loosened slightly to allow neutralization of oxygen within the tubes by the hydrogen gas generated within the system.

Cultivation of Anaerobic Bacteria

1. The relationship of O$_2$ to bacterial growth.

Drawing of
growth in
Eugon culture
tubes

P. aeruginosa *C. sporogenes* *E. coli*

Growth pattern _____ _____ _____

Cracking of agar due to gas (+ or −) _____ _____ _____

Classification according to growth pattern
 (figure 12.1) _____ _____ _____

2. Anaerobic broth cultures of bacteria.

Drawing of
culture in
anaerobic
broth culture
tubes

P. aeruginosa *C. sporogenes* *E. coli*

Comments _____ _____ _____

3. Drawings of plate cultures of aerobic and anaerobic bacteria.

Tryptic soy Brewer's plate GasPak plate OxyDish
agar plate

Comments _____ _____ _____ _____

4. Slant cultures of anaerobic bacteria.

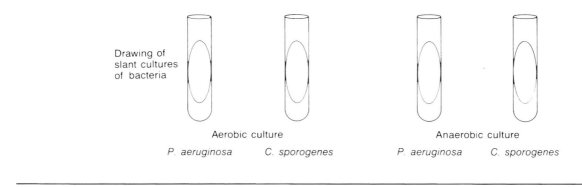

Drawing of
slant cultures
of bacteria

Aerobic culture

Anaerobic culture

P. aeruginosa C. sporogenes P. aeruginosa C. sporogenes

Comments

Review Questions

1. Explain how an anaerobic atmosphere can be created in a jar.

2. Explain what happens in a Wright's tube.

3. Differentiate between the following:
 a. an obligate anaerobe
 b. an obligate aerobe
 c. a facultative anaerobe
 d. an aerotolerant anaerobe
 e. a microaerophile

4. What are the ingredients in Brewer's anaerobic agar that remove O_2 from the medium? Briefly explain how an Oxyrase plate works.

5. Of the methods used in this exercise to create an anaerobic environment, which works the best, and why?

Determination of Bacterial Numbers

MATERIALS PER STUDENT

24-hour tryptic soy broth culture of *Escherichia coli* (ATCC 11229)
4 sterile 99-ml saline or phosphate buffer blanks
1-ml or 1.1-ml pipettes with pipettor
6 petri plates
6 agar pour tubes of tryptone glucose yeast agar (plate count agar)
48° to 50°C water bath
boiling water bath
Bunsen burner
cuvettes
spectrophotometer
4 tubes of tryptic soy broth

Learning Objectives

Each student should be able to

1. Describe several different ways to quantify the number of bacteria in a given sample
2. Determine quantitatively the number of viable cells in a bacterial culture by the standard plate count technique
3. Measure the turbidity of a culture with a spectrophotometer and relate this to the number (biomass) of bacteria

SUGGESTED READING IN TEXTBOOK

1. Measurement of Microbial Growth, section 6.2.
2. Measurement of Cell Numbers, section 6.2.
3. Measurement of Cell Mass, section 6.2.

Pronunciation Guide

Escherichia coli (esh-er-I-ke-a KOH-lee)

Why Is the Following Bacterium Used in This Exercise?

All of the learning objectives of this exercise are related to determining bacterial numbers. When working with large numbers and a short time frame, one of the most reliable microorganisms is one that has been used in previous experiments, namely, *Escherichia coli*. *E. coli* has a generation time at 37°C of 0.35 hours. Thus, it reproduces very rapidly and is easy to quantify (i.e., the number [biomass] of viable *E. coli* cells in a bacterial culture can be easily determined by spectrophotometry).

Medical Application

In the clinical laboratory and in research laboratories, it is frequently necessary to have an accurate count of living (viable) cells in a given culture. If done properly, counting procedures can produce very accurate results.

PRINCIPLES

Many studies require the quantitative determination of bacterial populations. The two most widely used methods for determining bacterial numbers are the **standard,** or **viable, plate count method** and **spectrophotometric (turbidimetric) analysis.** Although the two methods are somewhat similar in the results they yield, there are distinct differences. For example, the standard plate count method is an indirect measurement of cell density and reveals information related only to live bacteria. The spectrophotometric analysis is based on turbidity and indirectly measures all bacteria (cell biomass), dead or alive.

The standard plate count method consists of diluting a sample with sterile saline or phosphate buffer diluent until the bacteria are dilute enough to count accurately. That is, the final plates in the series should have between 25 and 250 colonies. Fewer than 25 colonies are not acceptable for statistical reasons, and more than 250 colonies on a plate are likely to produce colonies too close to each other to be distinguished as distinct **colony-forming units (CFUs)**. The assumption is that each viable bacterial cell is separate from all others and will develop into a single discrete colony (CFU). Thus, the number of colonies should give the number of live bacteria that can grow under the incubation conditions employed. A wide series of dilutions (e.g., 10^{-4} to 10^{-10}) is normally plated because the exact number of live bacteria in the sample is usually unknown. Greater precision is achieved by plating duplicates or triplicates of each dilution.

Increased turbidity in a culture is another index of bacterial growth and cell numbers (biomass). By using a spectrophotometer (*see appendix C*), the amount of transmitted light decreases as the cell population increases. The transmitted light is converted to electrical energy, and this is indicated on a galvanometer. The reading indirectly reflects the number of bacteria. This method is faster than the standard plate count but is limited because sensitivity is restricted to bacterial suspensions of 10^7 cells or greater.

DILUTION RATIOS

According to the *American Society for Microbiology Style Manual,* dilution ratios may be reported with either colons (:) or shills (/), but note there is a difference between them. A shill indicates the ratio of a part to a whole; for example, 1/2 means 1 of 2 parts, with a total of 2 parts. A colon indicates the ratio of 1 part to 2 parts, with a total of 3 parts. Thus, 1/2 equals 1:1, but 1:2 equals 1/3. (*See appendix A for a complete discussion of dilutions and sample problems.*)

Procedure

Standard Plate Count

1. With a wax pencil, label the bottom of six petri plates with the following dilutions: 10^{-4}, 10^{-5}, 10^{-6}, 10^{-7}, 10^{-8}, and 10^{-9}. Label four bottles of saline or phosphate buffer 10^{-2}, 10^{-4}, 10^{-6}, and 10^{-8}.
2. Using aseptic technique, the initial dilution is made by transferring 1.0 ml of liquid sample or 1 g of solid material to a 99-ml sterile saline blank (figure 19.1). This is a 1/100 or 10^{-2} dilution (*see appendix A*). Cap the bottle.
3. The 10^{-2} blank is then shaken vigorously 25 times by placing one's elbow on the bench and moving the

forearm rapidly in an arc from the bench surface and back. This serves to distribute the bacteria and break up any clumps of bacteria that may be present.

4. Immediately after the 10^{-2} blank has been shaken, uncap it and aseptically transfer 1.0 ml to a second 99-ml saline blank. Since this is a 10^{-2} dilution, this second blank represents a 10^{-4} dilution of the original sample. Cap the bottle.
5. Shake the 10^{-4} blank vigorously 25 times and transfer 1.0 ml to the third 99-ml blank. This third blank represents a 10^{-6} dilution of the original sample. Cap the bottle. Repeat the process once more to produce a 10^{-8} dilution.
6. Shake the 10^{-4} blank again and aseptically transfer 1.0 ml to one petri plate and 0.1 ml to another petri plate. Do the same for the 10^{-6} and the 10^{-8} blanks (figure 19.1).
7. Remove one agar pour tube from the 48° to 50°C water bath. Carefully remove the cover from the 10^{-4} petri plate and aseptically pour the agar into it. The agar and sample are immediately mixed by gently moving the plate in a figure-eight motion while it rests on the tabletop. Repeat this process for the remaining five plates.
8. After the pour plates have cooled and the agar has hardened, they are inverted and incubated at 35°C for 24 hours or 20°C for 48 hours.
9. At the end of the incubation period, select all of the petri plates containing between 25 and 250 colonies. Plates with more than 250 colonies cannot be counted and are designated **too numerous to count (TNTC)**. Plates with fewer than 25 colonies are designated **too few to count (TFTC)**. Count the colonies on each plate. If at all possible, a special counter such as a Quebec colony counter should be used (figure 19.2a). Your instructor will demonstrate how to use this counter or a handheld counter (figure 19.2b).
10. Calculate the number of bacteria (CFU) per milliliter or gram of sample by dividing the number of colonies by the dilution factor (*see appendix A*). The number of colonies per ml reported should reflect the precision of the method and should not include more than two significant figures. For example, suppose the plate of the 10^{-6} dilution yielded a count of 130 colonies. Then, the number of bacteria in 1 ml of the original sample can be calculated as follows:

Bacteria/ml = $(130) \div (10^{-6})$ = 1.3×10^8

or

130,000,000.

11. Record your results in the report for exercise 19.

Figure 19.1 Quantitative Plating Procedure.

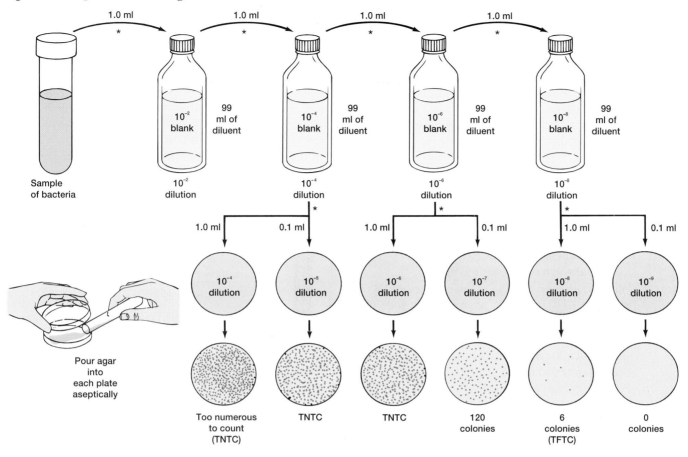

*Discard pipette after each transfer.

Figure 19.2 Colony Counters. (a) The Leica Dark-field Quebec Colony Counter. The counter illuminates the petri plate uniformly from the side, and the plate is magnified for easier counting of small colonies. The exclusive dark-field design provides even, glarefree illumination. Contrasted against dark-field background, colonies are bright—readily distinguished from other structures in agar. In this model, an electric probe (not shown) is touched to each colony to record the count automatically. **(b)** An electronic handheld colony counter. This model is a combination marker and counter. When the petri plate is touched, it is marked by the pen and counted at the same time. An electronic beep verifies each entry. A cumulative total appears on an easy-to-read LCD.

(a)

(b)

Turbidimetry Determination of Bacterial Numbers

1. Put one empty tube and four tubes of the sterile tryptic soy broth in a test-tube rack. With the exception of the empty tube, each tube contains 3 ml of sterile broth. Use four of these tubes (tubes 2 to 5) of broth to make four serial dilutions of the culture (figure 19.3).

2. Standardize and use the spectrophotometer (figure 19.4) as follows:

 a. Turn on the spectrophotometer by rotating knob *B* in figure 19.4 to the right.

 b. Set the monochromator dial (*D* in figure 19.4) so that the correct wavelength in nanometers (550 to 600 nm) is lined up with the indicator in the window adjacent to this dial. (Your instructor will inform you which wavelength to use.)

 c. When there is no cuvette in the cuvette holder (*A* in figure 19.4), the light source is blocked. The pointer should thus read zero transmittance or infinite absorbance. This is at the left end of the scale. Turn knob *B* until the pointer is aligned with the left end of the scale.

 d. Place in the cuvette holder (*A* in figure 19.4) the cuvette that contains just sterile broth. This tube is called the **blank** because it has a sample con-

centration equal to zero. It should therefore have an absorbance of zero (or a transmittance of 100%). This is at the right end of the scale. Set the pointer to the right end of the scale using knob *C* (figure 19.4).

 e. Place the other cuvettes, which contain the diluted bacterial suspension, in the cuvette chamber one at a time. Repeat steps c and d between experimental readings to confirm settings.

 f. Close the hatch and read the absorbance values of each bacterial dilution, and record your values. Remember to mix the bacterial suspension just before reading its absorbance.

 g. Record your values in the report for exercise 19. Using the plate count data, calculate the colony-forming units per milliliter for each dilution.

HINTS AND PRECAUTIONS

When mixing dilution tubes, do not use a vortex mixer.

Figure 19.4 Typical Spectrophotometers. (**a**) An analog model where (A) is the sample holder, (B) the power switch/zero control knob, (C) the 100% T control knob, and (D) the wavelength control knob. (**b**) A digital model that features direct concentration readout.

Figure 19.3 Twofold Serial Dilution for Standard Curve.

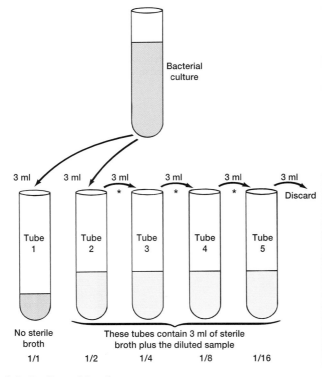

Bacterial culture

| 3 ml | 3 ml | 3 ml | 3 ml | 3 ml | 3 ml |

Discard

| Tube 1 | Tube 2 | Tube 3 | Tube 4 | Tube 5 |

No sterile broth

These tubes contain 3 ml of sterile broth plus the diluted sample

| 1/1 | 1/2 | 1/4 | 1/8 | 1/16 |

*Discard pipette after each transfer.

(a)

(b)

Laboratory Report 19

Name: _____

Date: _____

Lab Section: _____

Determination of Bacterial Numbers

1. Record your observations and calculated bacterial counts per milliliter in the following table.

Petri Plate	Dilution	ml of Dilution Plated	Number of Colonies	Bacterial Count per ml of Sample[a]
1	_____	_____	_____	_____
2	_____	_____	_____	_____
3	_____	_____	_____	_____
4	_____	_____	_____	_____

[a]This value is also expressed as colony-forming units per milliliter (CFU/ml).

2. Record your data from the turbidimetry experiment in the following table.

Turbidity-Absorbance Standard Curve

Dilution	CFU/ml	% T	Absorbance
Undiluted	_____	_____	_____
1/2 dilution	_____	_____	_____
1/4 dilution	_____	_____	_____
1/8 dilution	_____	_____	_____
1/16 dilution	_____	_____	_____

3. Construct a cell biomass standard curve by plotting the absorbance on the *y*-axis and the colony-forming units per milliliter on the *x*-axis.

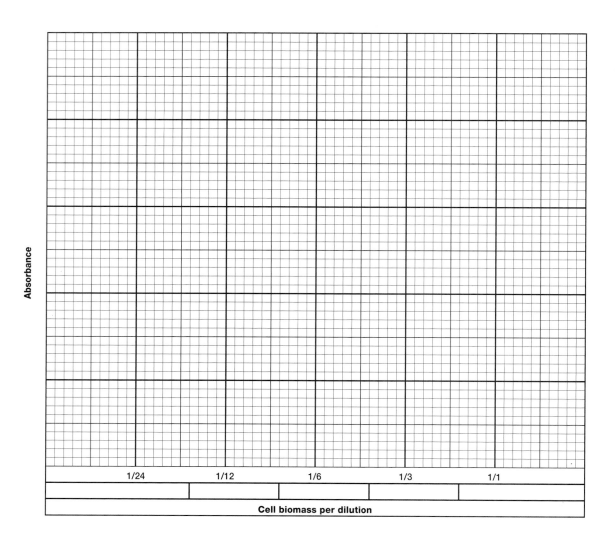

Review Questions

1. What is the difference between % T and absorbance?

2. Why is the viable plate count technique considered to be an indirect measurement of cell density, whereas the turbidimetry method is not a "count" at all?

3. Why is absorbance used in constructing a calibration curve instead of percent transmittance?

4. What is the purpose of constructing a calibration curve?

5. Why is it necessary to perform a plate count in conjunction with the turbidimetry procedure?

6. Give several reasons why it is necessary to shake the water blanks 25 times.

7. What is a CFU?

8. How would you prepare a series of dilutions to get a final dilution of 10^{-10}? Outline each step.

9. Why was 550 to 600 nm used in the spectroscopy portion of this experiment?

10. How would you define biomass?

11. What are several advantages to spectrophotometric determination of bacterial numbers? Several disadvantages?

Biochemical Activities of Bacteria

Life is driven by nothing else but electrons, by the energy given off by these electrons while cascading down from the high level to which they have been boosted up by photons. An electron going around is a little current. What drives life is thus a little current, kept up by the sunshine. All the complexes of intermediary metabolism are but lacework around this basic fact.
Albert Azent-Gyorgi (American physician and biochemist, Nobel recipient, 1893–1986)

Bacteria accomplish their various biochemical activities (growth and multiplication) using raw materials (nutrients) obtained from the environment. The biochemical transformations that occur both inside and outside of bacteria are governed by biological catalysts called **enzymes.**

This part of the laboratory manual presents exercises that have been designed to experimentally demonstrate or test for some of the biochemical activities of bacteria. This will be accomplished by observing the ability of bacteria to use enzymes and degrade carbohydrates, lipids, proteins, and amino acids. The metabolism, or use, of these organic molecules often produces by-products that can be used in the identification and characterization of bacteria.

After completing the exercises in part four, you will be able to demonstrate the ability to use appropriate microbiological media and test systems. This will meet the American Society for Microbiology Core Curriculum skill number 4 (*see pp. vi–viii*): (a) using biochemical test media and (b) accurately recording macroscopic observations.

Sir Alexander Fleming (1881–1955)

Fleming was a cowinner, with Ernst Boris Chain and Sir Howard Walter Florey, of the Nobel prize for medicine or physiology in 1945 for the discovery of penicillin.

In 1922, Alexander Fleming, a bacteriologist in London, had a cold. He allowed a few drops of his nasal mucus to fall on a culture plate containing bacteria. Some time later, he noticed that the bacteria near the mucus had been dissolved away, and he thought that the mucus might contain the "universal antibiotic" he was seeking. From this observation, Fleming showed that the antibacterial substance was an enzyme, which he named lysozyme—lyso- because of its ability to lyse bacteria and -zyme because it was an enzyme. He discovered a small coccus that was particularly susceptible to lysozyme and named it *Micrococcus lysodeikticus* because it exhibited lysis (Greek *deiktikos,* to show). Fleming also found that tears are a rich source of lysozyme. To continue his research, volunteers provided tears after they suffered a few squirts of lemon—an "ordeal by lemon." The *St. Mary's Hospital Gazette* published a cartoon showing children coming in to Fleming's laboratory for a few pennies where an attendant administered the beatings while another collected their tears. Fleming was disappointed to learn that lysozyme was not effective against harmful bacteria. However, seven years later, he discovered a highly effective antibiotic: penicillin—a striking illustration of Pasteur's famous comment, "Chance favors the prepared mind."

Carbohydrates I: Fermentation and β-Galactosidase Activity

MATERIALS PER STUDENT

20- to 24-hour tryptic soy broth cultures of *Escherichia coli* (ATCC 11229), *Alcaligenes faecalis* (ATCC 8750), *Salmonella cholerae-suis* (ATCC 29631), and a 7- to 10-day Sabouraud dextrose plate culture of *Saccharomyces cerevisiae* (ATCC 2366)

phenol red (or bromcresol purple) dextrose, lactose, and sucrose peptone broths with Durham tubes

Bunsen burner

inoculating loop and forceps

test-tube rack

incubator set at 35°C

5 tryptic agar base tubes

differentiation disks for dextrose, lactose, sucrose (Difco, BBL, or KEY Scientific Products, 1402 Chisholm Trail, Suite D, Round Rock, TX 78681; 800-843-1539; *www.keyscientific.com*)

ONPG disks (Difco) or ONPG tablets (KEY Scientific Products) for β-galactosidase

inoculating needle and forceps

100-ml beakers containing 70% ethanol

1-ml pipette and pipettor

sterile 0.85% NaCl (for β-galactosidase test)

3 sterile test tubes

Learning Objectives

Each student should be able to

1. Understand the biochemical process of fermentation

2. Describe how the carbohydrate fermentation patterns of some bacteria result in the production of an acid, or an acid and a gas

3. Explain how a Durham tube or a tryptic agar base tube and a differentiation disk can be used to detect acid and gas production

4. Perform a carbohydrate and a β-galactosidase fermentation test

5. Understand the function of β-galactosidase

SUGGESTED READING IN TEXTBOOK

1. Fermentations, section 9.3.

Pronunciation Guide

Alcaligenes faecalis (al-kah-LIJ-e-neez fee-KAL-iss)
Escherichia coli (esh-er-I-ke-a KOH-lee)
Salmonella cholerae-suis (sal-mon-EL-ah col-ER-AE su-is)
Saccharomyces cerevisiae (sak-a-row-MY-sees seri-VISS-ee-eye)
Staphylococcus aureus (staf-il-oh-KOK-kus ORE-ee-us)

Why Are the Above Microorganisms Used in This Exercise?

In this exercise, students will observe how microbial fermentation can yield acid, gas, or acid and a gas. Students also will test microorganisms for the presence of the enzyme β-galactosidase. To accomplish these objectives, the authors have chosen three bacteria and one yeast, each with a different fermentation pattern. *Escherichia coli* is a facultatively anaerobic gram-negative rod that produces acid and gas by catabolizing D-glucose (dextrose) and other carbohydrates. *E. coli* is β-galactosidase positive since it can also use lactose as its sole source of carbon.

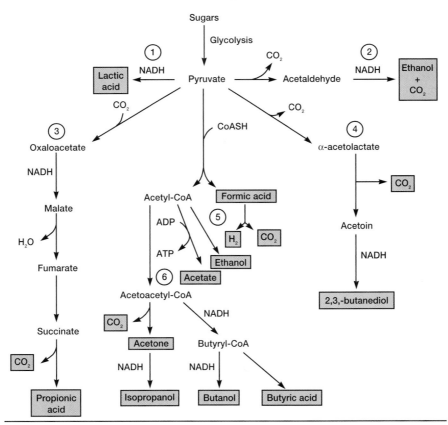

Lactose
(β-form)

β-galactosidase

H_2O

β-galactose + **β-glucose**

Alcaligenes faecalis is an obligately aerobic rod, coccal rod, or coccus that uses acetate, propionate, butyrate, and some other organic acids as a sole carbon and energy source; because carbohydrates are not used, it does not produce acid or gas. *A. faecalis* also is β-galactosidase negative. *Salmonella cholerae-suis* is a facultatively anaerobic gram-negative rod having both a respiratory and fermentative type of metabolism; thus, D-glucose and other types of carbohydrates are metabolized with the production of acid and sometimes gas. *S. cholerae-suis* is β-galactosidase positive. The yeast, *Saccharomyces cerevisiae*, will ferment glucose (but not sucrose) to produce gas, but no acid. *S. cerevisiae* is β-galactosidase negative.

PRINCIPLES

Fermentations are energy-producing biochemical reactions in which organic molecules serve both as electron acceptors and donors. The ability of microorganisms to ferment carbohydrates and the types of products formed are very useful in identification. A given carbohydrate may be fermented to a number of different end products depending upon the microorganism involved (figure 20.1). These end products (alcohols, acids, gases, or other organic molecules) are characteristic of the particular microorganisms. For example, if fermenting

Figure 20.1 Major Fermentation Pathways. Microorganisms produce various waste products when they ferment glucose. The by-products released (shaded boxes) are often characteristic of the microorganisms and can be used as identification tools.

1. Lactic acid fermentation. Lactic acid bacteria *(Streptococcus, Lactobacillus)*.
2. Alcoholic fermentation. *Zymomonas, Saccharomyces*.
3. Propionic acid fermentation. Propionic acid bacteria *(Propionibacterium)*.
4. 2,3,-butanediol fermentation. *Enterobacter, Serratia, Bacillus*.
5. Mixed acid fermentation. Enteric bacteria *(Escherichia, Enterobacter, Salmonella, Proteus)*.
6. Butyric acid fermentation. *Clostridium*.

Figure 20.2 Carbohydrate Fermentation. (**a**) Possible carbohydrate fermentation patterns of microorganisms, with phenol red as the pH indicator. (**b**) The tube on the left is the control. The next tube shows alcohol fermentation. Notice the gas bubble at the top. The third tube from the left shows no carbohydrate fermentation (negative). The tube on the right shows acid and gas production.

(a)

(b)

Figure 20.3 Carbohydrate Fermentation Using KEY Rapid Fermentation Tablets. The color change from red (right tube) to yellow (left tube) indicates acid production. Fermentation may be apparent within 30 minutes and usually is complete in 6 hours, but tubes should be held for 48 hours before being discarded. Gas production is shown by the appearance of bubbles near the bottom of the tube. Both of these tubes are negative for gas production.

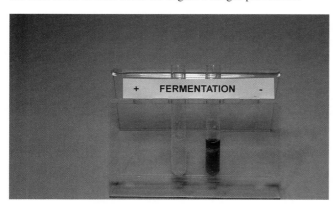

bacteria are grown in a liquid culture medium containing the carbohydrate glucose, they may produce organic acids as by-products of the fermentation. These acids are released into the medium and lower its pH. If a pH indicator such as phenol red or bromcresol purple is included in the medium, the acid production will change the medium from its original color to yellow (figures 20.2, 20.3; *see also appendix E*).

Gases produced during the fermentation process can be detected by using a small, inverted tube, called

a **Durham tube** (named after Herbert Edward Durham, English bacteriologist, 1866–1945), within the liquid culture medium. After the proper amount of broth is added, Durham tubes are inserted into each culture tube. During autoclaving, the air is expelled from the Durham tubes, and they become filled with the medium. If gas is produced, the liquid medium inside the Durham tube will be displaced, entrapping the gas in the form of a bubble (figure 20.2).

Some microorganisms, such as *E. coli,* can use lactose as their sole source of carbon. An essential enzyme in the metabolism of this sugar is **β-galactosidase.** β-galactosidase hydrolyzes lactose to galactose and glucose as shown in the diagram on p. 127.

Instead of lactose, the natural substrate of this enzyme, an artificial substrate, ONPG (o-nitro-phenyl-β-D-galactopyranoside), can be used. β-galactosidase catalyzes the hydrolysis of ONPG as follows:

$$ONPG \ + \ H_2O \ \xrightarrow{\text{β-galactosidase}}$$

galactose + o-nitrophenol.

ONPG is colorless but upon hydrolysis yields o-nitrophenol, which is yellow in an alkaline solution. If an ONPG disk or KEY tablet is incubated with a bacterial culture and the culture turns yellow, this is the positive test for β-galactosidase activity (figure 20.4).

Each of the above tests is important in the identification of certain bacteria.

Figure 20.4 ONPG Test Using Key Tablets. An ONPG positive reaction can be detected within 30 minutes by the appearance of a yellow color (tube on the left). A positive ONPG test shows that the bacteria being tested contain the enzyme beta-galactosidase, can ferment lactose, and may be properly classed as lactose fermenters. The tube on the right has not changed color; thus the bacteria would be ONPG negative.

Procedure: Durham Tube

First Period

1. Label five of the specified culture tubes with your name, date, and type of culture medium.
2. Label the first tube *E. coli;* the second, *S. cholerae-suis;* the third, *A. faecalis;* the fourth, *S. cerevisiae;* and the fifth, "control."
3. Using aseptic technique (*see figure 14.3*), inoculate each tube with the corresponding microbial culture. Leave the fifth tube uninoculated. Care should be taken during this step not to tip the fermentation tube, as this may accidentally force a bubble of air into the Durham tube and give a false-positive result. The tubes may be mixed by rolling them back and forth between the palms of the hands.
4. Place the five tubes in a test-tube rack and incubate at 35°C for 24 to 48 hours.

Second Period

1. Ideally, the tubes should be examined carefully between 2 to 4 hours, at 8 hours, and 18 hours in order to avoid false negatives due to reversal of the fermentation reactions that may occur with long incubations.
2. Examine all carbohydrate broth cultures for evidence of acid (A), or acid and gas (A/G) production. Use figure 20.2 as a reference. Acid production is detected by the medium turning yellow, and gas production by a gas bubble in the Durham tube.
3. The control tube should be negative for acid and gas production, and should have no turbidity.

4. Based on your observations, determine and record in the report for exercise 20 whether or not each microorganism was capable of fermenting the carbohydrate substrate with the production of acid, or acid and gas. Compare your results with other students who used other sugars.

Procedure: Sugar-Differentiation Disk

First Period

1. Label five tryptic agar base tubes with your name, date, and the sugar to be studied.
2. Label the first tube *E. coli;* the second, *S. cholerae-suis;* the third, *A. faecalis;* the fourth, *S. cerevisiae;* and the fifth, "control."
3. Using sterile forceps (or forceps that have been dipped in 70% ethanol and flamed), aseptically add one of your assigned sugar disks to each tube.
4. Inoculate each tube except the control in the following way. Heavily load the inoculating needle with microbial paste, then stab the deep until the needle reaches about one-half to two-thirds of the agar depth. The inoculating needle should strike the sugar-differentiation disk off-center and push it into the soft agar. The control tube should be stabbed with a sterile inoculating needle.
5. Incubate the tubes at 35°C for 24 hours.

Second Period

1. Examine the tubes after 2 to 4 hours, about 8 hours, and 18 hours.
2. Acid production is shown by yellowing of the medium around the disk, which often spreads throughout the agar. Gas production yields bubbles and splitting of the soft agar. Remember that positive-acid results can reverse with further incubation. Therefore, yellowing of the agar at 2 to 4 hours is a positive result, even if the tube turns red to violet upon longer incubation.
3. Compare your results with those of students who used different sugars.

ONPG Disks or Tablets

1. Dispense 0.2 ml of sterile 0.85% NaCl (Difco disks) or 1 ml (about 40 drops with a Pasteur pipette) of distilled water (KEY tablets) into four sterile test tubes.
2. Label each tube with the microorganism to be inoculated. Add your name and the date.
3. Suspend a loopful of paste (concentrated microorganisms) into each tube.

4. Place an ONPG disk or tablet into each tube and incubate at 35°C.
5. Check at 20 minutes and 4 hours (the Key tablet test can run up to 6 hours).
6. A positive β-galactosidase test is indicated by a yellow color; no color change indicates a negative test.
7. Record your results in the report for exercise 20.

HINTS AND PRECAUTIONS

(1) The amount of inoculum placed in each tube should be small. Too much inoculum may lead to overgrowth and neutralization of acidic fermentation products by the bacteria, thus resulting in false negatives.

(2) Do not vortex the fermentation tubes when inoculating because this can force air into the Durham tubes and result in false-positive recordings for gas production.

Name: _____

Date: _____

Lab Section: _____

Carbohydrates I: Fermentation and β-Galactosidase Activity

1. Complete the following table on carbohydrate fermentation.

Biochemical Results	Bacteria			
	E. coli	*S. cholerae-suis*	*A. faecalis*	*S. cerevisae*
Lactose (A), (A/G), (−)	_____	_____	_____	_____
Dextrose (A), (A/G), (−)	_____	_____	_____	_____
Sucrose (A), (A/G), (−)	_____	_____	_____	_____

A = Acid production.
A/G = Acid and gas production.
G = Gas production.
− = No growth or no change (alkaline).

2. Indicate whether the following bacteria have β-galactosidase activity.

a. *A. faecalis* _____

b. *E. coli* _____

c. *S. cholerae-suis* _____

d. *S. cerevisiae* _____

3. In summary, what is the major purpose of this exercise?

Review Questions

1. Define fermentation.

2. Do all microorganisms produce the same end product from pyruvate? Explain your answer.

3. What is the purpose of the phenol red or bromcresol purple in the fermentation tube?

4. What is the function of the Durham tube in the fermentation tube?

5. What are some of the metabolic end products produced by the different microorganisms used in this experiment?

6. What is the color of phenol red at an acid pH?

7. What is the function of β-galactosidase?

Carbohydrates II: Triple Sugar Iron Agar Test

MATERIALS PER STUDENT

24- to 48-hour tryptic soy broth cultures of *Alcaligenes faecalis* (ATCC 8750), *Escherichia coli* (ATCC 11229), *Proteus vulgaris* (ATCC 13315), *Pseudomonas aeruginosa* (ATCC 10145), and *Shigella flexneri* (ATCC 12661)
5 triple sugar iron agar slants
Bunsen burner
inoculating needle
incubator set at 35°C
test-tube rack

Learning Objectives

Each student should be able to

1. Understand the biochemical reactions involved in the triple sugar iron agar test
2. Differentiate among members of the family *Enterobacteriaceae*
3. Distinguish between the *Enterobacteriaceae* and other intestinal bacteria
4. Perform a TSI test

SUGGESTED READING IN TEXTBOOK

1. Carbohydrate catabolism, section 9.7.
2. The *Enterobacteriaceae,* section 22.3.

Pronunciation Guide

Alcaligenes faecalis (al-kah-LIJ-e-neez fee-KAL-iss)
Escherichia coli (esh-er-I-ke-a KOH-lee)
Proteus vulgaris (PRO-tee-us vul-GA-ris)
Pseudomonas aeruginosa (soo-do-MO-nas a-ruh-jin-OH-sah)
Shigella flexneri (shi-GEL-la flex-ner-i)

Why Are the Above Bacteria Used in This Exercise?

This exercise will provide the student experience in using the triple sugar iron agar test to differentiate among the members of the family *Enterobacteriaceae* and between *Enterobacteriaceae* and other intestinal bacteria. The authors have chosen three common bacteria in the family *Enterobacteriaceae: Escherichia coli, Proteus vulgaris,* and *Shigella flexneri.* All three are facultatively anaerobic gram-negative rods. In a TSI tube, *E. coli* produces an acid butt, an acid or alkaline slant, and no H_2S, but does produce gas. *P. vulgaris* produces an acid butt, an acid or alkaline slant, H_2S, and gas. *S. flexneri* produces an acid butt, an alkaline slant, no H_2S, and no gas. For the other intestinal bacteria, the authors have chosen *Alcaligenes faecalis* and *Pseudomonas aeruginosa.* Both of these intestinal bacteria are gram-negative aerobic rods. In a TSI tube, *A. faecalis* produces an alkaline butt, alkaline slant, H_2S, and gas; *P. aeruginosa,* an acid butt, alkaline slant, H_2S, and gas.

PRINCIPLES

As originally described in 1911 by F. F. Russell, the **triple sugar iron (TSI) agar test** is generally used for the identification of enteric bacteria (*Enterobacteriaceae*). It is also used to distinguish the *Enterobacteriaceae* from other gram-negative intestinal bacilli by their ability to catabolize glucose, lactose, or sucrose, and to liberate sulfides from ferrous ammonium sulfate or sodium thiosulfate. (*See exercise 24 for the biochemistry of H_2S production.*) TSI agar slants contain a 1% concentration of

Figure 21.1 Triple Sugar Iron Reactions (TSI-1) and Their Interpretation. (**a**) The tube on the left has a yellow butt (acid), red slant (alkaline), H_2S production as indicated by blackening of the agar, and no gas production. (**b**) The tube on the right shows no H_2S formation, a yellow slant (acid), gas production, and an acid butt. Note that the gas on the bottom has lifted the agar.

(a) (b)

	Tube a	Tube b
Slant	K	A
Butt	A	A
Gas	–	+
H_2S	+	–

lactose and sucrose, and a 0.1% glucose concentration. The pH indicator, phenol red, is also incorporated into the medium to detect acid production from carbohydrate fermentation (*see exercise 20*).

Often Kligler Iron Agar (named after I. J. Kligler in 1917), a differential medium similar to TSI, is used to obtain approximately the same information.

TSI slants are inoculated by streaking the slant surface using a zigzag streak pattern and then stabbing the agar deep with a straight inoculating needle (*see figure 14.5*). Incubation is for 18 to 24 hours in order to detect the presence of sugar fermentation, gas production, and H_2S production. The following reactions may occur in the TSI tube (figures 21.1–21.3):

1. **Yellow butt (A) and red slant (A)** due to the fermentation of glucose (phenol red indicator turns yellow due to the persisting acid formation in the butt). The **slant remains red (alkaline) (K)** because of the limited glucose in the medium and, therefore, limited acid formation, which does not persist.
2. A **yellow butt (A) and slant (A)** due to the fermentation of lactose and/or sucrose (yellow slant and butt due to the high concentration of these sugars) leading to excessive acid formation in the entire medium.

Figure 21.2 Triple Sugar Iron Reactions (TSI-2) and Their Interpretation. (**a**) The tube on the left has a red butt (alkaline), red slant (alkaline), and no acid or H_2S production. (**b**) The tube on the right has a yellow slant (acid), yellow butt (acid), and no gas or H_2S production.

(a) (b)

	Tube a	Tube b
Slant	K	A
Butt	K	A
Gas	–	–
H_2S	–	–

3. **Gas formation** noted by splitting of the agar.
4. **Gas formation (H_2S)** seen by blackening of the agar.
5. **Red butt (K) and slant (K)** indicates that none of the sugars were fermented and neither gas nor H_2S were produced.

Table 21.1 gives reactions usually expected from some of the more frequently encountered genera of the

Table 21.1			
Results of TSI Reaction			

	TSI REACTION			
Bacterium	**Butt**	**Slant**	**H_2S**	**Gas**
Enterobacter	A	A	–	+
Escherichia	A	A or K	–	+
Klebsiella	A	A	–	+
Citrobacter	A	K or A	V	+
Proteus vulgaris	A	A or K	+	+
Edwardsiella	A	K	–	+
Morganella	A	K	–	+
Serratia	A	K or A	–	V
Shigella	A	K	–	–
Salmonella typhi	A	K	+	–

A = acid, K = alkaline, V = varies between species.

Figure 21.3 **Triple Sugar Iron Reactions (TSI-3) and Their Interpretation.** (**a**) The tube on the left is an uninoculated control. Notice the red color. (**b**) The second tube from the left has a yellow slant (acid), yellow butt (acid), gas production at the bottom of the tube, and no H_2S production. This would indicate a weak lactose fermenter. (**c**) The third tube from the left has a red slant (alkaline), red butt (alkaline), and the black indicates H_2S production, but no gas. (**d**) The tube on the right has a red slant (alkaline), yellow butt (acid), H_2S production, but no gas production. This would indicate a nonlactose fermenter.

	Tube a	Tube b	Tube c	Tube d
Slant	–	A	K	K
Butt	–	A	K	A
Gas	–	+	–	–
H_2S	–	–	+	+

Enterobacteriaceae. Figure 21.4 summarizes the possible reactions and results in TSI for the various bacteria used in this experiment.

Procedure

First Period

1. Label each of the TSI agar slants with the name of the bacterium to be inoculated. Use one of the tubes as a control. Place your name and date on each tube.
2. Using aseptic technique (*see figure 14.3*), streak the slant with the appropriate bacterium and then stab the butt. Screw the caps on the tubes but do not tighten!
3. Incubate for only 18 to 24 hours at 35°C for changes in the butt and on the slant. Tubes should be incubated and checked daily for up to seven days in order to observe blackening.

Second Period

1. Examine all slant cultures for the color of the slant and butt, and for the presence or absence of blackening within the medium.
2. Record your results in the report for exercise 21.

HINTS AND PRECAUTIONS

(**1**) If screw-cap tubes are used, leave the caps loose about ¼ turn after inoculating the tubes to prevent excessive disruption of the agar should large amounts of gas be produced during incubation.

(**2**) Record the butt as acid production if the black color of FeS masks the color in the butt.

Figure 21.4 **The Possible Reactions and Results in TSI Agar for the Various Bacteria Used in This Experiment.**

No carbohydrate fermentation or hydrogen sulfide production

Example: *Alcaligenes faecalis*

glucose, lactose, sucrose → glucose, lactose, sucrose
(red slant/red butt) (**K; red slant/red butt**)

cysteine → cysteine
(**no black color**)

Glucose fermentation only

Example: *Shigella flexneri*

glucose → decrease in pH due to acid
(red butt) (**A; yellow butt**)

(continued)

Carbohydrates II: Triple Sugar Iron Agar Test 137

Figure 21.4 *(continued)*

lactose, sucrose → lactose, sucrose
(red slant) **(K; red slant)**

cysteine → cysteine
 (no black color)

Glucose fermentation only with hydrogen sulfide production

Example: *Pseudomonas aeruginosa*

glucose → decrease in pH due to acid
(red butt) **(A; yellow butt)**

lactose, sucrose → lactose, sucrose
 (red slant) **(K; red slant)**

cysteine → H_2S production

$H_2S + FeSO_4 \rightarrow FeS$
 (black color in media)

Lactose and/or sucrose and glucose fermentation

Example: *Escherichia coli*

glucose → decrease in pH due to acid
(red butt) **(A; yellow butt)**

lactose and/or sucrose → decrease in pH due to acid
 (red butt) **(A; yellow slant)**

cysteine → cysteine
 (no black color in media)

Lactose and/or sucrose and glucose fermentation with hydrogen sulfide (H_2S) production

Example: *Proteus vulgaris*

glucose → decrease in pH due to acid
(red butt) **(A; yellow butt)**

lactose and/or sucrose → decrease in pH due to acid
 (red slant) **(A; yellow butt)**

cysteine → H_2S production

$H_2S + FeSO_4 \rightarrow FeS$ production
 (black color in media)

Carbohydrates II: Triple Sugar Iron Agar Test

1. Complete the following table on the TSI test.

Bacterium	Carbohydrate Fermentation		H₂S Production	
	Butt Color	Slant Color	Black	H₂S
A. faecalis	_____	_____	_____	_____
E. coli	_____	_____	_____	_____
P. vulgaris	_____	_____	_____	_____
P. aeruginosa	_____	_____	_____	_____
S. flexneri	_____	_____	_____	_____

2. In summary, what is the major purpose of this exercise?

Review Questions

1. For what bacteria would you use the TSI test?

2. Why must TSI test observations be made between 18 to 24 hours after inoculation?

3. Distinguish between an acid and alkaline slant.

4. What is the purpose of thiosulfate in the TSI agar?

5. What is meant by a saccharolytic bacterium? What reaction would it give in a TSI tube?

6. Why is there more lactose and sucrose in TSI agar than glucose?

7. What is the pH indicator in TSI agar?

Carbohydrates III: Starch Hydrolysis

MATERIALS PER STUDENT

24- to 48-hour tryptic soy agar slant cultures of *Bacillus subtilis* (ATCC 6051), *Escherichia coli* (ATCC 11229), and *Proteus vulgaris* (ATCC 13315)
1 starch agar plate
Gram's iodine (1 g I_2, 2 g KI, 300 ml distilled water)
wax pencil
inoculating loop
Bunsen burner
Pasteur pipette with bulbs
incubator set at 35°C

Learning Objectives

Each student should be able to

1. Understand the biochemistry of starch hydrolysis
2. Perform a starch hydrolysis test

Pronunciation Guide

Bacillus subtilis (bah-SIL-lus SUB-til-is)
Escherichia coli (esh-er-I-ke-a KOH-lee)
Proteus vulgaris (PRO-te-us vul-GA-ris)

Why Are the Following Bacteria Used in This Exercise?

The major objective of this exercise is for the student to gain expertise in performing a starch hydrolysis test. If a bacterium produces α-amylase, it can hydrolyze starch; if α-amylase is not produced, the bacterium will not hydrolyze starch. The three bacteria the authors have chosen vary in their ability to produce α-amylase. *Bacillus subtilis* is amylase positive; *Escherichia coli* is amylase negative; and *Proteus vulgaris* is variable; it may be positive or negative.

PRINCIPLES

Many bacteria produce enzymes called **hydrolases.** Hydrolases catalyze the splitting of organic molecules into smaller molecules in the presence of water. This exercise will present the hydrolysis of the carbohydrate starch.

The starch molecule consists of two constituents: **amylose,** an unbranched glucose polymer (200 to 300 units), and **amylopectin,** a large branched polymer. Both amylopectin and amylose are rapidly hydrolyzed by certain bacteria, using their **α-amylases,** to yield dextrins, maltose, and glucose, as follows:

$$\begin{array}{c} \text{Starch} \\ [\text{Amylose} + \text{Amylopectin}] \\ (\text{Large polysaccharide}) \end{array} \xrightarrow[\text{H}_2\text{O}]{\alpha\text{-amylase}}$$

$$\begin{array}{c} \text{Dextrins} \\ (\text{Intermediate} \\ \text{polysaccharides}) \end{array} + \begin{array}{c} \text{Maltose} \\ (\text{Disaccharide}) \end{array} + \begin{array}{c} \text{Glucose} \\ (\text{Monosaccharide}) \end{array}$$

Figure 22.1 **Test for Starch Hydrolysis after Adding Gram's Iodine.** (**a,c**) Positive hydrolysis. The complete breakdown of all starch is shown by the clear (white) halo. (**b**) Negative hydrolysis. Starch remains intact—no color change as indicated by the purple to brown color around the streak.

(a) (b)

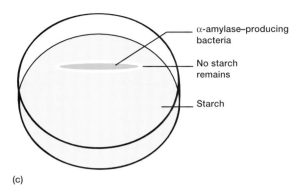

- α-amylase–producing bacteria
- No starch remains
- Starch

(c)

Gram's iodine can be used to indicate the presence of starch. When it contacts starch, it forms a blue to brown complex. Hydrolyzed starch does not produce a color change. If a clear area appears after adding Gram's iodine to a medium containing starch and bacterial growth, α-amylase has been produced by the bacteria (figure 22.1). If there is no clearing, starch has not been hydrolyzed.

Procedure

First Period: Starch Hydrolysis Test

1. With a wax pencil, divide a starch agar plate into three straight sections as indicated. Label each with the bacterium to be inoculated. Add your name and date to the plate.

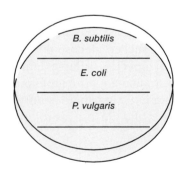

B. subtilis

E. coli

P. vulgaris

2. Using aseptic technique (*see figure 14.3*), streak the respective bacteria onto the plate in a **straight line** within the section.
3. Incubate the plate for 24 to 48 hours at 35°C.

Second Period

1. Place several drops of Gram's iodine on each of the line streaks on the starch agar plate. If the area around the line of growth is clear, starch has been hydrolyzed, and the test is positive; if it is not clear or the entire medium turns blue, starch has not been hydrolyzed, and the test is negative.
2. If the results are difficult to read, an alternative procedure is to invert the plate (after removing the lid) over a beaker containing iodine crystals. The rising vapor will react with the starch without the interference of the red-brown color of the unreacted iodine.
3. Record your results in the report for exercise 22.

HINTS AND PRECAUTIONS

(**1**) Carefully adding iodine to only a small part of the growth at one end of the streak does not contaminate the plate, and it may be reincubated and subsequently retested if necessary.

(**2**) Upon addition of iodine, record the presence or absence of blue color immediately.

(**3**) Test bacteria giving a red-violet color with iodine (partial hydrolysis) should be retested after an additional incubation period (see no. 1 above).

Name: _____

Date: _____

Lab Section: _____

Carbohydrates III: Starch Hydrolysis

1. In the following plate, sketch the presence or absence of starch hydrolysis.

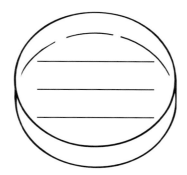

2. In summary, what is the major purpose of this exercise?

Review Questions

1. Describe the function of hydrolases.

2. Describe the chemistry of starch hydrolysis.

3. The chemical used to detect microbial starch hydrolysis on starch plates is _____.

4. What does starch hydrolysis by a bacterium indicate?

5. Amylase is an enzyme that attacks starch. The smallest product of this hydrolysis is called _____.

6. How is it possible that bacteria may grow heavily on starch agar but not necessarily produce α-amylase?

7. What are the ingredients of starch agar?

Lipids: Lipid Hydrolysis

MATERIALS PER STUDENT

tryptic soy broth cultures of *Proteus mirabilis* (ATCC 14273) and *Staphylococcus epidermidis* (ATCC 14990)
petri plate containing spirit blue agar with 3% Bacto lipase reagent (Difco)
inoculating loop
incubator set at 35°C
wax pencil
Bunsen burner

Learning Objectives

Each student should be able to

1. Understand the biochemical process of lipid hydrolysis
2. Determine the ability of bacteria to hydrolyze lipids by producing specific lipases
3. Explain how it is possible to detect the hydrolysis of lipids by a color change reaction
4. Perform a lipid hydrolysis test

SUGGESTED READING IN TEXTBOOK

1. Lipid Catabolism, section 9.8; see also figures 9.21 and 9.22.

Pronunciation Guide

Proteus mirabilis (PRO-te-us meh-RA-bill-iss)
Staphylococcus epidermidis (staf-il-oh-KOK-kuse-pee-DER-meh-diss)

Why Are the Following Bacteria Used in This Exercise?

After this exercise, the student will be able to differentiate between bacteria that can produce lipases and those that cannot. The authors have thus chosen a lipase positive and a lipase negative bacterium to demonstrate this difference. *Proteus mirabilis* (L. adj. wonderful, surprising) is a facultatively anaerobic gram-negative rod that produces lipase. *P. mirabilis* occurs in the intestines of humans and a wide variety of animals; it also occurs in manure, soil, and polluted waters. *Staphylococcus epidermidis* (Gr. *epidermidis*, outer skin) is a gram-positive coccus that does not produce lipase. It is mostly associated with the skin and mucous membranes of warm-blooded vertebrates but is often isolated from food products, dust, and water.

PRINCIPLES

Lipids are high-molecular-weight compounds possessing large amounts of stored energy. The two common lipids catabolized by bacteria are the triglycerides (triacylglycerols) and phospholipids. Triglycerides are hydrolyzed by the enzymes called **lipases** into glycerol and free fatty acid molecules as indicated in the following diagram. Glycerol and free fatty acid molecules can then be taken up by the bacterial cell and further metabolized through reactions of glycolysis, β-oxidation pathway, and the citric acid cycle. These lipids can also enter other metabolic pathways where they are used for the synthesis of cell membrane phospholipids. Since phospholipids are functional components of all cells, the ability of bacteria to hydrolyze host-cell phospholipids is an important factor in the spread of pathogenic bacteria. In addition, when lipase-producing

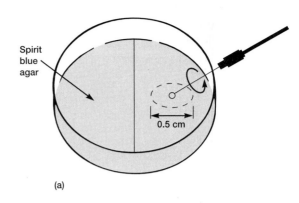

$$\begin{array}{ccccc}
\text{CH}_2-\text{O}-\overset{\overset{\text{O}}{\|}}{\text{C}}-\text{R}_1 & & & \text{CH}_2-\text{OH} & + \text{R}_1\text{COOH} \\[2mm]
\text{CH}-\text{O}-\overset{\overset{\text{O}}{\|}}{\text{C}}-\text{R}_2 & + 3\text{H}_2\text{O} & \xrightarrow{\text{lipases}} & \text{CH}-\text{OH} & + \text{R}_2\text{COOH} \\[2mm]
\text{CH}_2-\text{O}-\overset{\overset{\text{O}}{\|}}{\text{C}}-\text{R}_3 & & & \text{CH}_2-\text{OH} & + \text{R}_3\text{COOH} \\[2mm]
\text{Triglyceride} & \text{Water} & & \text{Glycerol} & \text{Fatty acids}
\end{array}$$

bacteria contaminate food products, the lipolytic bacteria hydrolyze the lipids, causing spoilage termed **rancidity.**

When these same lipids are added to an agar-solidified culture medium and are cultured with lipolytic bacteria, the surrounding medium becomes acidic due to the release of fatty acids. By adding a pH indicator to the culture medium, it is possible to detect the hydrolysis of lipids by a color change. For example, spirit blue agar with Bacto lipase reagent has a lavender color. It turns royal blue around lipolytic bacterial colonies due to the acid pH.

Procedure

First Period

1. With a wax pencil, divide the bottom of a spirit blue agar plate in half and label half the plate *P. mirabilis* and the other half *S. epidermidis.* Place your name and date on the plate.
2. Spot-inoculate (figure 23.1*a*) the spirit blue agar plates with the respective bacteria.
3. Incubate the plate in an inverted position for 24 to 48 hours at 35°C.

Second Period

1. Examine the plate for evidence of lipid hydrolysis (figure 23.1*b*). Hydrolysis is evidenced by a blue zone around the bacterial growth. If no lipid hydrolysis has taken place, the zone around the colony will remain lavender.
2. Measure the zone of hydrolysis and record your results in the report for exercise 23.

Figure 23.1 Lipid Hydrolysis. (**a**) Procedure for spot inoculating a spirit blue agar plate. (**b**) Positive and negative reactions.

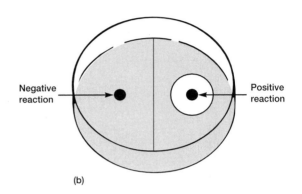

Laboratory Report 23

Name: _____

Date: _____

Lab Section: _____

Lipids: Lipid Hydrolysis

1. Based on your observations, complete the following table on lipid hydrolysis.

Bacterium	Lipid Hydrolysis	Zone of Hydrolysis
P. mirabilis	_____	_____
S. epidermidis	_____	_____

2. Sketch and describe what is happening on the petri plate with respect to lipid hydrolysis.

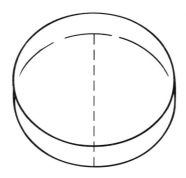

3. In summary, what is the major purpose of this exercise?

Review Questions

1. What is the function of lipases?

2. How can one determine whether a bacterium is lipolytic?

3. What are two functions of lipids in bacterial cells?

4. Give some examples of foods that might be spoiled by lipolytic bacteria.

5. How is the ability of certain bacteria to attack phospholipids related to pathogenicity?

6. What is the difference between a triglyceride (triacylglycerol) and a phospholipid?

7. What are several pathways that bacteria use to metabolize lipids?

Proteins, Amino Acids, and Enzymes I: Hydrogen Sulfide Production and Motility

MATERIALS PER STUDENT

24- to 48-hour tryptic soy broth cultures of *Klebsiella pneumoniae* (ATCC e13883), *Proteus vulgaris* (ATCC 13315), and *Salmonella typhimurium* (ATCC 29631)
Bunsen burner
inoculating needle
test-tube rack
3 SIM (sulfide-indole-motility) agar deeps
3 motility test medium deeps
Kovacs' reagent
incubator set at 35°C
wax pencil

Learning Objectives

Each student should be able to

1. Understand the biochemical process of hydrogen sulfide production by bacteria
2. Describe two ways hydrogen sulfide production can be detected
3. Describe how motility can be detected
4. Perform hydrogen sulfide and motility tests

SUGGESTED READING IN TEXTBOOK

1. Requirements for Nitrogen, Phosphorus, and Sulfur, section 5.4.
2. Oxidation of Inorganic Molecules, section 9.10.

Pronunciation Guide

Klebsiella pneumoniae (kleb-se-EL-lah nu-MO-ne-ah)
Proteus vulgaris (PRO-tee-us vul-GA-ris)
Salmonella typhimurium (sal-mon-EL-ah tie-fee-MUR-ee-um)

Why Are the Following Bacteria Used in This Exercise?

In this exercise, the student will gain experience in differentiating between bacteria based on hydrogen sulfide production (H_2S) and motility. Thus, the authors have chosen two bacteria that are H_2S producers and motile, and one bacterium that is nonmotile and does not produce H_2S. *Salmonella typhimurium* (Gr. *typhus,* a stupor) is a facultatively anaerobic gram-negative rod that occurs in humans, warm- and cold-blooded animals, food, and the environment. It is H_2S positive and motile. *Proteus vulgaris* (L. *vulgaris,* common) is a gram-negative facultatively anaerobic rod that occurs in the intestines of both humans and a wide variety of animals, manure, and polluted waters. It is motile and produces H_2S. *Klebsiella pneumonia* (Gr. *pneumonia,* pneumonia, inflammation of the lungs) is a facultatively anaerobic gram-negative rod that occurs in human feces, clinical specimens, soil, water, grain, fruits, and vegetables. It is nonmotile and does not produce H_2S.

PRINCIPLES

Many proteins are rich in sulfur-containing amino acids such as cysteine. When these proteins are hydrolyzed by some bacteria, the amino acids are released and taken up as nutrients. Cysteine, in the presence of **cysteine desulfurase,** loses its sulfur atom through the addition of hydrogen from water to form hydrogen sulfide gas (figure 24.1*a*).

Gaseous hydrogen sulfide may also be produced by the reduction of inorganic sulfur-containing compounds such as thiosulfate ($S_2O_3^{2-}$), sulfate (SO_4^{2-}), or sulfite (SO_3^{2-}). For example, when certain bacteria take up sodium thiosulfate, they can reduce it to sulfite using the enzyme thiosulfate reductase, with the release of hydrogen sulfide gas (figure 24.1*b*).

In this exercise, the SIM medium (named after J. S. Simmons in 1926) contains peptones and sodium thiosulfate as substrates, and ferrous ammonium sulfate, $Fe(NH_4)SO_4$, as the H_2S indicator. Cysteine is a component of the peptones used in SIM medium. Sufficient agar is present to make the medium semisolid. Once H_2S is produced, it combines with the ferrous ammonium sulfate, forming an insoluble, black ferrous sulfide precipitate that can be seen along the line of the stab inoculation. If the organism is also motile, the entire tube may turn black. This black line or tube indicates a positive H_2S reaction; absence of a black precipitate indicates a negative reaction (figure 24.1*c*).

SIM agar may also be used to detect the presence or absence of motility in bacteria as well as indole production. (*See exercise 25 for a discussion of indole production.*) Motility is present when the growth of the culture is not restricted to the stab line of the inoculation. Growth of nonmotile bacteria is confined to the line of inoculation.

One can also use semisolid media (motility test medium deeps) to determine whether a bacterial strain is motile. During growth, motile bacteria will migrate from the line of inoculation to form a dense turbidity in the surrounding medium; nonmotile bacteria will grow only along the line of the inoculation.

Procedure

First Period

1. Label each of the SIM agar deep tubes with the name of the bacterium to be inoculated, your name, and date.
2. Using aseptic technique (*see figure 14.3*), inoculate each tube with the appropriate bacterium by stab-

bing the medium three-fourths of the way to the bottom of the tube. Do the same for the three motility test medium deeps.
3. Incubate the cultures for 24 to 48 hours at 35°C.

Second Period

1. Examine the SIM cultures for the presence or absence of a black precipitate along the line of the stab inoculation. A black precipitate of FeS indicates the presence of H_2S.
2. Based on your observations, determine and record in the report for exercise 27 whether or not each bacterium was capable of H_2S production, and the presence (+) or absence (−) of motility.
3. If desired, one can also test for indole production by adding 5 drops of Kovacs' (named after the German bacteriologist, Nikolaus Kovacs, in the early 1900s) reagent to the SIM cultures and looking for the development of a red color at the top of the deeps (*see exercise 25*).

HINTS AND PRECAUTIONS

(1) Be careful when inoculating the deeps to withdraw the needle from the agar in a line as close as possible to the line used when entering the agar.

(2) Another aid in visualizing motility is to slowly rotate questionable tubes containing small amounts of growth around the stab line. When this is done, the growth appears much wider on the two opposite sides and narrower on the other two sides on which the bacteria is not motile.

(3) To observe motility, make sure the outside of your tubes are clean by wiping them with a Kimwipe.

(4) Any blackening of the medium is considered a positive test for H_2S.

Proteins, Amino Acids, and Enzymes I:
Hydrogen Sulfide Production and Motility

1. Complete the following table on hydrogen sulfide production and motility.

Bacterium	H_2S Production (+) or (−)	Motility (+) or (−)	
		SIM	Motility Test Medium
K. pneumoniae	_____	_____	_____
S. typhimurium	_____	_____	_____
P. vulgaris	_____	_____	_____

2. Sketch and describe what is happening in each tube with respect to H_2S production and motility.

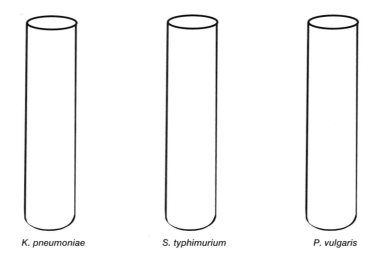

K. pneumoniae *S. typhimurium* *P. vulgaris*

3. In summary, what is the major purpose of this exercise?

Review Questions

1. Of what use to bacteria is the ability to produce H_2S?

2. How is SIM medium used to detect motility?

3. What substrates are acted on in SIM medium in order for H_2S to be produced?

4. In addition to H_2S production and motility, for what other test can SIM medium be used?

5. How does a black precipitate of FeS indicate the production of H_2S?

6. What does cysteine desulfurase catalyze? Show the reaction.

7. What does thiosulfate reductase catalyze? Show the reaction.

Proteins, Amino Acids, and Enzymes II: The IMViC Tests

SAFETY CONSIDERATIONS

Be careful with the Bunsen burner flame. No mouth pipetting. Barritt's reagent contains naphthol, which is toxic and may cause peeling of the skin; thus, wear gloves when using this reagent. Kovacs' reagent is also caustic to the skin and mucous membranes due to the concentrated HCl and p-dimethylaminobenzaldehyde. In case of contact with either reagent, immediately flush eyes or skin with plenty of water for at least 15 minutes. Keep all culture tubes upright in a test-tube rack or can.

MATERIALS PER STUDENT

24- to 48-hour tryptic soy broth cultures of *Enterobacter aerogenes* (ATCC 13048), *Escherichia coli* (ATCC 11229), *Klebsiella oxytoca* (ATCC 13182), and *Proteus vulgaris* (ATCC 13315)

4 SIM agar deep tubes

Kovacs' reagent, Key Indole Test Tablets, or Difco's SpotTest Indole Reagent Kovacs

Bunsen burner

inoculating loop and needle

4 MR-VP broth tubes each containing 5 ml of medium

methyl red indicator

Barritt's reagent (solutions A and B) or Difco's SpotTest Voges-Proskauer reagents A and B

4 Simmons citrate agar slants

4 empty test tubes

4-ml pipettes with pipettor

wax pencil

disposable gloves

Learning Objectives

Each student should be able to

1. Understand how some bacteria degrade the amino acid tryptophan

2. Determine the ability of some bacteria to oxidize glucose with the production of acid end products

3. Differentiate between glucose-fermenting enteric bacteria

4. Explain the purpose of the Voges-Proskauer test

5. Differentiate among enteric bacteria on the basis of their ability to ferment citrate

6. Perform the IMViC series of tests

SUGGESTED READING IN TEXTBOOK

1. Catabolism of Carbohydrates and Intracellular Reserve Polymers, section 9.7, see figure 9.10.

2. The *Enterobacteriaceae,* section 22.3, see table 22.7.

Pronunciation Guide

Enterobacter aerogenes (en-ter-oh-BAK-ter a-RAH-jen-eez)
Escherichia coli (esh-er-I-ke-a KOH-lee)
Klebsiella oxytoca (kleb-se-EL-lah ok-se-TO-se-ah)
Proteus vulgaris (PRO-te-us vul-GA-ris)
Salmonella (sal-mon-EL-ah)
Shigella (shi-GEL-la)
Enterobacteriaceae

Enterobacter aerogenes
Escherichia coli } Lactose fermenters
Klebsiella oxytoca

Alcaligenes faecalis
Proteus vulgaris } Lactose nonfermenter
Salmonella typhimurium
Shigella dysentariae

Some enteric (intestinal) bacteria

Why Are the Following Bacteria Used in This Exercise?

In this exercise, the student will learn how to perform the IMViC series of tests that distinguish between different enteric (pertaining to the small intestine) bacteria. To illustrate the various IMViC reactions, the authors have chosen four enteric bacteria. *Enterobacter aerogenes* (Gr. *aer,* air) is a facultatively anaerobic gram-negative rod that has peritrichous flagella. It is a motile lactose fermenter. *E. aerogenes* is widely distributed in nature, occurring in fresh water, soil, sewage, plants, vegetables, and animal and human feces. It is indole negative, MR negative, VP positive, and Simmons citrate positive. *Escherichia coli* (Gr. *colon,* large intestine) is a facultatively anaerobic gram-negative rod that is motile with peritrichous flagella or nonmotile. It is a lactose fermenter. *E. coli* occurs as normal flora in the lower part of the intestine of warm-blooded animals. It is indole positive, MR positive, VP negative, and Simmons citrate negative. *Klebsiella oxytoca* is a facultatively anaerobic gram-negative rod. It is nonmotile and a lactose fermenter. *K. oxytoca* occurs in human feces and clinical specimens, soil, water, grain, fruits, and vegetables. It is indole positive, often MR negative, VP positive, and Simmons citrate positive. *Proteus vulgaris* (L. *vulgaris,* common) is a gram-negative facultatively anaerobic rod that occurs in the intestines of humans and a wide variety of animals, in manure, and in polluted waters. It has peritrichous flagella, is motile, and does not ferment lactose. *P. vulgaris* is indole positive, MR positive, VP negative, and sometimes Simmons citrate positive.

Medical Application

The following medically important bacteria are MR+: *Escherichia coli* (opportunistic urinary tract infections), *Salmonella typhi* (typhoid fever), *Shigella dysenteriae* (bacterial dysentery), and *Yersinia pestis* (plague). The following is MR−: *Enterobacter aerogenes* (urinary tract infections).

Bordetella pertussis (whooping cough) is citrate negative whereas all other *Bordetella* species are citrate positive. The enteric bacteria such as *Klebsiella pneumoniae* (pneumonia) and *Enterobacter* are citrate positive and can be distinguished in the clinical laboratory from the opportunistic pathogen *Escherichia coli* (urinary tract infections), which is citrate negative.

PRINCIPLES

The identification of enteric (intestinal) bacteria is of prime importance in determining certain food-borne and water-borne diseases. Many of the bacteria that are found in the intestines of humans and other mammals belong to the family *Enterobacteriaceae.* These bacteria are short, gram-negative, nonsporing bacilli. They can be subdivided into lactose fermenters and nonfermenters. Examples include pathogens (*Salmonella* and *Shigella,* lactose nonfermenters), occasional pathogens (*Klebsiella* and *Escherichia,* lactose fermenters; and *Proteus,* lactose nonfermenter), and normal intestinal microbiota (*Enterobacter,* lactose fermenter).

The differentiation and identification of these enteric bacteria can be accomplished by using the **IMViC test** (**i**ndole, **m**ethyl red, **V**oges-Proskauer, and **c**itrate; the "i" is for ease of pronunciation).

Indole Production

The amino acid **tryptophan** is found in nearly all proteins. Bacteria that contain the enzyme **tryptophanase** can hydrolyze tryptophan to its metabolic products, namely, indole, pyruvic acid, and ammonia. The bacteria use the pyruvic acid and ammonia to satisfy nutritional needs; indole is not used and accumulates in the medium. The presence of indole can be detected by the addition of **Kovacs' reagent.** Kovacs' reagent reacts with the indole, producing a bright red compound on the surface of the medium (figures 25.1, 25.2). Bacteria producing a red layer following addition of Kovacs' reagent are **indole positive;** the absence of a red color indicates tryptophan was not hydrolyzed, and the bacteria are **indole negative.**

Methyl Red Test

All enteric bacteria catabolize glucose for their energy needs; however, the end products vary depending on the enzyme pathways present in the bacteria. The pH indicator **methyl red** (*see appendix E*) detects a pH change to the acid range as a result of acidic end products such as lactic, acetic, and formic acids. This test is of value in distinguishing between *E. coli* (a mixed acid fermenter) and *E. aerogenes* (a butanediol fermenter). **Mixed acid fermenters** such as *E. coli* produce a mixture of fermentation acids and thus acidify the medium. **Butanediol fermenters** such as *E. aerogenes* form butanediol, acetoin, and fewer organic acids. The pH of the medium does not fall as low as during mixed acid fermentation. As illustrated in figure 25.3, at a pH of 4, the methyl red indicator turns red—a **positive methyl red test.** At a pH of 6, the indicator turns yellow—a **negative methyl red test.**

Voges-Proskauer Test

The **Voges-Proskauer test** (named after Daniel Voges, German physician, and Bernhard Proskauer, German hygienist, in the early twentieth century) identifies bacteria that ferment glucose, leading to **2,3-butanediol** accumulation in the medium. The

Figure 25.2 **Indole Test with KEY Tablets.** The development of a red color within a few minutes after adding 10 drops of Kovacs' reagent—a positive test (tube on the left). Negative reactions are colorless (tube on the right).

addition of 40% KOH and a 5% solution of alpha-naphthol in absolute ethanol (Barritt's reagent) will detect the presence of **acetoin**—a precursor in the synthesis of 2,3-butanediol. In the presence of the reagents and acetoin, a cherry-red color develops. Development of a red color in the culture medium 15 minutes following the addition of Barritt's reagent represents a **positive VP test;** absence of a red color is a **negative VP test** (figure 25.4).

Citrate Utilization Test

The **citrate utilization test** determines the ability of bacteria to use **citrate** as a sole carbon source for their energy needs. This ability depends on the presence of a

citrate permease that facilitates transport of citrate into the bacterium. Once inside the bacterium, citrate is converted to pyruvic acid and CO_2. Simmons citrate agar slants contain sodium citrate as the carbon source, NH_4^+ as a nitrogen source, and the pH indicator (*see appendix D*) bromothymol blue. This test is done on slants since O_2 is necessary for citrate utilization. When bacteria oxidize citrate, they remove it from the medium and liberate CO_2. CO_2 combines with sodium (supplied by sodium citrate) and water to form sodium carbonate—an alkaline product. This raises the pH, turns the pH indicator to a blue color, and represents a **positive citrate test;** absence of a color change is a **negative citrate test** (figure 25.5). Citrate-negative cultures will also show no growth in the medium.

Procedure

Indole Production Test

First Period

1. Label each of the SIM deep tubes with the name of the bacterium to be inoculated (*E. coli, P. vulgaris,* and *E. aerogenes*), your name, and date. (If SIM medium is unavailable, tryptic soy broth is a good substitute for testing indole production.)

2. Using aseptic technique (*see figure 14.3*), inoculate each tube by a stab inoculation or with a loopful of culture.

3. Incubate the tubes for about 24 hours at 35°C.

Figure 25.3 Methyl Red Test. (a) *Escherichia coli,* MR+. **(b)** *Enterobacter aerogenes,* MR−.

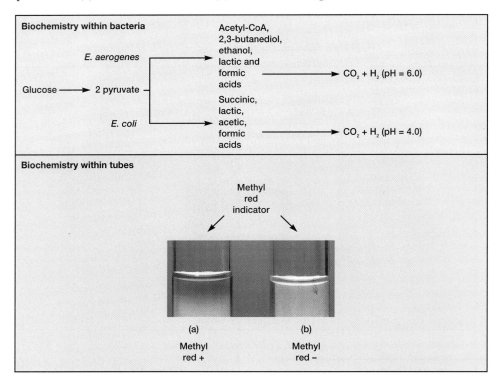

Biochemistry within bacteria

Glucose ⟶ 2 pyruvate

E. aerogenes ⟶ Acetyl-CoA, 2,3-butanediol, ethanol, lactic and formic acids ⟶ $CO_2 + H_2$ (pH = 6.0)

E. coli ⟶ Succinic, lactic, acetic, formic acids ⟶ $CO_2 + H_2$ (pH = 4.0)

Biochemistry within tubes

Methyl red indicator

(a) Methyl red +

(b) Methyl red −

Figure 25.4 Voges-Proskauer Test. (a) *Enterobacter aerogenes,* VP+. **(b)** *Escherichia coli,* VP−.

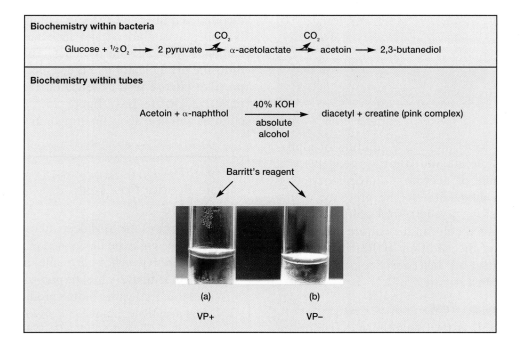

Biochemistry within bacteria

Glucose + ½O_2 ⟶ 2 pyruvate ⟶(CO_2) α-acetolactate ⟶(CO_2) acetoin ⟶ 2,3-butanediol

Biochemistry within tubes

Acetoin + α-naphthol ⟶ (40% KOH / absolute alcohol) ⟶ diacetyl + creatine (pink complex)

Barritt's reagent

(a) VP+

(b) VP−

Figure 25.5 Citrate Test. (a) *Enterobacter aerogenes;* blue color is positive. **(b)** *Escherichia coli;* green color is negative.

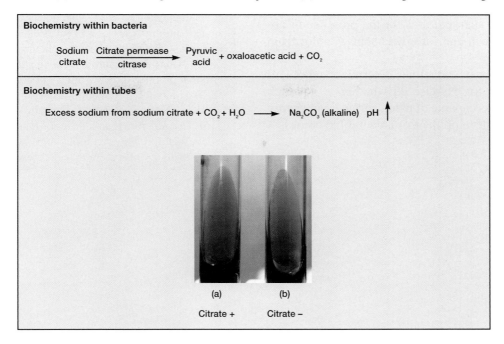

Biochemistry within bacteria

$$\text{Sodium citrate} \xrightarrow[\text{citrase}]{\text{Citrate permease}} \text{Pyruvic acid} + \text{oxaloacetic acid} + CO_2$$

Biochemistry within tubes

$$\text{Excess sodium from sodium citrate} + CO_2 + H_2O \longrightarrow Na_2CO_3 \text{ (alkaline)} \quad pH \uparrow$$

(a) (b)

Citrate + Citrate –

Second Period

1. Remove the tubes from the incubator and, while wearing disposable gloves, add 0.5 ml (about 10 drops) of Kovacs' reagent to each tube, and shake the tube gently. A deep red develops in the presence of indole. Negative reactions remain colorless or light yellow.

2. Based on your observations, determine and record in the report for exercise 25 whether or not each bacterium was capable of hydrolyzing the tryptophan.

Methyl Red Test
First Period

1. Label each of the MR-VP broth media tubes with the name of the bacterium (*E. coli, E. aerogenes,* and *K. oxytoca*) to be inoculated, your name, and date.

2. Using aseptic technique, inoculate each tube with the appropriate bacterium by means of a loop inoculation.

3. Incubate all tubes at 35°C for 24 to 48 hours. For slow fermenters, it may take four to five days.

Second Period

1. Transfer one-third of each culture into an empty test tube and set these aside for the Voges-Proskauer test.

2. To the two-thirds of the culture remaining in each tube, add 0.2 ml (about 4 to 5 drops) of methyl red indicator. Carefully note any color change (a red color is positive).

3. Based on your observations, determine and record in the report for exercise 25 whether or not each bacterium was capable of fermenting glucose, lowering the pH of the medium.

Voges-Proskauer Test
Second Period

1. Use the one-third aliquot from the methyl red test. While wearing disposable gloves, add 0.6 ml of Barritt's solution A and 0.2 ml of solution B to each culture, and shake vigorously to aerate. (Alternatively, about 15 drops of reagent A followed by 5 drops of reagent B works fairly well and avoids pipetting.) Positive reactions occur at once or within 20 minutes and are indicated by the presence of a red color.

2. Based on your observations, determine and record in the report for exercise 25 whether or not each bacterium was capable of fermenting glucose, with the production of acetylmethylcarbinol.

Citrate Utilization Test
First Period

1. Label each of the Simmons citrate agar slants with the name of the bacterium (*E. coli, E. aerogenes,* and *K. oxytoca*) to be inoculated, your name, and date.

2. Using aseptic technique, inoculate each bacterium into its proper tube by means of a stab-and-streak inoculation.

3. Incubate these cultures for 24 to 48 hours at 35°C.

Second Period

1. Examine the slant cultures for the presence or absence of growth and for any change in color from green to blue.
2. Based on your observations, determine and record in the report for exercise 25 whether or not each bacterium was capable of using citrate as an energy source. The development of a deep blue color is a positive test.

HINTS AND PRECAUTIONS

(1) Incubate the SIM agar deeps for only 24 hours prior to adding Kovacs' reagent because the indole may be further metabolized if the incubation time is prolonged. This could result in false negatives for some bacteria that produce indole from tryptophan.

(2) The indole production test also can be performed by adding Kovacs' reagent to a tryptic soy broth culture.

(3) Use no more than 5 drops of methyl red. If more is added, this may impart a red color to the medium that is unrelated to specific metabolic end products.

Laboratory Report 25

Name: _____

Date: _____

Lab Section: _____

Proteins, Amino Acids, and Enzymes II: The IMViC Tests

1. Based on your observations, record whether or not each bacterium was capable of hydrolyzing the tryptophan to produce indole.

Bacterium	Color of Reagent Layer	Indole + or −
E. coli	_____	_____
E. aerogenes	_____	_____
P. vulgaris	_____	_____

2. Based on your observations, record whether or not each bacterium was capable of fermenting glucose, with the production of either acids or acetylmethylcarbinol.

	Methyl Red Test		Voges-Proskauer Test	
Bacterium	Medium Color	+ or −	Medium Color	+ or −
E. coli	_____	_____	_____	_____
E. aerogenes	_____	_____	_____	_____
K. oxytoca	_____	_____	_____	_____

3. Based on your observations, record whether or not each bacterium was capable of using citrate as an energy source.

Bacterium	Presence or Absence of Growth	Color of Medium	Citrate Use + or −
E. coli	_____	_____	_____
E. aerogenes	_____	_____	_____
K. oxytoca	_____	_____	_____

4. In summary, what is the major purpose of this exercise?

Review Questions

1. What is the component in the SIM deep tubes that makes this medium suitable to detect the production of indole by bacteria?

2. What organic molecule is necessary to detect mixed acid fermentation by bacteria?

3. Why did you shake the MR-VP culture?

4. Can a bacterium that ferments using the 2,3-butanediol pathway also use the mixed acid route? Explain your answer.

5. Why is a chemically defined medium necessary for the detection of citrate utilization by bacteria?

6. Complete the following table.

Test	Medium	Significant Ingredients
Indole	_____	_____
Methyl red	_____	_____
Voges-Proskauer	_____	_____
Citrate	_____	_____

Exercise **26**

Proteins, Amino Acids, and Enzymes III: Casein Hydrolysis

SAFETY CONSIDERATIONS

Be careful with the Bunsen burner flame and the water baths. No mouth pipetting. Keep all culture tubes upright in a test-tube rack or in a can.

MATERIALS PER STUDENT

- 24- to 48-hour tryptic soy broth cultures of *Escherichia coli* (ATCC 11229), *Bacillus subtilis* (ATCC 6051), and *Pseudomonas aeruginosa* (ATCC 10145)
- tubes of plate count agar (tryptone glucose yeast agar)
- inoculating loop
- boiling water bath
- sterile petri plates
- water bath set at 48° to 50°C
- water bath set at 35°C
- Bunsen burner
- test-tube rack
- wax pencil
- sterile skim milk
- 5-ml pipette with pipettor

Learning Objectives

Each student should be able to

1. Understand the biochemical process of deamination
2. Determine the ability of some bacteria to secrete proteolytic enzymes capable of hydrolyzing the protein casein by performing a casein hydrolysis test
3. Explain what a zone of proteolysis indicates

SUGGESTED READING IN TEXTBOOK

1. Protein and Amino Acid Catabolism, section 9.9.
2. Dairy Products, section 41.6.

Pronunciation Guide

Bacillus subtilis (bah-SIL-lus SUB-til-us)
Escherichia coli (esh-er-I-ke-a KOH-lee)
Pseudomonas aeruginosa (soo-do-MO-nas a-ruh-jin-OH-sa)

Why Are the Above Bacteria Used in This Exercise?

In this exercise, the student will learn how to perform a casein hydrolysis test to detect the presence of proteolytic enzymes. Thus, the authors have chosen three bacteria that have been used in prior exercises. *Escherichia coli* will produce a negative reaction; *Bacillus subtilis* and *Pseudomonas aeruginosa* will produce positive reactions.

PRINCIPLES

Casein is a large milk protein incapable of permeating the plasma membrane of bacteria. (Its presence is the reason milk is white.) Therefore, before casein can be used by some bacteria as their source of carbon and energy, it must be degraded into amino acids. Bacteria accomplish this by secreting **proteolytic enzymes** that catalyze the hydrolysis of casein to yield amino acids (figure 26.1), which are then transported into the cell and catabolized.

When milk is mixed with plate count agar, the casein in the milk makes the agar cloudy. Following inoculation of the plate count agar, bacteria that liberate proteases (e.g., caseinase) will produce a **zone of proteolysis** (a clear area surrounding the colony). Clearing of the cloudy agar (a positive reaction) is the result of a hydrolytic reaction that yields soluble amino acids

163

Figure 26.1 **Proteolytic Hydrolysis.**

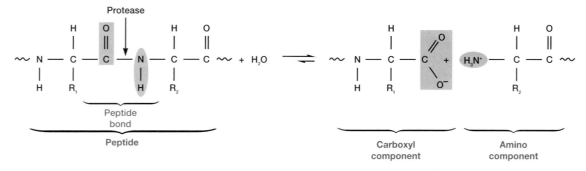

(figure 26.2b). In a negative reaction, there is no protease activity, and the medium surrounding the bacterial colony remains opaque (figure 26.2b).

Procedure

First Period

1. Melt the tubes of plate count agar by placing them in the boiling water bath. After melting, place the tubes in the 48° to 50°C water bath for 10 minutes.

Figure 26.2 Procedure for Determining Casein Hydrolysis. (**a**) Spot inoculation of a plate count agar with milk plate. (**b**) Plate exhibiting two zones of proteolysis.

(a)

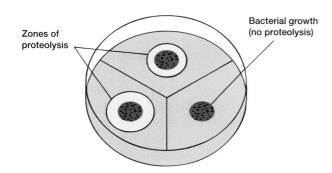

(b)

2. With a wax pencil, mark the bottom of a petri plate into three sections: label one *E. coli,* the second, *B. subtilis,* and the third, *P. aeruginosa.* Add your name and date to the plate.
3. Pipette 2 ml of warm (48° to 50°C) sterile skim milk into the petri plate. Add the melted agar and mix thoroughly by moving the plate in a circular motion. Allow this medium to gel on a cool, level surface.
4. As shown in figure 14.3, aseptically spot-inoculate (figure 26.2a) each third of the petri plate with the appropriate bacterium as per the label. Place a loopful of culture on the center of each section and spread it in a circular fashion to cover an area about the size of a dime or less (5 to 18 mm in diameter).
5. Incubate the plate in an inverted position at 35°C for 24 to 48 hours.

Second Period

1. Examine the plate count agar for the presence or absence of a clear zone (zone of proteolysis) surrounding the growth of each of the bacterial test organisms. You can see the clear zones best against a black background.
2. Based on your observations, determine and record in the report for exercise 26 which of the bacteria were capable of hydrolyzing the casein. Also, measure the zone of hydrolysis for each colony.

HINTS AND PRECAUTIONS

Aseptic technique must be followed because possible contaminating microorganisms from the air or body might be capable of hydrolyzing casein and thus lead to erroneous results.

Name: _____

Date: _____

Lab Section: _____

Proteins, Amino Acids, and Enzymes III: Casein Hydrolysis

1. Complete the following table on casein hydrolysis.

Bacterium	Casein Hydrolysis (+ or −)	Diameter of Hydrolysis Zone
E. coli	_____	_____
B. subtilis	_____	_____
P. aeruginosa	_____	_____

2. Draw your skim milk agar plates showing the growth patterns of the above three bacteria. Label accordingly.

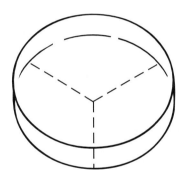

3. In summary, what is the major purpose of this exercise?

Review Questions

1. Define the following terms:

 a. protein

 b. hydrolysis

 c. casein

 d. protease

 e. amino acid

 f. peptide bond

 g. proteolysis

2. How can plate count agar that contains milk be used to demonstrate proteolysis?

3. Why are some bacteria able to grow on plate count agar that contains milk even though they do not produce any proteases?

4. Draw the chemical reaction for proteolytic hydrolysis.

5. Why was sterile skim milk used in this experiment?

6. Why is milk white?

Proteins, Amino Acids, and Enzymes IV: Gelatin Hydrolysis

SAFETY CONSIDERATIONS

Be careful with the Bunsen burner flame. No mouth pipetting. Keep all culture tubes upright in a test-tube rack.

MATERIALS PER STUDENT

24- to 48-hour tryptic soy broth cultures of
 Enterobacter aerogenes (ATCC 13048, gel. +),
 Escherichia coli (ATCC 11229, gel. −), and
 Proteus vulgaris (ATCC 13315, gel. +)
4 nutrient gelatin deep tubes
Bunsen burner
inoculating loop
1-ml pipettes with pipettor
refrigerator or ice-water bath
test-tube rack
incubator set at 35°C
3 KEY Rapid Gelatin Test Strips (KEY Scientific Products, 1402 Chisholm Trail, Suite D, Round Rock, TX 78681; 800-843-1539)
3 sterile test tubes
wax pencil

Learning Objectives

Each student should be able to

1. Understand how the proteolytic enzyme gelatinase liquefies gelatin
2. Explain why some bacteria hydrolyze gelatin
3. Describe how gelatin liquefaction is tested for in the laboratory, and perform a gelatin hydrolysis test

SUGGESTED READING IN TEXTBOOK

1. Complex Media, section 5.7; see also box 5.1 and table 5.5.

Pronunciation Guide

Enterobacter aerogenes (en-ter-oh-BAK-ter a-RAH-jen-eez)
Escherichia coli (esh-er-I-ke-a KOH-lee)
Proteus vulgaris (PRO-te-us vul-GA-ris)

Why Are the Above Bacteria Used in This Exercise?

This exercise shows the student how to differentiate between bacteria based on their ability to produce the enzyme gelatinase. The authors have chosen the following three bacteria to accomplish this objective. *Enterobacter aerogenes* is gelatinase positive, but gelatin is very slowly liquefied by most strains. In contrast, *Proteus vulgaris* also is gelatinase positive and liquefies gelatin very rapidly. *Escherichia coli* is gelatinase negative.

 Medical Application

In the clinical laboratory, gelatin hydrolysis is used to distinguish between the pathogenic *Staphylococcus aureus* (+) and the nonpathogenic *S. epidermidis* (slow−). It can also be used to distinguish *Listeria monocytogenes* (− for gelatin hydrolysis and one cause of bacterial meningitis) from some species of *Corynebacterium.*

PRINCIPLES

When boiled in water, the connective tissue collagen (which is stringy, insoluble, and indigestible) changes into **gelatin,** a soluble mixture of polypeptides. Certain bacteria are able to hydrolyze gelatin by secreting a proteolytic enzyme called **gelatinase.** The resulting

amino acids can then be used as nutrients by the bacteria. Since hydrolyzed gelatin is no longer able to gel, it is a liquid. The ability of some bacteria to digest gelatin is an important characteristic in their differentiation. For example, when grown on a gelatin medium (Thiogel), *Clostridium perfringens* causes liquefaction, whereas *Bacteroides fragilis* does not. Gelatin hydrolysis can also be used to assess the pathogenicity of certain bacteria. The production of gelatinase can often be correlated with the ability of a bacterium to break down tissue collagen and spread throughout the body of a host.

Gelatin liquefaction (the formation of a liquid) can be tested for by stabbing nutrient gelatin deep tubes. Following incubation, the cultures are placed in a refrigerator or ice bath at 4°C until the bottom resolidifies. If gelatin has been hydrolyzed, the medium will remain liquid after refrigeration. If gelatin has not been hydrolyzed, the medium will resolidify during the time it is in the refrigerator (figure 27.1). Nutrient gelatin may require up to a 14-day incubation period for positive results.

Another way to test for gelatinase is by the use of KEY Rapid Test Strips. These strips are used in the rapid test (within 24 to 48 hours) for gelatin liquefaction. Liquefaction is demonstrated by the bacterium's ability to remove, with gelatinase, the outer layer of the strip when the gelatin test strip is immersed in a suspension of bacterial cells. If gelatin is removed, the strip changes to a blue color, and the test is positive; if there is no color change, the test is negative.

Procedure

First Period

1. Label three nutrient gelatin deeps with your name, date, and the bacterium to be inoculated. Label the fourth tube "control."
2. Using aseptic technique (*see figure 14.3*), inoculate three of the deeps with the appropriate bacterium by stabbing the medium three-fourths of the way to the bottom of the tube.
3. Incubate the four tubes for 24 to 48 hours or longer at 35°C. The incubation time depends on the species of bacteria; some may require incubation for up to 2 weeks. If the latter is the case, observe on days 7 and 14.

KEY Rapid Test

1. Into three small test tubes, pipette 1/2 ml (or about 20–25 drops with a Pasteur pipette) of a heavy bacterial suspension or suspend paste in 1/2 to 1 ml of water. With a wax pencil, label each of the tubes with the appropriate bacterium, your name, and date.
2. Drop one gelatin test strip into each tube.
3. Incubate for 24 to 48 hours at 35°C.

Second Period

1. Remove the nutrient gelatin deep tubes from the incubator and place them in the refrigerator at 4°C for 30 minutes or in an ice bath for 3 to 5 minutes.
2. When the bottom resolidifies, remove the tubes and gently slant them. Notice whether or not the surface of the medium is fluid or liquid. If the nutrient gelatin is liquid, this indicates that gelatin has been hydrolyzed by the bacterium. If no hydrolysis occurred, the medium will remain a gel. The uninoculated control should also be negative.

KEY Rapid Test

1. Observe the color of the three gelatin test strips. Liquefaction will appear first along the surface of the suspension. A blue color is a positive test; no color change is a negative test.
2. Based on your observations, determine and record in the report for exercise 27 which of the three bacteria were capable of hydrolyzing gelatin.

Figure 27.1 Hydrolysis of Gelatin. If gelatin is hydrolyzed by the enzyme gelatinase, it does not gel when cooled but remains a liquid. Thus it flows when the culture is tilted backward (right tube). A negative control tube is on the left. Notice that the solid gelatin does not flow when the tube is tilted.

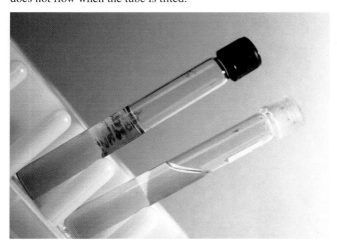

HINTS AND PRECAUTIONS

Do not shake the tubes when moving them to a refrigerator; gelatin digestion may have occurred only at the surface.

Name: _____

Date: _____

Lab Section: _____

Proteins, Amino Acids, and Enzymes IV: Gelatin Hydrolysis

1. Complete the following table on gelatin hydrolysis.

Bacterium	Gelatin Hydrolysis (+ or −)	
	Tube Results	**KEY Strips**
E. aerogenes	_____	_____
E. coli	_____	_____
P. vulgaris	_____	_____

2. Sketch and describe what is happening in each tube with respect to gelatin hydrolysis.

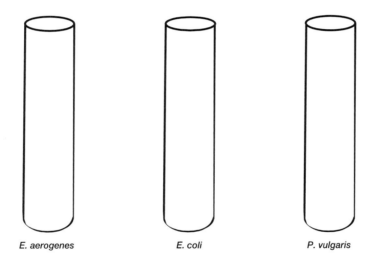

E. aerogenes E. coli P. vulgaris

3. In summary, what is the major purpose of this exercise?

Review Questions

1. How can gelatin hydrolysis be beneficial to certain bacteria?

2. What is gelatin?

3. What is unique about gelatin at 35°C versus 5°C?

4. Why did you refrigerate the gelatin cultures before observing them for liquefaction?

5. Can gelatin hydrolysis be correlated with the pathogenicity of a bacterium? Explain your answer.

6. Why is gelatin liquefied in the presence of gelatinase?

7. How does a KEY Rapid Gelatin Test Strip work?

Proteins, Amino Acids, and Enzymes V: Catalase Activity

SAFETY CONSIDERATIONS

Be careful with the Bunsen burner flame. No mouth pipetting. Three percent hydrogen peroxide is caustic to the skin and mucous membranes. Keep all culture tubes upright in a test-tube rack.

MATERIALS PER STUDENT

18- to 24-hour tryptic soy broth cultures of *Staphylococcus aureus* (ATCC 25923), *Enterococcus faecalis* (ATCC 19433), and *Micrococcus luteus* (ATCC 9341)
tryptic soy agar slants
3% hydrogen peroxide (H_2O_2)(caustic) or Difco's SpotTest Catalase Reagent
Bunsen burner
inoculating loop
Pasteur pipette with pipettor
incubator set at 35°C
test-tube rack
wax pencil
clean glass slides
wooden applicator stick (or nichrome wire loop)

Learning Objectives

Each student should be able to

1. Understand the biochemical process of hydrogen peroxide detoxification by aerobic bacteria through the production of the enzyme catalase
2. Describe how catalase production can be determined
3. Perform a catalase test

SUGGESTED READING IN TEXTBOOK

1. Oxygen Concentration, section 6.4.

Pronunciation Guide

Enterococcus faecalis (en-te-ro-KOK-kus fee-KAL-iss)
Micrococcus luteus (my-kro-KOK-us LOO-tee-us)
Staphylococcus aureus (staf-il-oh-KOK-kus ORE-ee-us)

Why Are the Above Bacteria Used in This Exercise?

In this exercise, the student will learn to perform the catalase test. The catalase test is very useful in differentiating between groups of bacteria. The authors have chosen the following three bacteria to accomplish the above objective. *Staphylococcus aureus* (L. *aureus,* golden) is a gram-positive coccus that is catalase positive when grown in an aerobic environment. *S. aureus* is mainly associated with the human skin and mucous membranes of warm-blooded vertebrates, but is often isolated from food products, dust, and water. *Enterococcus faecalis* (L. *faecium,* of the dregs, of feces) is a catalase negative, gram-positive coccus. *E. faecalis* occurs widely in the environment, particularly in feces of vertebrates. *Micrococcus luteus* (L. *luteus,* golden yellow) is another gram-positive coccus that also is catalase positive. *M. luteus* occurs primarily on mammalian skin and in soil, but commonly can be isolated from food products and air.

PRINCIPLES

Some bacteria contain flavoproteins that reduce O_2, resulting in the production of **hydrogen peroxide** (H_2O_2) or **superoxide** (O_2^-). These are extremely toxic because they are powerful oxidizing agents and destroy cellular constituents very rapidly. A bacterium must be able to protect itself against such O_2 products or it will be killed.

Many bacteria possess enzymes that afford protection against toxic O_2 products. Obligate aerobes and facultative anaerobes usually contain the enzyme **superoxide dismutase,** which catalyzes the destruction of superoxide, and either **catalase** or **peroxidase,** which catalyzes the destruction of hydrogen peroxide as follows:

$$2O_2^- + 2H^+ \xrightarrow[\text{dismutase}]{\text{superoxide}} \underset{\text{Oxygen}}{O_2} + \underset{\text{Hydrogen peroxide}}{H_2O_2}$$

$$2H_2O_2 \xrightarrow[\text{peroxidase}]{\text{catalase or}} \underset{\text{Water}}{2H_2O} + \underset{\text{Free oxygen}}{O_2}$$

Most strict anaerobes lack both enzymes and therefore cannot tolerate O_2.

Catalase production and activity can be detected by adding the substrate H_2O_2 to an appropriately incubated (18- to 24-hour) tryptic soy agar slant culture. If catalase was produced by the bacteria, the above chemical reaction will liberate free O_2 gas. Bubbles of O_2 represent a positive catalase test; the absence of bubble formation is a negative catalase test.

Catalase activity is very useful in differentiating between groups of bacteria. For example, the morphologically similar *Enterococcus* (catalase negative) and *Staphylococcus* (catalase positive) can be differentiated using the catalase test (figure28.1).

Procedure

First Period

1. Label each of the tryptic soy agar slants with the name of the bacterium to be inoculated, your name, and date.
2. Using aseptic technique (*figure 14.3*), heavily inoculate each experimental bacterium into its appropriately labeled tube by means of a streak inoculation.
3. Incubate the slants at 35°C for 18 to 24 hours.

Second Period

1. To test for catalase, set the slant in an inclined position and pipette several drops of a 3% solution of H_2O_2 over the growth on the slant or use 3 to 5 drops of Difco's SpotTest Catalase Reagent.
2. The appearance of gas bubbles (figure 28.1*a*) indicates a positive test; the absence of gas bubbles is a negative test (figure 28.1*b*).
3. Based on your observations, determine and record in the report for exercise 28 whether or not each bacterium was capable of catalase activity.
4. Note: An alternative procedure for doing the catalase test is to remove growth from a slant using a

Figure 28.1 Catalase Test on Slants. (**a**) *Staphylococcus aureus,* catalase positive. Notice the bubbles of oxygen (tube on the left). (**b**) *Enterococcus faecalis,* catalase negative; note the absence of bubbles (tube on the right).

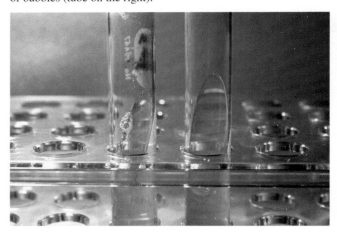

Figure 28.2 Catalase Test on Slides. A positive catalase reaction (left slide) shows gas bubbles; a negative catalase reaction reveals an absence of gas bubbles (right slide).

wooden applicator stick or nichrome wire loop and place the growth on a glass slide. The cells are then mixed in a drop of 3% H_2O_2 or a drop of Difco's SpotTest Catalase Reagent. Immediate bubbling indicates a positive catalase test (figure 28.2).

HINTS AND PRECAUTIONS

(**1**) Dispose of the hydrogen peroxide slides in the appropriate container filled with disinfectant.

(**2**) When using a slant for other purposes in the same laboratory period, it is possible to save material by adding H_2O_2 to the slant after finishing with it.

(**3**) Extreme care must be exercised if a colony is taken from a blood agar plate. Erythrocytes contain catalase, and a transfer of only a few blood cells can give a false-positive reaction.

(**4**) Always use fresh hydrogen peroxide, since it is unstable.

Name: _____

Date: _____

Lab Section: _____

Proteins, Amino Acids, and Enzymes V: Catalase Activity

1. Complete the following table on catalase activity.

Bacterium	Presence of Bubbling (catalase positive)	Absence of Bubbling (catalase negative)
E. faecalis	_____	_____
S. aureus	_____	_____
M. luteus	_____	_____

2. In summary, what is the major purpose of this exercise?

Review Questions

1. What is the importance of catalase to certain bacteria?

2. Do anaerobic bacteria require catalase? Explain your answer.

3. Write a balanced equation for the degradation of H_2O_2 in the presence of catalase.

4. What two groups of bacteria can be differentiated with the catalase test?

5. What are three products that result when flavoproteins reduce O_2?

 a.

 b.

 c.

6. What are several bacteria that produce catalase?

7. What is the substrate of the catalase reaction?

Exercise 29

Proteins, Amino Acids, and Enzymes VI: Coagulase and DNase Activity

SAFETY CONSIDERATIONS

Be careful with the Bunsen burner flame and water bath. One normal HCl can cause severe skin burns and irritate and burn the respiratory and digestive systems. No mouth pipetting. Keep all culture tubes upright in a test-tube rack or in a can.

MATERIALS PER STUDENT

24- to 48-hour tryptic slant cultures of
 Staphylococcus aureus (ATCC 25923) and
 Staphylococcus epidermidis (ATCC 14990)
citrated rabbit plasma
DNase test agar plates
water bath at 35°C
inoculating loop
Bunsen burner
1 N hydrochloric acid (caustic)
small test tubes
1-ml Pasteur pipettes with pipettor
wax pencil
incubator set at 35°C
test-tube rack

Learning Objectives

Each student should be able to

1. Understand the biochemistry of the enzymes coagulase and DNase
2. Explain how coagulase and DNase confer a survival advantage to bacteria that produce these enzymes
3. Describe how pathogenic species of staphylococci can be differentiated from nonpathogenic species
4. Perform coagulase and DNase tests

SUGGESTED READING IN TEXTBOOK

1. The *Staphylococcaceae*, section 23.4.

Pronunciation Guide

Staphylococcus aureus (staf-il-oh-KOK-kus ORE-ee-us)
S. epidermidis (e-pee-DER-meh-diss)

 Why Are the Above Bacteria Used in This Exercise?

This exercise shows the student how to differentiate pathogenic species of staphylococci from nonpathogenic species by performing catalase and DNase tests. The two most commonly encountered staphylococci will be used: *Staphylococcus aureus* and *S. epidermidis*. *S. aureus* is coagulase and DNase positive, whereas *S. epidermidis* is coagulase and DNase negative.

⚕ Medical Application

In the clinical laboratory, coagulase activity is used to distinguish between pathogenic (coagulase +) and nonpathogenic (coagulase −) staphylococci.

PRINCIPLES

Coagulases are enzymes that clot blood plasma by a mechanism that is similar to normal clotting. Although coagulase activity is not required for pathogenicity, this enzyme is a good indicator of the pathogenic potential of *S. aureus*. Coagulase-producing staphylococci (termed **coagulase positive**) form a fibrin clot around themselves and avoid attack by the host's defenses. In the coagulase test (figure 29.1), coagulase-positive staphylococci will cause the plasma to clot by using coagulase to initiate the clotting cascade. Citrate and EDTA are usually added to act as anticoagulants

175

(a)

(b)

and prevent false-positive results. Cultures should be considered **coagulase negative** if they are unclotted after 4 hours.

In addition to coagulase production, most pathogenic strains of staphylococci produce a nuclease enzyme called **DNase.** DNase degrades host DNA and increases the pathogenicity of staphylococci that possess it. To demonstrate the presence of DNase, agar containing dissolved DNA is spot-inoculated with staphylococci. A zone of clearing around the colony indicates a positive DNase test. This clearing occurs because the large DNA molecule has been degraded by the enzyme, and the end products dissolve in the added

acid. Intact DNA does not dissolve in weak acid but rather is precipitated by it; thus, the medium around colonies that do not produce DNase becomes opaque.

Procedure

First Period

Coagulase Test

1. Add 0.5 ml of citrated rabbit plasma to two small test tubes. With the wax pencil, label the tubes with the respective bacteria, your name, and date.
2. Inoculate one tube with enough *S. aureus* paste to make a cloudy suspension. Inoculate the other tube with *S. epidermidis*. Alternatively, one can add about 5 drops of thick 18- to 24-hour broth culture to each tube.
3. Incubate both tubes at 35°C for 1 to 4 hours in a water bath. Afterward, examine both tubes for the presence or absence of clouding and clots.

 A positive coagulase test is represented by any degree of clotting, from a loose clot suspended in plasma to a solid clot (figure 29.1*a*).

DNase Test

1. Divide a DNase test agar plate in half by marking the bottom with a wax pencil. Add your name and date.
2. On half of the plate, heavily spot-inoculate *S. aureus* over a 0.5-cm area (figure 29.2*a*). Do the same with *S. epidermidis* on the other half of the plate.
3. Incubate for 18 to 24 hours at 35°C.

Second Period

1. Using a Pasteur pipette with bulb, flood the DNase test agar plate with 1 N HCl. A zone of clearing around the colony indicates a positive DNase test (i.e., the DNA in the medium has been degraded) (figure 29.2*b*).
2. Based on your observations, determine and record in the report for exercise 29 whether or not each bacterium was coagulase and DNase positive or negative.

HINTS AND PRECAUTIONS

(1) Do not shake or vortex the citrated rabbit plasma when inoculating it with the bacteria.

(2) In a positive DNA test, the zone of clearing should appear immediately.

Figure 29.2 DNase Test. (a) Spot inoculate half of a DNase test agar plate with *S. aureus* and the other half with *S. epidermidis*. **(b)** DNase degradation (*S. aureus,* right side of plate) and no degradation (*S. epidermidis,* left side of plate).

(a)

(b)

Proteins, Amino Acids, and Enzymes VI:
Coagulase and DNase Activity

1. Complete the following table on coagulase and DNase activity.

Biochemical Results	S. aureus	S. epidermidis
Coagulase test (+ or −)	_____	_____
DNase test (+ or −)	_____	_____

2. Make a drawing of your DNase test agar plate observations.

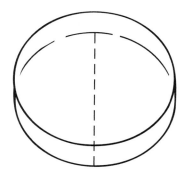

3. In summary, what is the major purpose of this exercise?

Review Questions

1. How does the enzyme coagulase function?

2. How does the enzyme DNase function?

3. False-positive coagulase tests have been reported for some bacteria that can metabolize citrate. Explain.

4. Does a coagulase-positive staphylococcus also have to be DNase positive? Explain why or why not.

5. What is the function of the 1 N HCl added to the DNase plates after incubation?

6. Describe a positive DNase test.

7. Describe a negative DNase test.

Proteins, Amino Acids, and Enzymes VII: Oxidase Test

MATERIALS PER STUDENT

young 24-hour tryptic soy broth cultures of *Alcaligenes faecalis* (ATCC 8750), *Escherichia coli* (ATCC 25922), and *Pseudomonas aeruginosa* (ATCC 27853)

tryptic soy agar plates

tetramethyl-p-phenylenediamine dihydrochloride (oxidase reagent)

Bunsen burner

platinum or plastic loops

wax pencil

Pasteur pipette with pipettor

Oxidase Disks or Dry Slides (Difco); Oxidase Test Strips (KEY Scientific Products); SpotTest Oxidase Reagent (Difco)

wooden applicator sticks

Whatman No. 2 filter paper

Learning Objectives

Each student should be able to

1. Understand the biochemistry underlying oxidase enzymes
2. Describe the experimental procedure that enables one to distinguish between groups of bacteria based on cytochrome oxidase activity
3. Give examples of oxidase-positive and oxidase-negative bacteria
4. Perform an oxidase test

SUGGESTED READING IN TEXTBOOK

1. The Electron Transport Chain, section 9.5; see also figures 9.13–9.15.
2. Rapid Methods of Identification, section 36.2; see also table 36.3.

Pronunciation Guide

Alcaligenes faecalis (al-kah-LIJ-e-neez fee-KAL-iss)
Escherichia coli (esh-er-I-ke-a KOH-lee)
Pseudomonas aeruginosa (soo-do-MO-nas a-ruh-jin-OH-sah)

Why Are the Above Bacteria Used in This Exercise?

This exercise gives the student experience in performing the oxidase test. The oxidase test distinguishes between groups of bacteria based on cytochrome oxidase activity. Three bacteria will be used. *Alcaligenes faecalis* (L. *faecium,* of the dregs, of feces) is a gram-negative, aerobic rod (coccal rod or coccus) that possesses a strictly respiratory type of metabolism with oxygen as the terminal electron acceptor. It is thus oxidase positive. *Escherichia coli* is a facultatively anaerobic gram-negative rod that has both respiratory and fermentative types of metabolism and is oxidase negative. *Pseudomonas aeruginosa* is a gram-negative, aerobic rod having a strictly respiratory type of metabolism with oxygen as the terminal electron acceptor and thus is oxidase positive.

Medical Application

The oxidase test is a useful procedure in the clinical laboratory because some gram-negative pathogenic species of bacteria (such as *Neisseria gonorrhoeae, P. aeruginosa,*

and *Vibrio* species) are oxidase positive, in contrast to species in the family *Enterobacteriaceae,* which are oxidase negative.

PRINCIPLES

Oxidase enzymes play an important role in the operation of the electron transport system during aerobic respiration. **Cytochrome oxidase** (aa_3 type) uses O_2 as an electron acceptor during the oxidation of reduced cytochrome *c* to form water and oxidized cytochrome *c*.

The ability of bacteria to produce cytochrome oxidase can be determined by the addition of the oxidase test reagent or test strip (tetramethyl-*p*-phenylenediamine dihydrochloride or an Oxidase Disk, p-minodimethylaniline) to colonies that have grown on a plate medium. Or, using a wooden applicator stick, a bacterial sample can either be rubbed on a Dry Slide Oxidase reaction area, on a KEY test strip, or filter paper moistened with the oxidase reagent. The light pink oxidase test reagent (disk, strip, or slide) serves as an artificial substrate, donating electrons to cytochrome oxidase and in the process becoming oxidized to a purple and then dark purple (figure 30.1) compound in the presence of free O_2 and the oxidase. The presence of this dark purple coloration represents a positive test. No color change or a light pink coloration on the colonies indicates the absence of oxidase and is a negative test.

Procedure

First Period

1. With a wax pencil, divide the bottom of a tryptic soy agar plate into three sections and label each with the name of the bacterium to be inoculated, your name, and date.
2. Using aseptic technique (*see figure 14.3*), make a single streak-line inoculation on the agar surface with the appropriate bacterium.

3. Incubate the plate in an inverted position for 24 to 47 hours at 35°C.

Second Period

1. Add 2 to 3 drops of the oxidase reagent to the surface of the growth of several isolated colonies of each test bacterium or to some paste that has been transferred to a piece of filter paper. Using another colony, place an Oxidase Disk on it. Add a drop of sterile water. If Dry Slides or test strips are available, use a wooden applicator stick to transfer a sample to the slide, test strip, or filter paper moistened with oxidase reagent. Alternatively, drop a KEY Oxidase Test Strip onto the surface of a slant culture and moisten it with water if necessary.
2. Observe the colony or sample for the presence or absence of a color change from pink to purple, and finally to dark purple. This color change will occur within 20 to 30 seconds. Color changes after 20 to 30 seconds are usually disregarded since the reagent begins to change color with time due to auto-oxidation. Oxidase-negative bacteria will not produce a color change or will produce a light pink color.
3. Based on your observations, determine and record in the report for exercise 30 whether or not each bacterium was capable of producing oxidase.

<div style="border:1px solid">

HINTS AND PRECAUTIONS

(1) Students should note the color change immediately following the addition of oxidase reagent. Color changes after 20 seconds are not valid.

(2) Using nichrome or other iron-containing inoculating devices may cause false-positive reactions.

(3) If bacterial paste is transferred with an applicator stick, put the stick in a jar of disinfectant or a Biohazard bag immediately after use.

</div>

Figure 30.1 **Oxidase Test.** Note the purple to dark purple color after the colonies have been added to filter paper moistened with oxidase reagent.

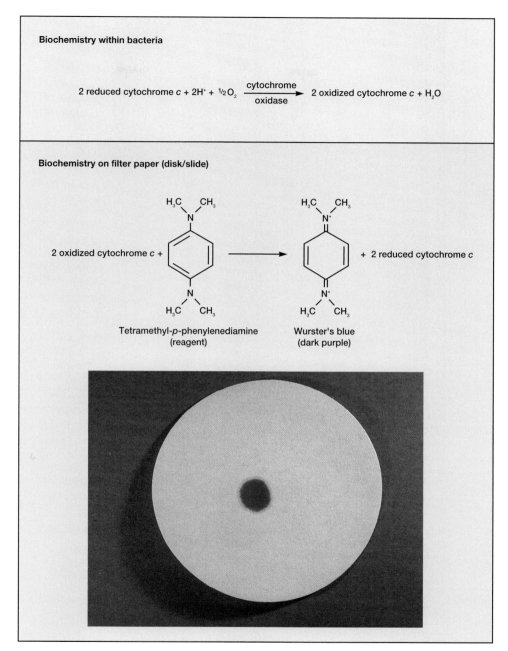

Laboratory Report 30

Name: _____

Date: _____

Lab Section: _____

Proteins, Amino Acids, and Enzymes VII: Oxidase Test

1. Complete the following table on the oxidase test.

Bacterium	Color of Colonies after Adding		Oxidase Production (+ or −)	
	Reagent	Disk or Slide	Reagent	Disk or Slide
A. faecalis	_____	_____	_____	_____
E. coli	_____	_____	_____	_____
P. aeruginosa	_____	_____	_____	_____

2. In summary, what is the major purpose of this exercise?

Review Questions

1. What metabolic property characterizes bacteria that possess oxidase activity?

2. What is the importance of cytochrome oxidase to bacteria that possess it?

3. Do anaerobic bacteria require oxidase? Explain your answer.

4. What is the function of the test reagent in the oxidase test?

5. The oxidase test is used to differentiate among which groups of bacteria?

6. Why should nichrome or other iron-containing inoculating devices not be used in the oxidase test?

7. Are there limitations to the oxidase test?

Proteins, Amino Acids, and Enzymes VIII: Urease Activity

MATERIALS PER STUDENT

24- to 48-hour tryptic soy agar slants of *Escherichia coli* (ATCC 11229), *Klebsiella pneumoniae* (ATCC e13883), *Proteus vulgaris* (ATCC 13315), and *Salmonella cholerae-suis* (ATCC 29631)
5 urea broth tubes
Bunsen burner
test-tube rack
inoculating loop
incubator set at 35°C
urea disks (Difco) or urease test tablets (KEY Scientific Products)
4 sterile test tubes
wax pencil
sterile forceps

Learning Objectives

Each student should be able to

1. Understand the biochemical process of urea hydrolysis
2. Determine the ability of bacteria to degrade urea by means of the enzyme urease
3. Tell when the urease test is used
4. Perform a urease test

SUGGESTED READING IN TEXTBOOK

1. *Pseudomonas* and the *Enterobacteriaceae,* section 22.3; see also figure 22.8 and tables 22.6, 22.7.

Pronunciation Guide

Escherichia coli (esh-er-I-ke-a KOH-lee)
Klebsiella pneumoniae (kleb-se-EL-lah nu-mo-ne-ah)
Proteus vulgaris (PRO-tee-us vul-GA-ris)
Salmonella cholerae-suis (sal-mon-EL-ah coler-ah SU-is)

Why Are the Above Bacteria Used in This Exercise?

In this exercise, the student will perform a urease test to determine the ability of bacteria to degrade urea by means of the enzyme urease. The authors have chosen two urease-positive bacteria (*Klebsiella pneumoniae* and *Proteus vulgaris*) and two urease-negative bacteria (*Escherichia coli* and *Salmonella cholerae-suis*).

Medical Application

In the clinical laboratory, members of the genus *Proteus* can be distinguished from other enteric nonlactose-fermenting bacteria (*Salmonella, Shigella*) by their fast urease activity. *P. mirabilis* is a major cause of human urinary tract infections.

PRINCIPLES

Some bacteria are able to produce an enzyme called **urease** that attacks the nitrogen and carbon bond in amide compounds such as urea, forming the end products ammonia, CO_2, and water (figure 31.1).

Urease activity (the **urease test**) is detected by growing bacteria in a medium containing urea and using

Figure 31.1 **Urea Hydrolysis.** (**a**) Uninoculated control. (**b**) Weakly positive reaction (delayed positive). (**c**) Very rapid positive reaction. (**d**) Negative reaction. (*© Raymond B. Otero/Visuals Unlimited*)

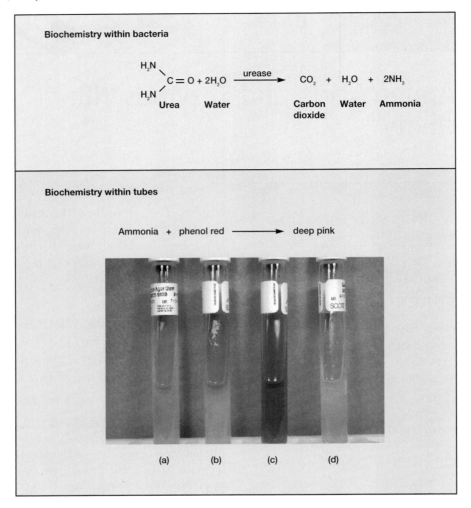

a pH indicator such as phenol red (*see appendix E*). When urea is hydrolyzed, ammonia accumulates in the medium and makes it alkaline. This increase in pH causes the indicator to change from orange-red to deep pink or purplish red (cerise) and is a positive test for urea hydrolysis. Failure to develop a deep pink color is a negative test.

Procedure

First Period

1. Label each of the urea broth tubes with the name of the bacterium to be inoculated, your name, and date.

2. Using aseptic technique (*see figure 14.3*), inoculate each tube with the appropriate bacterium by means of a loop inoculation.

3. Incubate the tubes for 24 to 48 hours at 35°C.

Urea Disks or Tablets

1. Add 0.5 ml (about 20 drops) of sterile distilled water to four sterile test tubes for the Difco disk or 1 ml distilled water for the KEY tablet.

2. Transfer one or two loopfuls of bacterial paste to each tube. Label with your name and date.

3. Using sterile forceps, add one urea or urease disk tablet to each tube.

4. Incubate up to 4 hours at 35°C. Check for a color change each hour. (The KEY test may be incubated up to 24 hours if necessary.)

Second Period

1. Examine all of the urea broth cultures and urea disk or urease tablet tubes to determine their color (figures 31.1 and 31.2).
2. Based on your observations, determine and record in the report for exercise 31 whether each bacterium was capable of hydrolyzing urea.

HINTS AND PRECAUTIONS

Some bacteria have a delayed urease reaction that may require an incubation period longer than 48 hours.

Figure 31.2 KEY Test for Urea. After incubation, a pink to red color constitutes a positive test (tube on the left). If the original straw color persists, the test is negative (tube on the right).

Name: _____

Date: _____

Lab Section: _____

Proteins, Amino Acids, and Enzymes VIII: Urease Activity

1. Complete the following table on urease activity.

| Bacterium | Color of | | Urea Hydrolysis (+ or −) |
	Urea Broth	Disks	
E. coli	_____	_____	_____
K. pneumoniae	_____	_____	_____
P. vulgaris	_____	_____	_____
S. cholerae-suis	_____	_____	_____

2. In summary, what is the major purpose of this exercise?

Review Questions

1. Explain the biochemistry of the urease reaction.

2. What is the purpose of the phenol red in the urea broth medium?

3. When would you use the urease test?

4. Why does the urea disk change color?

5. What is the main advantage of the urea disk over the broth tubes with respect to the detection of urease?

6. What is in urea broth?

7. What color is cerise?

Proteins, Amino Acids, and Enzymes IX: Lysine and Ornithine Decarboxylase Test

MATERIALS PER STUDENT

24- to 48-hour tryptic soy broth cultures of *Enterobacter aerogenes* (ATCC 13048), *Citrobacter freundii* (ATCC 8090), *Klebsiella pneumoniae* (ATCC e13883), and *Proteus vulgaris* (ATCC 13315)

4 Moeller's lysine decarboxylase broth with lysine (LDC)

4 lysine iron agar slants (LIA)

4 Moeller's ornithine decarboxylase broth with ornithine (ODC)

1 Moeller's lysine decarboxylase broth without lysine (DC), which will serve as the control

1 Moeller's ornithine decarboxylase broth without ornithine (OD), which will serve as the control

Pasteur pipettes with pipettor

inoculating loop

test-tube rack

sterile distilled water

sterile mineral oil

incubator set at 35°C

8 sterile test tubes

ornithine, lysine, and decarboxylase KEY Rapid Substrate Tablets and strips (KEY Scientific Products, 1402 Chisholm Trail, Suite D, Round Rock, TX 78681; 800-843-1539; *www.keyscientific.com*)

Bunsen burner

ninhydrin in chloroform (Dissolve 50 mg ninhydrin in 0.4 ml of dimethylsulfoxide [DMSO], then add 25 ml of chloroform to the DMSO solution.)

10% KOH

Learning Objectives

Each student should be able to

1. Understand the biochemical process of decarboxylation
2. Tell why decarboxylases are important to some bacteria
3. Explain how the decarboxylation of lysine can be detected in culture
4. Perform lysine and ornithine decarboxylase tests

SUGGESTED READING IN TEXTBOOK

1. Protein and Amino Acid Catabolism, section 9.9; see also figure 9.23.

Pronunciation Guide

Citrobacter freundii (SIT-ro-bac-ter FRUN-dee)
Enterobacter aerogenes (en-ter-oh-BAK-ter a-RAH-jen-eez)
Klebsiella pneumoniae (kleb-se-EL-lah nu-MO-ne-ah)
Proteus vulgaris (PRO-te-us vul-GA-ris)

Why Are the Above Bacteria Used in This Exercise?

This exercise gives the student experience using the lysine and ornithine decarboxylase test to differentiate between bacteria. Two lysine decarboxylase-positive (*Enterobacter aerogenes* and *Klebsiella pneumoniae*) and two lysine decarboxylase-negative (*Proteus vulgaris* and *Citrobacter freundii*) bacteria, and two ornithine decarboxylase-positive (*E. aerogenes* and *Citrobacter freundii*) and two ornithine decarboxylase-negative (*K. pneumoniae* and *P. vulgaris*) bacteria were chosen to demonstrate the lysine and ornithine decarboxylase tests.

Medical Application

In the clinical laboratory, decarboxylase differential tests are used to differentiate between organisms in the *Enterobacteriacea E.*

PRINCIPLES

Decarboxylation is the removal of a carboxyl group from an organic molecule. Bacteria growing in liquid media decarboxylate amino acids most actively when conditions are anaerobic and slightly acidic. Decarboxylation of amino acids, such as lysine and ornithine, results in the production of an amine and CO_2 as illustrated below.

$$R - \underset{\underset{\text{An amino acid}}{\underset{|}{NH_2}}}{CH} - \boxed{COOH} \xrightarrow{\text{decarboxylase}} R - CH_2 - NH_2 + \boxed{CO_2}$$

An amine Carbon dioxide gas

Bacteria that are able to produce the enzymes **lysine decarboxylase** and **ornithine decarboxylase** can decarboxylate lysine and ornithine and use the amines as precursors for the synthesis of other needed molecules. In addition, when certain bacteria carry out fermentation, acidic waste products are produced, making the medium acidic and inhospitable. Many decarboxylases are activated by a low pH. They remove the acid groups from amino acids, producing alkaline amines, which raise the pH of the medium making it more hospitable.

Decarboxylation of lysine or ornithine can be detected by culturing bacteria in a medium containing the desired amino acid, glucose, and a pH indicator (bromcresol purple, *see appendix E*). Before incubation, sterile mineral oil is layered onto the broth to prevent oxygen from reaching the bacteria and inhibiting the reaction. The acids produced by the bacteria from the fermentation of glucose will initially lower the pH of the medium and cause the pH indicator to change from purple to yellow. The acid pH activates the enzyme that causes decarboxylation of lysine or ornithine to amines and the subsequent neutralization of the medium. This results in another color change from yellow back to purple (figure 32.1).

Lysine iron agar (LIA) is also used for the cultivation and differentiation of members of the *Enterobacteriaceae* based on their ability to decarboxylate lysine and to form H_2S. Bacteria that decarboxylate lysine turn the medium purple. Bacteria that produce H_2S appear as black colonies.

The lysine decarboxylase test is useful in differentiating *Pseudomonas* (L. −), *Klebsiella* (L. +), *Enterobacter* (L. +), and *Citrobacter* (L. −) species. The ornithine decarboxylase test is helpful in distinguishing between *Klebsiella* (O. −) and *Proteus* (O. −), and *Enterobacter* (O. +) bacteria.

A quick test for ornithine or lysine decarboxylase is to use the KEY Rapid Substrate Tablets and strips. These tablets contain the respective amino acids in a mixture of salts correctly buffered for each test. In addition, a pH indicator is present in the tablet, which changes color as the decarboxylation reaction progresses. In the lysine decarboxylase test tablet, the indicator is bromcresol purple, which turns purple as the test becomes positive (figure 32.2). The indicator in the ornithine decarboxylase test tablet is phenol red, which turns red in a positive test.

Procedure

First Period (Standard Method)

1. Label four LDC tubes and/or LIA slants with the names of the respective bacteria (*K. pneumoniae, E. aerogenes, P. vulgaris,* and *C. freundii*) to be inoculated. Do the same for one control DC tube. Add your name and date to the tubes.
2. Do the same with the four ODC and one OD tubes.
3. As shown in figure 14.3, aseptically inoculate the tubes with the proper bacteria.
4. With a sterile Pasteur pipette, layer about 1 ml of sterile mineral oil on top of the inoculated media. LIA slants do not need mineral oil.
5. Incubate the cultures for 24 to 48 hours at 35°C.

KEY Test Tablet/Strip Method

1. Label eight sterile test tubes with the respective bacteria, your name, and date.
2. Pipette 1 ml of sterile distilled water in each tube for regular tablets and 0.5 ml for ODC test strips.
3. Add a loopful of cell paste or 0.1 ml of thick bacterial culture to each tube.
4. Add four ornithine test strips to the first four tubes and four lysine tablets to the other four tubes.
5. Incubate the LDC tubes at 35°C for 24 hours and the ODC test strips for 4 to 6 hours.
6. A color change to purple (LDC) or red (ODC) constitutes a positive test; no color change is a negative test.

Figure 32.1 Ornithine Decarboxylase Test. (a) The tube on the left is the uninoculated control. It is purple due to the pH indicator brom-cresol purple. **(b)** The second tube from the left (yellow) is negative for ornithine decarboxylase; weak acid production (pH less than 5.2) from glucose fermentation has turned it yellow due to the accumulation of acidic end products (e.g., *Proteus vulgaris*). If the bacterium is only capable of glucose fermentation, the medium will remain yellow. **(c)** The third tube from the left (light purple) is slightly positive for ornithine decarboxylase due to the accumulation of alkaline end products. **(d)** The fourth tube from the left is more positive for the enzyme since it is a darker purple. **(e)** The tube on the right is strongly positive for ornithine decarboxylase (e.g., *Klebsiella pneumoniae*).

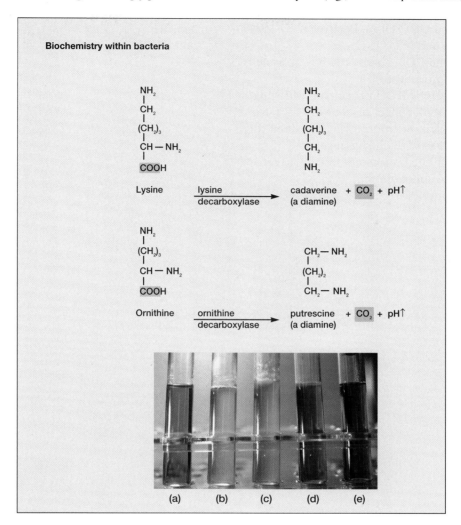

Biochemistry within bacteria

$$\text{Lysine} \xrightarrow[\text{decarboxylase}]{\text{lysine}} \text{cadaverine (a diamine)} + CO_2 + pH\uparrow$$

$$\text{Ornithine} \xrightarrow[\text{decarboxylase}]{\text{ornithine}} \text{putrescine (a diamine)} + CO_2 + pH\uparrow$$

(a) (b) (c) (d) (e)

Figure 32.2 Lysine Decarboxylase KEY Test. The purple color in the tube on the left is a positive reaction to lysine. No color change (the tube on the right) is a negative reaction.

+ LYSINE -

Second Period

1. Examine the cultures for color changes in the medium and record your results in the report for exercise 32. Enzymatic activity is indicated by an alkaline (dark purple) reaction when compared with the inoculated control medium (light slate color) in the LDC, LIA, and ODC tubes. Positive KEY tests are purple (LDC) and red (ODC).

2. The KEY ODC and LDC results can be confirmed by the ninhydrin procedure.
 a. Add 1 drop of 10% KOH to each tube and mix.
 b. Add either 1.0 ml (tablet test) or 0.5 ml (strip test) of ninhydrin in chloroform. Let stand for 10 to 15 minutes without shaking.
 c. Purple color in the bottom chloroform layer is positive for decarboxylation.

HINTS AND PRECAUTIONS

(1) In biochemical tests involving visual evaluation of color changes that are sometimes minimal, it is often useful to hold the control and experimental tubes next to each other to discern any color differences.

(2) In decarboxylase tests, any trace of purple, from light to dark purple, is considered a positive test.

Proteins, Amino Acids, and Enzymes IX: Lysine and Ornithine Decarboxylase Test

1. Results from the decarboxylase tests.

Bacterium	Color of LIA	Color of LDC	Color of ODC	LD Tablets	OD Tablets
C. freundii	_____	_____	_____	_____	_____
E. aerogenes	_____	_____	_____	_____	_____
K. pneumoniae	_____	_____	_____	_____	_____
P. vulgaris	_____	_____	_____	_____	_____

2. Tabulate the significant ingredients of the following broths.

Medium	Ingredients
Moeller's lysine decarboxylase broth	_____

Moeller's ornithine decarboxylase broth	_____

Lysine iron agar	_____

3. In summary, what is the major purpose of this exercise?

Review Questions

1. Explain what occurs during decarboxylation.

2. Why does the LDC broth or lysine iron agar turn purple when lysine is decarboxylated?

3. Why does the LDC medium always turn yellow regardless of the ability of the bacteria to produce lysine decarboxylase?

4. Why is the lysine decarboxylase test negative if both LDC and DC broths turn purple?

5. Why is sterile mineral oil added to LDC test media?

6. What is the basis for the quick KEY test for ornithine or lysine decarboxylase?

7. How does the pH indicator bromcresol purple indicate a change in pH?

Proteins, Amino Acids, and Enzymes X: Phenylalanine Deamination

MATERIALS PER STUDENT

24- to 48-hour tryptic soy broth cultures of *Escherichia coli* (ATCC 11229) and *Proteus vulgaris* (ATCC 13315)

3 phenylalanine deaminase agar slants or phenylalanine deaminase test tablets (KEY Scientific Products)

10% aqueous ferric chloride solution (or 10% $FeCl_3$ in 50% HCl)

inoculating loop

Pasteur pipette with pipettor

test-tube rack

incubator set at 35°C

wax pencil

Learning Objectives

Each student should be able to

1. Understand the biochemical process of phenylalanine deamination
2. Describe how to perform the phenylalanine deamination test
3. Perform a phenylalanine test

SUGGESTED READING IN TEXTBOOK

1. Protein and Amino Acid Catabolism, section 9.9; see also figure 9.23.

Pronunciation Guide

Escherichia coli (esh-er-I-ke-a KOH-lee)
Proteus vulgaris (PRO-tee-us vul-GA-ris)

Why Are the Following Bacteria Used in This Exercise?

In this exercise, the student will learn how to perform the phenylalanine deaminase test to differentiate between various enteric bacteria. The ability of certain bacteria to oxidatively degrade phenylalanine is of taxonomic importance. The two enteric bacteria chosen to show this differentiation are *Escherichia coli* and *Proteus vulgaris*. *P. vulgaris* produces the enzyme phenylalanine deaminase whereas *E. coli* does not.

Medical Application

In the clinical laboratory, phenylalanine deamination can be used to differentiate the genera *Morganella, Proteus,* and *Providencia* (+) from the *Enterobacteriaceae* (−). Bacteria in these genera can cause urinary tract infections and are capable of causing opportunistic infections elsewhere in the body. It is of interest to note that *Proteus* is named for the Greek god Proteus who changed shape to avoid detection.

PRINCIPLES

Phenylalanine deaminase catalyzes the removal of the amino group (NH_3^+) from phenylalanine (figure 33.1). The resulting products include organic acids, water, and ammonia. Certain enteric bacteria (e.g., *Proteus, Morganella,* and *Providencia*) can use the organic acids in biosynthesis reactions. In addition, the deamination detoxifies inhibitory amines.

Figure 33.1 Phenylalanine Deamination. (a) Uninoculated control. **(b)** Phenylalanine negative. **(c)** Phenylalanine positive.

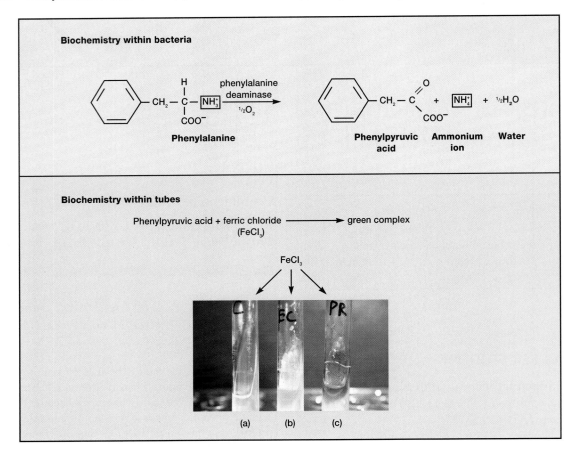

The **phenylalanine deaminase test** can be used to differentiate among enteric bacteria such as *E. coli* and *P. vulgaris*. *P. vulgaris* produces the enzyme phenylalanine deaminase, which deaminates phenylalanine, producing phenylpyruvic acid. When ferric chloride is added to the medium, it reacts with phenylpyruvic acid, forming a green compound. Since *E. coli* does not produce the enzyme, it cannot deaminate phenylalanine. When ferric chloride is added to an *E. coli* culture, there is no color change.

Procedure

First Period

1. Label two slants of phenylalanine deaminase agar with the name of the bacterium to be tested. Use another slant as a control. Add your name and date to each slant.
2. Using aseptic technique (*see figure 14.3*), inoculate each of the slants with the respective bacteria.
3. Incubate aerobically at 35°C for 18 to 24 hours.

4. Alternatively, the cultures can be directly tested by the addition of KEY test tablets. Add a tablet to 1 ml distilled water, inoculate heavily with paste, and incubate for about 20 to 24 hours at 35°C.

 Add 1 or 2 drops of 10% $FeCl_3$ reagent. A yellowish green color that develops within 1 to 5 minutes is a positive test (figure 33.2).

Second Period

1. With the Pasteur pipette, add a few drops of the 10% $FeCl_3$ to the growth on the slant. Rotate each tube between your palms to wet and loosen the bacterial growth. The presence of phenylpyruvic acid is indicated by the development of a green color within 5 minutes and indicates a positive test for phenylalanine deamination. If there is no color change after adding the reagent, the test is negative, and no deamination has occurred.
2. Based on your observations, determine and record in the report for exercise 33 which of the bacteria were able to deaminate phenylalanine.

Figure 33.2 KEY Test for Phenylalanine. A greenish-yellow color developing in 1 to 5 minutes (tube on the left) is a positive test for phenylalanine deaminase. No color change (the tube on the right) is a negative reaction.

Name: _____

Date: _____

Lab Section: _____

Proteins, Amino Acids, and Enzymes X: Phenylalanine Deamination

1. Complete the following table on phenylalanine deamination.

Bacterium	Color of the Slant	Deamination (+ or −)
E. coli	_____	_____
P. vulgaris	_____	_____

2. Describe the phenylalanine deamination reaction.

3. In summary, what is the major purpose of this exercise?

Review Questions

1. What are two ways that phenylalanine can be used by *P. vulgaris?*

2. What is the purpose of the ferric chloride in the phenylalanine deamination test?

3. When would you use the phenylalanine deamination test?

4. Name some bacteria that can deaminate phenylalanine.

5. Describe the process of deamination.

6. Why must the phenylalanine test be determined within 5 minutes?

7. Describe the color of an uninoculated tube of phenylalanine agar.

Exercise 34

Proteins, Amino Acids, and Enzymes XI: Nitrate Reduction

MATERIALS PER STUDENT

24- to 48-hour tryptic soy broth cultures of
 Escherichia coli (ATCC 11229), *Pseudomonas fluorescens* (ATCC 13525), and
 Staphylococcus epidermidis (ATCC 14990)
garden soil
Bunsen burner
inoculating loop
1-ml pipette with pipettor
nitrate broth tubes or nitrate agar slants
nitrite test reagent A or Difco's SpotTest Nitrate
 Reagent A
nitrite test reagent B or Difco's SpotTest Nitrate
 Reagent B
zinc powder or dust or Difco's SpotTest Nitrate
 Reagent C
test-tube rack
incubator set at 35°C
5 sterile test tubes
wax pencil
disposable gloves

Learning Objectives

Each student should be able to

1. Understand the biochemical process of nitrate reduction by bacteria

2. Describe how nitrate reduction can be determined from bacterial cultures
3. Perform a nitrate reduction test

SUGGESTED READING IN TEXTBOOK

1. Anaerobic Respiration, section 9.6.

Pronunciation Guide

Escherichia coli (esh-er-I-ke-a KOH-lee)
Pseudomonas fluorescens (soo-do-MO-nas floor-es-shens)
Staphylococcus epidermidis (staf-il-oh-KOK-kus e-pee-DER-meh-diss)

Why Are the Above Bacteria Used in This Exercise?

In this exercise, the student will learn how to perform the nitrate reduction test in order to differentiate between bacteria. Three different bacteria that give three different nitrate reduction results will be used. *Staphylococcus epidermidis* is unable to use nitrate as a terminal electron acceptor; therefore, it cannot reduce nitrate. *Escherichia coli* can reduce nitrate only to nitrite. *Pseudomonas fluorescens* (M. L. *fluoresco,* fluoresce; the fluorescent *Pseudomonas* species are characterized by excretion of diffusible yellow-green pigments that fluoresce in ultraviolet light) often reduces nitrate completely to molecular nitrogen.

Medical Application

Most enteric bacteria are nitrate reducers. Pathogenic examples include *Escherichia coli* (opportunistic urinary tract infections), *Klebsiella pneumoniae* (bacterial pneumonia),

Morganella morganii and *Proteus mirabilis* (nosocomial infections). Nonenteric nitrogen reducing pathogens include *Staphylococcus aureus* (staphylococcal food poisoning, bacteremia, various abscesses) and *Bacillus anthracis* (anthrax).

PRINCIPLES

Chemolithoautotrophic bacteria (bacteria that obtain energy through chemical oxidation; they use inorganic compounds as electron donors and CO_2 as their primary source) and many chemoorganoheterotrophs (bacteria that require organic compounds for growth; the organic compounds serve as sources of carbon and energy) can use nitrate (NO_3^-) as a terminal electron acceptor during anaerobic respiration. In this process, nitrate is reduced to nitrite (NO_2^-) by **nitrate reductase** as illustrated in figure 34.1. Some of these bacteria possess the enzymes to further reduce the nitrite to either the ammonium ion or molecular nitrogen as also illustrated in figure 34.1.

The ability of some bacteria to reduce nitrate can be used in their identification and isolation. For example, *E. coli* can reduce nitrate only to nitrite, *P. fluorescens* reduces it completely to molecular nitrogen, and *S. epidermidis* is unable to use nitrate as a terminal electron acceptor.

The **nitrate reduction test** is performed by growing bacteria in a culture tube with a nitrate broth medium containing 0.5% potassium nitrate (KNO_3). After incubation, the culture is examined for the presence of gas and nitrite ions in the medium. The gas (a mixture of CO_2 and N_2) is released from the reduction of nitrate (NO_3^-) and from the citric acid cycle (CO_2) (figure 34.1). The nitrite ions are detected by the addition of sulfanilic acid and N,N-dimethyl-1-naphthylamine to the culture. Any nitrite in the medium will react with these reagents to produce a pink or red color.

If a culture does not produce a color change, several possibilities exist: (1) the bacteria possess nitrate reductase and also reduce nitrite further to ammonia or molecular nitrogen; (2) they possess other enzymes that reduce nitrite to ammonia; or (3) nitrates were not reduced by the bacteria. To determine if nitrates were reduced past nitrite, a small amount of zinc powder or 5 to 10 drops of SpotTest Nitrate Reagent C is added to the culture containing the reagents. Since zinc reduces nitrates to nitrites, a pink or red color will appear and verifies the fact that nitrates were not reduced to nitrites by the bacteria. If a red color does not appear, the nitrates in the medium were reduced past the nitrite stage to either ammonia or nitrogen gas.

Procedure

First Period

1. Label three tubes of nitrate broth or nitrate agar slants with the three respective bacteria (*E. coli*, *P. fluorescens*, and *S. epidermidis*); label the fourth tube "garden soil" and the fifth tube "control." Add your name and date to each tube. The control tube serves two purposes: (1) to determine if the medium is sterile and (2) to determine if any O_2 comes out of the medium instead of out of the gas produced by the bacteria.
2. Using aseptic technique (*see figure 14.3*), inoculate three tubes with the respective bacteria, and the fourth with about a gram of garden soil.
3. Incubate all five tubes for 24 to 48 hours at 35°C.

Second Period

1. Observe the tubes for the presence of growth, and the absence of growth in the control tube.
2. With a pipette and pipettor, while wearing disposable gloves, add 0.5 ml of nitrate test reagent A and 0.5 ml of test reagent B to each of the culture tubes and mix. (Alternatively, about 5 to 10 drops of each reagent works well.) A distinct pink or red color indicates a positive test, provided the uninoculated control medium is negative.
3. Negative tests should be confirmed by adding several grains of zinc powder or 5 to 10 drops of Difco's nitrate reagent C and gently shaking the tube. If nitrate is present in the medium, it will turn red within 5 to 10 minutes; if it is absent, there will be no color change.
4. Record your results in the report for exercise 34.

HINTS AND PRECAUTIONS

(1) Although disposable gloves should be worn when using nitrite reagents A and B, if these solutions get on your hands, wash them immediately with soap and water for at least 15 minutes.

(2) Bubbles indicate a positive test for nonfermenters only; fermenters may also produce gas from carbohydrates.

(3) Even a small amount of gas or bubble production is a positive test for nonfermenters.

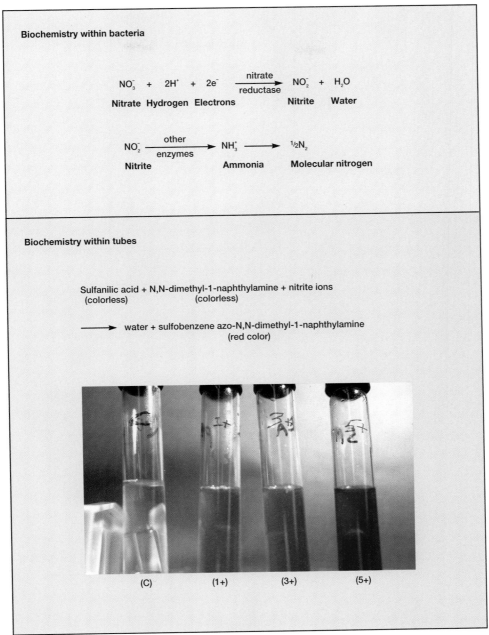

Biochemistry within bacteria

$$NO_3^- + 2H^+ + 2e^- \xrightarrow{\text{nitrate reductase}} NO_2^- + H_2O$$

Nitrate Hydrogen Electrons Nitrite Water

$$NO_2^- \xrightarrow{\text{other enzymes}} NH_3^+ \longrightarrow \tfrac{1}{2}N_2$$

Nitrite Ammonia Molecular nitrogen

Biochemistry within tubes

Sulfanilic acid + N,N-dimethyl-1-naphthylamine + nitrite ions
 (colorless) (colorless)

\longrightarrow water + sulfobenzene azo-N,N-dimethyl-1-naphthylamine
 (red color)

(C) (1+) (3+) (5+)

Laboratory Report *34*

Name: _____

Date: _____

Lab Section: _____

Proteins, Amino Acids, and Enzymes XI: Nitrate Reduction

1. On the basis of your observations, complete the following table.

Bacterium or Soil	Color with Reagents	Color with Zinc	Nitrate Reduction (+ or −)	End Products	Gas
E. coli	_____	_____	_____	_____	_____
P. fluorescens	_____	_____	_____	_____	_____
S. epidermidis	_____	_____	_____	_____	_____
Soil	_____	_____	_____	_____	_____
Control tube	_____	_____	_____	_____	_____

2. Illustrate or outline a complete test for the presence of nitrate reductase.

3. In summary, what is the major purpose of this exercise?

Review Questions

1. From your results, which bacteria are negative for nitrate reduction? Which are positive?

2. How do you explain the results from the soil sample?

3. Why is the development of a red color a negative test when zinc is added?

4. What are the end products that may result from the action of bacteria with nitrate-reducing enzymes?

5. What is the purpose of a control tube in this exercise?

6. How would you perform a complete test for the presence of nitrate reduction?

Rapid Multitest Systems

Man is a tool-using animal . . . without tools he is nothing, with tools he is all.
Thomas Carlyle (Scottish critic, historian, 1795–1881)

Staining, morphology, motility, culture data, enzymatic, and biochemical activities can be used to identify microorganisms. These characteristics were the basis for parts two through four of this manual. In determining each of these characteristics, a large amount of media, test tubes, culture plates, and time were involved. In order to reduce these factors, several rapid multitest systems have been developed for the identification of medically important microorganisms in a relatively short period. To facilitate rapid identification, some of these systems use a uniform coding procedure that can be computerized.

The purpose of part five of this manual is to introduce you to two of these systems: API 20E and Enterotube II. The API 20E and Enterotube II Systems are used to identify primarily the gram-negative, oxidase-negative *Enterobacteriaceae*.

After completing either of these two exercises, you will, at the minimum, be able to demonstrate the ability to use appropriate microbiological test systems and accurately record macroscopic observations. This will meet the American Society for Microbiology Core Curriculum skill number 4 (*see pp. vi–viii*).

The API 20E System

MATERIALS PER GROUP OF STUDENTS

API 20E System strip, incubator tray, and cover (bioMérieux Vitek, P.O. Box 42016, Hazelwood, MO 63042-2395; 800-638-4835)
tryptic soy agar slant or plate culture of unknown member of the *Enterobacteriaceae*
5 ml of sterile 0.85% saline in test tube with cap
50-ml plastic squeeze bottle containing tap water
5-ml Pasteur pipettes with pipettor
oxidase test reagent or disk/strip
petri plate
TSA plate
Whatman No. 2 filter paper
10% ferric chloride
Barritt's reagents A and B
sterile mineral oil
tube containing McFarland No. 3 (BaSO$_4$) reference standard
platinum inoculating loop (nichrome wire loops should NOT be used) or a wooden applicator stick
Bunsen burner
Kovacs' reagent
nitrite test reagents A and B solution
zinc dust or 20-mesh granular zinc
1.5% hydrogen peroxide
incubator set at 35° to 37°C
API 20E Quick Index Booklet
test-tube rack

wax pencil
disposable gloves

Learning Objectives

Each student should be able to

1. Understand why an oxidase test is first performed in the API 20E System
2. Correctly inoculate the API 20E System
3. Read and record the various biochemical reactions that occur after 18 to 24 hours of incubation
4. Determine a seven-digit profile number
5. Look up the unknown bacterium in the API 20E Quick Index Booklet

SUGGESTED READING IN TEXTBOOK

1. The *Enterobacteriaceae,* section 22.3.
2. Rapid Methods of Identification, section 36.2; see also figures 36.7 and 36.8.

 Medical Application

The API 20E System and Enterotube II System are used in the clinical laboratory for the identification of enteric bacteria. Among them are *Escherichia coli* (opportunistic urinary tract infections), *Proteus mirabilis* (opportunistic urinary tract infections), *Shigella dysenteriae* (bacillary dysentery), *Salmonella typhi* (typhoid fever), and *Yersinia pestis* (plague).

PRINCIPLES

The **API 20E System** is a standardized, miniaturized version of conventional biochemical procedures used in the identification of *Enterobacteriaceae* and other gram-negative bacteria. A total of 127 taxa can be identified with this system. It is a ready-to-use, microtube system

(a) All tests: positive

(b) All tests: negative

that performs 22 standard biochemical tests on pure bacterial cultures from appropriate, primary isolation media.

This system consists of a strip containing 20 chambers (figure 35.1), each consisting of a microtube and a depression called a **cupule.** The tubes contain dehydrated substrates. The substrates are rehydrated by

Cupule

Tube

adding a bacterial saline suspension. To create anaerobic conditions, sterile mineral oil is added to several of the microtubes. The strip of microtubes is then incubated for 18 to 24 hours at 35° to 37°C so that the bacterium can act on the substrates. The strip is read by noting color changes after the various indicator systems have been affected by the metabolites or added reagents (table 35.1). The identification of the unknown bacterium is achieved by determining a seven-digit **profile index number** and consulting the **API 20E Profile Recognition System** or the **API 20E Profile Index Booklet.** Charts can also be used to determine the unknown bacterium (*see appendix G*).

Procedure

First Period

1. Using aseptic technique (*see figure 14.3*) and an applicator stick or inoculating loop, select a well-isolated colony from the unknown streak plate or from the pure culture slant. Smear a small amount of the colony over a small area of filter paper. Add several drops of the oxidase reagent. Note the color change (*see exercise 30, figure 30.1*). You may also use oxidase test strips or disks as described in exercise 30.

2. Transfer another loopful of bacteria to the test tube containing 5 ml sterile saline. (If an applicator stick is used, load the tip with bacteria from the colony.) Recap and shake the tube. The turbidity should match that of the McFarland No. 3 ($BaSO_4$) standard. Add more bacteria if necessary.

3. Label the elongated flap of the incubation tray with your name and date. From the wash bottle, add 5 ml of water to the bottom of the tray. This provides a humid atmosphere during incubation.

4. Remove the plastic API strip from the sealed pouch and place it in the incubation tray. Reseal the pouch to protect the remaining strips.

5. Shake the 5 ml of bacterial suspension. Remove the cap and fill the 5-ml Pasteur pipette with the bacterial suspension. (Tilt the strip to avoid forming bubbles.) Inoculate the strip as follows:

 a. Tilt the API 20E incubation tray and place the pipette tip against the side of each cupule. Fill the tubes of the ONPG, TDA, IND, GLU, MAN, INO, SOR, RHA, SAC, MEL, AMY, and ARA microtubes.

 b. Slightly underfill the ADH, LDC, ODC, H_2S, and URE microtubes. Note that these microtubes are underlined on the test strip!

 c. Fill both the tube and cupule section of the |CIT|, |VP|, and |GEL| tubes. Note that these microtubes have brackets on the test strip!

 d. After inoculation, completely fill the cupule section of the ADH, LDC, ODC, H_2S, and URE tubes with mineral oil. This creates anaerobic conditions.

6. Place the lid on the incubation tray and incubate at 35°C for 18 to 24 hours. If the strip cannot be read after 24 hours, place it in a refrigerator at 2° to 8°C until the reactions can be read.

7. Make an isolation streak on a TSA plate with a portion of the bacterial suspension to ascertain the purity of the suspension.

Second Period

1. After 18 hours and before 24 hours incubation, record all reactions not requiring the addition of reagents (do not read TDA, VP, and IND). Table 35.1 summarizes these reactions. Record these results by placing a + for a positive reaction and a − for a negative reaction in the 24-hour box for the report for exercise 35. Also record the colors.

2. If the GLU microtube is negative (blue or green), do not add reagents. This negative glucose test

Table 35.1

Chart I. Summary of Results—18- to 24-Hour Procedure

Tube	Positive		Negative	Comments
	Interpretation of Reactions			
	Positive		Negative	Comments
ONPG	Yellow		Colorless	(1) Any shade of yellow is a positive reaction. (2) VP tube, before the addition of reagents, can be used as a negative control.
ADH	Incubation 18–24 h: Red or orange 36–48 h: Red		Yellow Yellow or orange	Orange reactions occurring at 36 to 48 hours should be interpreted as negative.
LDC	18–24 h: Red or orange 36–48 h: Red		Yellow Yellow or orange	Any shade of orange within 18 to 24 hours is a positive reaction. At 36 to 48 hours, orange decarboxylase reactions should be interpreted as negative.
ODC	18–24 h: Red or orange 36–48 h: Red		Yellow Yellow or orange	Orange reactions occurring at 36 to 48 hours should be interpreted as negative.
CIT	Turquoise or dark blue		Light green or yellow	(1) Both the tube and cupule should be filled. (2) Reaction is read in the aerobic (cupule) area.
H$_2$S	Black deposit		No black deposit	(1) H$_2$S production may range from a heavy black deposit to a very thin black line around the tube bottom. Carefully examine the bottom of the tube before considering the reaction negative. (2) A "browning" of the medium is a negative reaction unless a black deposit is present. "Browning" occurs with TDA positive organisms.
URE	18–24 h: Red or orange 36–48 h: Red		Yellow Yellow or orange	A method of lower sensitivity has been chosen. *Klebsiella, Proteus,* and *Yersinia* routinely give positive reactions.
TDA	Add 1 drop 10% ferric chloride — Brown-red		Yellow	(1) Immediate reaction. (2) Indole positive organisms may produce a golden orange color due to indole production. This is a negative reaction.
IND	Add 1 drop Kovacs' reagent — Red ring		Yellow	(1) The reaction should be read within 2 minutes after the addition of the Kovacs' reagent and the results recorded. (2) After several minutes, the HCl present in Kovacs' reagent may react with the plastic of the cupule resulting in a change from a negative (yellow) color to a brownish-red. This is a negative reaction.
VP	Add 1 drop of 40% potassium hydroxide, then 1 drop of 6% α-naphthol — Red		Colorless	(1) Wait 10 minutes before considering the reaction negative. (2) A pale pink color (after 10 minutes) should be interpreted as negative. A pale pink color that appears immediately after the addition of reagents but turns dark pink or red after 10 minutes should be interpreted as positive. Motility may be observed by hanging drop or wet-mount preparation.
GEL	Diffusion of the pigment		No diffusion	(1) The solid gelatin particles may spread throughout the tube after inoculation. Unless diffusion occurs, the reaction is negative. (2) Any degree of diffusion is a positive reaction.
GLU	Yellow or gray		Blue or blue-green	**Fermentation** (*Enterobacteriaceae, Aeromonas, Vibrio*) (1) Fermentation of the carbohydrates begins in the most anaerobic portion (bottom) of the tube. Therefore, these reactions should be read from the bottom of the tube to the top. (2) A yellow color at the bottom of the tube only indicates a weak or delayed positive reaction.
MAN INO SOR RHA SAC MEL AMY ARA	Yellow		Blue or blue-green	**Oxidation** (Other gram negatives) (1) Oxidative utilization of the carbohydrates begins in the most aerobic portion (top) of the tube. Therefore, these reactions should be read from the top to the bottom of the tube. (2) A yellow color in the upper portion of the tube and a blue color in the bottom of the tube indicate oxidative utilization of the sugar. This reaction should be considered positive **only** for non *Enterobacteriaceae* gram-negative rods. This is a negative reaction for fermentative organisms such as *Enterobacteriaceae*.
GLU Nitrate reduction	After reading GLU reaction, add 2 drops 0.8% sulfanilic acid and 2 drops 0.5% N,N-dimethyl-α-naphthylamine NO$_2$: Red N$_2$ gas: Bubbles; Yellow after reagents and zinc		Yellow Orange after reagents and zinc	(1) Before addition of reagents, observe GLU tube (positive or negative) for bubbles. Bubbles are indicative of reduction of nitrate to the nitrogenous (N$_2$) state. (2) A positive reaction may take 2 to 3 minutes for the red color to appear. (3) Confirm a negative test by adding zinc dust or 20-mesh granular zinc. A pink-orange color after 10 minutes confirms a negative reaction. A yellow color indicates reduction of nitrates to nitrogenous (N$_2$) state.
MAN INO SOR Catalase	After reading carbohydrate reaction, add 1 drop 1.5% H$_2$O$_2$ — Bubbles		No bubbles	(1) Bubbles may take 1 to 2 minutes to appear. (2) Best results will be obtained if the test is run in tubes that have no gas from fermentation.

Note: The column between GLU and MAN/INO/SOR/RHA/SAC/MEL/AMY/ARA rows reads vertically: "Comments for all carbohydrates"

Source: Copyright bioMérieux Vitek, Inc., Hazelwood, MO. Reprinted by permission.
Note: The following abbreviations are used: ONPG (β-galactosidase), ADH (arginine dihydrolase), LDC (lysine decarboxylase), ODC (ornithine decarboxylase), CIT (citrate), H$_2$S (hydrogen sulfide), URE (urease), TDA (tryptophan deaminase), IND (indole), VP (Voges-Proskauer), GEL (gelatin), GLU (glucose), MAN (mannitol), INO (inositol), SOR (sorbitol), RHA (rhamnose), SAC (sucrose), MEL (melibiose), AMY (amygdalin), ARA (arabinose).

Chart I. Summary of Results—18- to 24-Hour Procedure

Tube	Chemical/Physical Principles	Components	
		Reactive Ingredients	Quantity
ONPG	Hydrolysis of ONPG by beta-galactosidase releases yellow orthonitrophenol from the colorless ONPG; ITPG (isopropylthiogalactopyranoside) is used as inducer.	ONPG ITPG	0.2 mg 8.0 µg
ADH	Arginine dihydrolase transforms arginine into ornithine, ammonia, and carbon dioxide. This causes a pH rise in the acid-buffered system and a change in the indicator from yellow to red.	Arginine	2.0 mg
LDC	Lysine decarboxylase transforms lysine into a basic primary amine, cadaverine. This amine causes a pH rise in the acid-buffered system and a change in the indicator from yellow to red.	Lysine	2.0 mg
ODC	Ornithine decarboxylase transforms ornithine into a basic primary amine, putrescine. This amine causes a pH rise in the acid-buffered system and a change in the indicator from yellow to red.	Ornithine	2.0 mg
CIT	Citrate is the sole carbon source. Citrate utilization results in a pH rise and a change in the indicator from green to blue.	Sodium Citrate	0.8 mg
H_2S	Hydrogen sulfide is produced from thiosulfate. The hydrogen sulfide reacts with iron salts to produce a black precipitate.	Sodium Thiosulfate	80.0 µg
URE	Urease releases ammonia from urea; ammonia causes the pH to rise and changes the indicator from yellow to red.	Urea	0.8 mg
TDA	Tryptophan deaminase forms indolepyruvic acid from tryptophane. Indolepyruvic acid produces a brownish-red color in the presence of ferric chloride.	Tryptophane	0.4 mg
IND	Metabolism of tryptophane results in the formation of indole. Kovacs' reagent forms a colored complex (pink to red) with indole.	Tryptophane	0.2 mg
VP	Acetoin, an intermediary glucose metabolite, is produced from sodium pyruvate and indicated by the formation of a colored complex. Conventional VP tests may take up to four days, but by using sodium pyruvate, API has shortened the required test time. Creatine intensifies the color when tests are positive.	Sodium Pyruvate Creatine	2.0 mg 0.9 mg
GEL	Liquefaction of gelatin by proteolytic enzymes releases a black pigment that diffuses throughout the tube.	Kobn Charcoal Gelatin	0.6 mg
GLU MAN INO SOR RHA SAC MEL AMY ARA	Utilization of the carbohydrate results in acid formation and a consequent pH drop. The indicator changes from blue to yellow.	Glucose Mannitol Inositol Sorbitol Rhamnose Sucrose Melibiose Amygdalin (L +) Arabinose	2.0 mg 2.0 mg 2.0 mg 2.0 mg 2.0 mg 2.0 mg 2.0 mg 2.0 mg 2.0 mg
GLU Nitrate Reduction	Nitrites form a red complex with sulfanilic acid and N,N-dimethylalphanaphthylamine. In case of negative reaction, addition of zinc confirms the presence of unreduced nitrates by reducing them to nitrites (pink-orange color). If there is no color change after the addition of zinc, this is indicative of the complete reduction of nitrates through nitrites to nitrogen gas or to an anaerogenic amine.	Potassium Nitrate	80.0 µg
MAN INO SOR Catalase	Catalase releases oxygen gas from hydrogen peroxide.		

Source: Copyright bioMérieux Vitek, Inc., Hazelwood, MO. Reprinted by permission.

indicates that your unknown is not a member of the *Enterobacteriaceae* and the testing cannot proceed any further as outlined in this exercise.

3. If the GLU tube is positive (yellow with or without gas bubbles), add the following test reagents in the order listed. In all cases, read the results immediately after adding the reagents and waiting the proper length of time. Do not replace the lid on the tray until all results have been collected. Record the results obtained on the report for exercise 35 as per step 1.

 a. Add 1 drop of 10% ferric chloride to the TDA microtube. A positive reaction is indicated by a brown-red color. A negative reaction is yellow.

 b. Add 1 drop each of Barritt's A (α-naphthol) and B (40% KOH) solutions to the VP (Voges-Proskauer) microtube. The KOH should be added first. It may take 10 minutes for this reaction to occur. A positive reaction is pink to red. No color change is a negative reaction.

 c. Add 1 drop of Kovacs' reagent to the IND (indole) microtube. A positive test is indicated by a red ring within 2 minutes. A yellow ring is a negative reaction.

 d. Examine the GLU (glucose) tube (positive or negative) for bubbles. Bubbles indicate the reduction of nitrate and the production of nitrogen gas (*see exercise 34*). Record the presence or absence of gas in the report for exercise 35.

 e. Add 2 drops of each nitrite test reagent to the GLU microtube. If nitrates are reduced, a red color (with or without gas bubbles) occurs within 2 to 3 minutes. This is a positive test, while a yellow color is a negative test.

 f. If the test is negative, add a speck of zinc dust. A pink-orange color within 10 minutes indicates that nitrate reduction did not occur. A yellow color indicates that nitrogen gas was produced.

 g. Add 1 drop of hydrogen peroxide to the MAN, INO, and SOR cupules. If catalase is produced (*see exercise 28*), gas bubbles appear within 2 minutes and the test is positive. No bubbles indicates a negative test.

4. After all reactions have been recorded on the report sheet, determine the seven-digit profile number as follows:

 a. Within each test section, add the numbers containing only the tests that are positive.

 b. Enter the sum of the positive tests for each section in the square labeled "profile number." The seven digits represent the profile number for your enteric unknown. See no. 2 in the report for exercise 35.

 c. Identify the unknown by looking up the profile number in the API 20E Quick Index Booklet (*also see appendix G*).

5. When you are finished with the API 20E System, the entire unit must be autoclaved, incinerated, or immersed in a germicide prior to disposal.

HINTS AND PRECAUTIONS

(1) Make sure that the culture is adequately mixed and checked against the McFarland standard to ensure an even inoculation of the system.

(2) Be wary of any air bubbles in the cupules.

(3) Do not cover the incubation tray with its lid after adding reagents until all changes have been observed.

The API 20E System

1. Tabulation of API 20E System results.

Test	ONPG	ADH	LDC	ODC	CIT	H$_2$S	URE	TDA	IND	VP	GEL	GLU	MAN	INO	SOR	RHA	SAC	MEL	AMY	ARA	OXI
Test value	1	+ 2	+ 4	1	+ 2	+ 4	1	+ 2	+ 4	1	+ 2	+ 4	1	+ 2	+ 4	1	+ 2	+ 4	1	+ 2	+ 4
Results 24 hr																					
Profile number																					

Catalase: _____

Nitrate reduction: _____

2. Construction of the seven-digit profile number. For example, *E. coli* = 5 144 572.

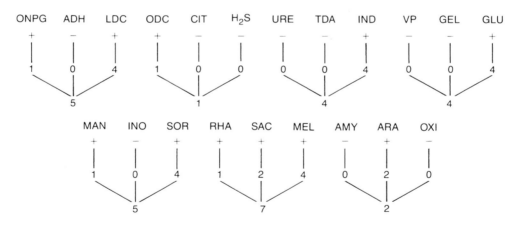

3. Number of unknown: _____. Look up the unknown in the API 20E Quick Index Booklet.

 Name of unknown: _____.

4. In summary, what is the major purpose of this exercise?

Review Questions

1. What is the purpose of the sterile mineral oil in the API 20E System?

2. What group of pathogens is the API 20E System used to detect in a clinical laboratory?

3. Why is the recommended incubation procedure for the API 20E System 18 to 24 hours?

4. Can gram-negative, oxidase-positive bacteria be identified with the API 20E System? Explain your answer.

5. What is a cupule?

6. What is a profile index number?

7. What is found in the API 20E Profile Index Booklet?

The Enterotube II System

MATERIALS PER STUDENT

1 BBL Enterotube II System (Becton Dickinson Microbiology Systems, Sparks, MD 21152 [1-800-638-8663] also available from Fisher Scientific)

plate culture(s) of unknown *Enterobacteriaceae* from EMB, MacConkey, SS, Hektoen enteric, or TSA

TSA plate

Kovacs' reagent

10% KOH with 3% creatine solution

5% α-naphthol in absolute ethanol

3 syringes with needles or Pasteur pipettes with pipettor

test-tube rack

platinum inoculating loop (nichrome wire should NOT be used)

wax pencil

Bunsen burner

Roche Enterotube II Computer Coding Manual

Learning Objectives

Each student should be able to

1. Correctly inoculate the Enterotube II System
2. Read and record the various biochemical reactions that occur after 18 to 24 hours of incubation
3. Determine a five-digit numerical code

4. Identify unknown *Enterobacteriaceae* using the Enterotube II System and the Computer Coding Manual

SUGGESTED READING IN TEXTBOOK

1. The *Enterobacteriaceae,* section 22.3.
2. Rapid Methods of Identification, section 36.2.

PRINCIPLES

The basic philosophy of the **Enterotube II System** is the same as the API 20E System (*see exercise 35*)—speed, ease, and low cost in the identification of gram-negative, glucose-fermenting, oxidase-negative *Enterobacteriaceae.* Some microbiologists prefer this system over the API 20E System. Both systems are presented in this manual although they accomplish the same end.

The Enterotube II System consists of a single tube containing 12 compartments (figure 36.1), each containing a different agar-solidified culture medium. Compartments that require aerobic conditions have small openings that allow air in; those requiring anaerobic conditions have a layer of paraffin wax on the top of the media. There is a self-enclosed inoculating needle (wire). This needle can touch an isolated bacterial colony and then in one movement can be drawn through the 12 compartments, inoculating the test media. Fifteen standard tests are performed. After 18 to 24 hours of incubation, the color changes that occur in each of the compartments are recorded and interpreted according to the manufacturer's instructions. This is done by determining a five-digit code from the results and consulting a coding manual, or by comparing the results obtained with those outlined in a differential chart provided by the manufacturer (*see appendix G*). This quick multitest system also has a computer-assisted program called **ENCISE (Enterobacteriaceae Numerical Coding and Identification System for Enterotube).**

Figure 36.1 **Enterotube II Multitest System.** (**a**) Uninoculated colors. (**b**) Inoculated colors. *(Becton-Dickinson Microbiology Systems, Cockeysville, MD. Reprinted by permission.)*

*Certain microorganisms will not always produce the ideal "strong" positive A. Reactions similar to B should also be considered positive.

Procedure

First Period

1. Using aseptic technique (*see figure 14.3*), select a well-isolated colony from the unknown plate and do an isolation streak on a TSA plate. Incubate at 35°C in the inverted position for 24 to 48 hours. This plate serves as a control to make sure your unknown is a pure culture. Alternately, it can serve as the source colony for the Enterotube II System.

2. Select a well-isolated colony from either the TSA plate or a plate provided by your instructor. Perform an oxidase test on the bacteria either by placing a small amount of the bacteria on the filter paper and adding several drops of oxidase reagent or by using a special test disk/strip. Note the color change (*see exercise 30, figure 30.1*). If oxidase is positive, stop. The Enterotube II System cannot be used with oxidase-positive bacteria. If oxidase is negative, proceed as follows:

 a. Place your name and date on the white paper label on the side of the Enterotube above the VP. Remove the caps from both ends of the Enterotube. The tip of the wire is sterile and does not need to be flamed.

 b. Touch a well-isolated colony with the tip of the wire.

 c. Inoculate the Enterotube with the bacterial culture by drawing, and at the same time rotating, the wire through the 12 compartments. **Do not pull the wire out of the Enterotube!**

 d. Push the wire back through the Enterotube so that the 12 chambers are reinoculated.

 e. Withdraw the wire once again until the tip is in the H₂S/indole compartment and then break the wire at the notch by bending it back and forth. Discard the needle remnant.

 f. Replace the caps but do not tighten. This creates an aerobic environment where the wire is (chambers GLU-GAS, LYS, ORN), which is necessary for dextrose fermentation.

 g. Remove the blue plastic strip to expose the aeration holes (for chambers ADON, LAC, ARAB, SORB, VP, DUL-PA, UREA, and CIT) and slide the clear plastic band over the glucose compartment. This later step will contain any small amount of sterile wax that may escape due to excessive gas production by some bacteria.

 h. Incubate the Enterotube for 18 to 24 hours at 35° to 37°C. Make sure the Enterotube is placed

on a flat surface and that there is space between adjacent tubes to allow for air circulation.

Second Period

1. After 18 to 24 hours of incubation, examine the Enterotube and notice the color changes that have occurred in each compartment. Use table 36.1 and appendix G as a guide for comparing and interpreting your results. Record both positive and negative results on the white label on the side of the Enterotube and in the report for exercise 36.

2. Perform the indole test by placing the Enterotube in a test-tube rack with the GLU-GAS compartment facing downward. Add 1 or 2 drops of Kovacs' reagent to the H_2S compartment using either a Pasteur pipette with pipettor or a syringe equipped with a needle. If a pipette is used, first make a small hole in the plastic film with a hot inoculating loop or needle. Results are read after 10 seconds. If the fluid or film turns red, this is a positive indole test; if there is no color change, the test is negative. Record your results in the report for exercise 36.

3. To the Voges-Proskauer compartment, add 2 drops of 20% KOH and 3 drops of α-naphthol. Use a new syringe and needle or pipette for each solution. Results should be read within 20 minutes after adding the solutions. Record your results in the report for exercise 36.

4. Determine the five-digit identification number as follows:
 a. Use only the tests that are positive. Add the numbers under the results within each test section.
 b. Enter the sum of the positive tests for each test section in the square labeled "ID value." These five digits represent the identification number for your unknown. For example, in the illustration, the circles represent positive tests.

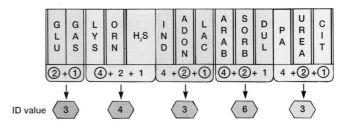

5. Determine the identity of your enteric unknown by comparing the five-digit identification number with the Enterotube II Computer Coding Manual.

HINTS AND PRECAUTIONS

All results for the Enterotube II system should be recorded before injecting the indole and Voges-Proskauer test reagents because discoloration of other test chambers can occur when these reagents are applied.

Table 36.1

Biochemical Reactions of Enterotube II

Symbol	Uninoculated color	Reacted color	Type of reaction
ADON			**Adonitol** Bacterial fermentation of adonitol, which results in the formation of acidic end products, is indicated by a change in color of the indicator present in the medium from red (alkaline) to yellow (acidic). Any sign of yellow should be interpreted as a positive reaction; orange should be considered negative.
LAC			**Lactose** Bacterial fermentation of lactose, which results in the formation of acidic end products, is indicated by a change in color of the indicator present in the medium from red (alkaline) to yellow (acidic). Any sign of yellow should be interpreted as a positive reaction; orange should be considered negative.
ARAB			**Arabinose** Bacterial fermentation of arabinose, which results in the formation of acidic end products, is indicated by a change in color from red (alkaline) to yellow (acidic). Any sign of yellow should be interpreted as a positive reaction; orange should be considered negative.
SORB			**Sorbitol** Bacterial fermentation of sorbitol, which results in the formation of acidic end products, is indicated by a change in color from red (alkaline) to yellow (acidic). Any sign of yellow should be interpreted as a positive reaction; orange should be considered negative.
V.P.			**Voges-Proskauer** Acetylmethylcarbinol (acetoin) is an intermediate in the production of butylene glycol from glucose fermentation. The presence of acetoin is indicated by the development of a red color within 20 minutes. Most positive reactions are evident within 10 minutes.
DUL-PA		A* B*	**Dulcitol** Bacterial fermentation of dulcitol, which results in the formation of acidic end products, is indicated by a change in color of the indicator present in the medium from green (alkaline) to yellow or pale yellow (acidic). **Phenylalanine Deaminase** This test detects the formation of pyruvic acid from the deamination of phenylalanine. The pyruvic acid formed reacts with a ferric salt in the medium to produce a characteristic black to smoky gray color.
UREA			**Urea** The production of urease by some bacteria hydrolyzes urea in this to produce ammonia, which causes a shift in pH from yellow (acidic) to reddish-purple (alkaline). This test is strongly positive for *Proteus* in 6 hours and weakly positive for *Klebsiella* and some *Enterobacter* species in 24 hours.
CIT		A* B*	**Citrate** Organisms that are able to utilize the citrate in this medium as their sole source of carbon produce alkaline metabolites that change the color of the indicator from green (acidic) to deep blue (alkaline). Any degree of blue should be considered positive.

Biochemical Reactions of Enterotube II

Symbol	Uninoculated color	Reacted color	Type of reaction
GLU-GAS			**Glucose (GLU)** The end products of bacterial fermentation of glucose are either acid or acid and gas. The shift in pH due to the production of acid is indicated by a color change from red (alkaline) to yellow (acidic). Any degree of yellow should be interpreted as a positive reaction; orange should be considered negative. **Gas Production (GAS)** Complete separation of the wax overlay from the surface of the glucose medium occurs when gas is produced. The amount of separation between the medium and overlay will vary with the strain of bacteria.
LYS			**Lysine Decarboxylase** Bacterial decarboxylation of lysine, which results in the formation of the alkaline end product cadaverine, is indicated by a change in color of the indicator from pale yellow (acidic) to purple (alkaline). Any degree of purple should be interpreted as a positive reaction. The medium remains yellow if decarboxylation of lysine does not occur.
ORN			**Ornithine Decarboxylase** Bacterial decarboxylation of ornithine causes the alkaline end product putrescine to be produced. The acidic (yellow) nature of the medium is converted to purple as alkalinity occurs. Any degree of purple should be interpreted as a positive reaction. The medium remains yellow if decarboxylation of ornithine does not occur.
H₂S/IND			**H₂S Production** Hydrogen sulfide, liberated by bacteria that reduce sulfur-containing compounds such as peptones and sodium thiosulfate, reacts with the iron salts in the medium to form a black precipitate of ferric sulfide usually along the line of inoculation. Some *Proteus* and *Providencia* strains may produce a diffuse brown coloration in this medium, which should not be confused with true H₂S production. **Indole Formation** The production of indole from the metabolism of tryptophan by the bacterial enzyme tryptophanase is detected by the development of a pink to red color after the addition of Kovac's reagent.

Source: Copyright © Becton Dickinson Microbiology Systems. Reprinted by permission.

The Enterotube II System

1. Record the results of your unknown in the following chart. Use a + for a positive test and a − for a negative test.

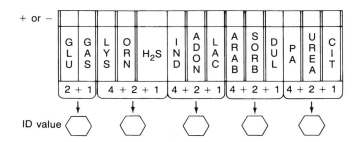

2. Using the Enterotube II Computer Coding Manual, determine the five-digit code number by encircling the numbers (4, 2, or 1) only under the tests that are positive. Total these numbers within each group to form a digit for that group.
3. In summary, what is the major purpose of this exercise?

Review Questions

1. What are some advantages of the Enterotube II System over the API 20E System? Some disadvantages?

2. Why can the Enterotube II System be used to identify only gram-negative, oxidase-negative *Enterobacteriaceae?*

3. What is the medical significance of members of the *Enterobacteriaceae?*

4. Why is part of the wire left in the Enterotube II? Why is part of the wire broken off?

5. What is a coding manual?

6. What are some biochemical characteristics shared by the family *Enterobacteriaceae?*

7. What is the advantage of ENCISE over a coding manual?

Unknown Identification

Exercises 6–12 and 20–36 are designed to enable students to gather morphological, staining, cultural, and biochemical data on the characteristics of different microorganisms. The tests used are those that are employed in bacterial identification in all microbiology laboratories. The approach is to gather as much information about the microorganism as possible. After this information has been collected, it can be used to identify a microorganism.

The first edition of *Bergey's Manual of Systematic Bacteriology* consists of four volumes of material that can be used to identify microorganisms. (The second edition of this manual with a revised classification system is currently being prepared. Since this laboratory manual will be published before the entire second edition of *Bergey's Manual* is available to the majority of higher education institutions, we will continue to use the first edition.) For example, in *Bergey's Manual,* bacteria are grouped according to Gram-stain reaction, cell shape, cell arrangement, oxygen requirements, motility, and nutritional and metabolic properties.

In 1994, the ninth edition of *Bergey's Manual of Determinative Bacteriology* was published. It is a departure from past editions that attempted, usually inadequately, to combine systematic and determinative information. Systematic information continues to be found in *Bergey's Manual of Systematic Bacteriology,* with the *Determinative* manual serving as a reference aid in the identification of unknown bacteria. This ninth edition is intended solely for the identification of those bacteria that have been described and cultured. They are a small number and represent only a fraction of those existing in nature. Much work remains to be done, and future editions of both Bergey's manuals will only become larger.

The purpose of part six of this manual (chapters 37 and 38) is to first introduce the student to the proper use of *Bergey's Manual of Systematic Bacteriology* and then to employ it in the identification of a general unknown mixture of gram-positive and gram-negative bacteria.

After completing the two exercises in part six, you will, at the minimum, be able to demonstrate an increased level of laboratory thinking skills, including:

(a) following an experimental protocol;

(b) collecting and organizing data in a systematic fashion;

(c) assessing the validity of data (including integrity and significance); and

(d) drawing appropriate conclusions based on the results.

This will meet the American Society for Microbiology Core Curriculum Laboratory Cognitive Skill number 1 and Thinking Skill number 2 (*see pp. vi–viii*).

David Hendricks Bergey (1860–1937)
(National Library of Medicine)

Bergey was the American bacteriologist who first proposed the system of bacterial classification in which bacteria are grouped according to Gram reaction, metabolism, and morphology, with each group being further subdivided into orders, families, genera, and species.

The first edition of *Bergey's Manual of Determinative* [*Systematic*] *Bacteriology* was published in 1923. Throughout the years, *Bergey's Manual* has become a widely used international reference work for bacterial taxonomy. At first, Dr. Bergey was the nominal owner of *Bergey's Manual.* In 1936, he executed a trust indenture whereby any publication income is used solely for preparing, editing, and publishing revisions and successive editions of *Bergey's Manual,* as well as for any research that may be necessary or desirable in such activities. Today, *Bergey's Manual* is a worldwide collaborative effort. It has an editorial board of 13 trustees, and over two hundred specialists from 19 countries are listed as contributors to the first two volumes.

Using the First Edition of *Bergey's Manual of Systematic Bacteriology* to Identify Bacteria

MATERIALS PER GROUP OF STUDENTS

Krieg, N. R., ed. 1984. *Bergey's Manual of Systematic Bacteriology,* editor-in-chief J. G. Holt, vol. 1. Baltimore, Maryland: Williams & Wilkins.

Sneath, P. H. A., Mair, N. S., and Sharpe, M. E. 1986. *Bergey's Manual of Systematic Bacteriology,* editor-in-chief J. G. Holt, vol. 2. Baltimore, Maryland: Williams & Wilkins.

Learning Objectives

Each student should be able to

1. Use *Bergey's Manual* to identify an unknown bacterial strain
2. Use *Bergey's Manual* as a source of further information about a group of bacteria

SUGGESTED READING IN TEXTBOOK

1. Microbial Taxonomy, Chapters 19.
2. The bacterial survey chapters, Chapters 20–24. These should be consulted for specific information about particular volumes of the manual and groups of bacteria. Chapters 21 and 22 cover the bacteria in volume 1 of *Bergey's Manual* and chapters 23 and 24 describe the bacteria in volume 2.
3. Bacterial Classification according to the first edition of *Bergey's Manual of Systematic Bacteriology,* appendix III.

Pronunciation Guide

Streptococcus (strep-to-KOK-us)

PRINCIPLES

Bergey's Manual of Systematic Bacteriology is intended to aid microbiologists in identifying unknown bacteria and also to provide information about the biology of each bacterial group. The first edition of the manual is divided into four volumes and 33 sections. Each section contains bacteria sharing a few easily determined characteristics and bears a title that either describes these properties or provides the vernacular names of the bacteria included. The characteristics used to define sections are normally features such as general shape and morphology, Gram-staining properties, oxygen dependence, motility, the presence of endospores, and the mode of energy production. Each volume focuses on a different collection of bacteria.

Volume 1. The gram-negative bacteria of general, medical, or industrial importance

Volume 2. Gram-positive bacteria other than actinomycetes

Volume 3. Gram-negative bacteria with distinctive properties, cyanobacteria, and archaeobacteria

Volume 4. Actinomycetes (gram-positive filamentous bacteria)

Several useful introductory articles on various aspects of bacterial taxonomy are located at the beginning of each volume. These articles cover such topics as using the manual, bacterial classification and nomenclature, the identification of bacteria, reference collections of bacteria, and higher bacterial taxa. The volumes are numbered sequentially. For example, volume 1 ends on page 964, and the introduction to bacterial classification in volume 2 begins on page 965.

The sections in the first edition of *Bergey's Manual* vary considerably in content. Some contain only one major bacterial taxon. For example, Section 1 has only the order *Spirochetales.* Other sections contain more than one family (section 5, the facultatively anaerobic gram-negative rods) or more than one order (section 9, the Rickettsias and Chlamydiae). Many sections lack any taxa above the genus level (section 13, endospore-forming, gram-positive rods and cocci). Several sections also have genera grouped under the heading "Other Genera" or "Other Organisms." These groups belong in the section, but cannot yet be placed in a recognized family.

Each article on a genus of bacteria is organized in a specific sequence to provide essential information about the genus as efficiently as possible. Some of the more important parts of each article are the following:

1. **Name of the Genus.** The accepted name is given in boldface.
2. **Capsule Description.** A brief summary of the major characteristics of the genus, with the most important properties in boldface, is provided for convenience.
3. **Further Descriptive Information.** A more extensive description of the biology of the genus is then given. Although this description is not exhaustive, it provides a great deal of useful information about such items as morphology, growth conditions and nutrition, physiology and metabolism, genetics, pathogenicity, and ecology.
4. **Enrichment and Isolation.** Selected methods for growth of the bacteria are presented.
5. **Differentiation of the Genus from Other Genera.** Characteristics that are particularly useful in distinguishing related genera are given, often in the form of a table.
6. **Taxonomic Comment.** This part summarizes the reasons for placing the genus in its current taxonomic position. The placement of species in the genus is also discussed.
7. **Differentiation of the Species of the Genus.** The characteristics most useful for distinguishing between species are described and usually summarized in table form.
8. **List of the Species of the Genus.** Descriptions of each species in the genus are given. Where possible, much of the essential information is summarized in tables.

In articles on the bacterial genera, three types of tables are used to summarize a great deal of information: (1) tables that differentiate between related genera; (2) tables that differentiate between species within a genus; and (3) tables that provide extra information about particular species. A variety of symbols are used in these tables. The most important are defined as follows:

+ 90% or more strains are positive for the characteristic
− 90% or more strains are negative
d 11% to 89% of the strains are positive
v strain instability (this is not equivalent to "d")
D different reactions in different taxa (species of a genus or genera of a family)

Footnotes to the tables give additional symbols and exceptions to the above symbols.

One can use the manual for unknown identification in the following way. First, determine which volume to employ based on the Gram-staining properties of the unknown and its general shape. For example, most ordinary gram-negative bacteria are found in volume 1, while ordinary gram-positive bacteria (excluding the actinomycetes) are found in volume 2. Most unknowns provided for student laboratory exercises will be found in the first two volumes. Next, proceed to the table of contents in the appropriate volume and carefully study the descriptive title for each section. The titles should provide sufficient information to place the unknown within a section. Determine the genus of the unknown using the keys or tables that are normally found at the beginning of each section. Once the proper genus has been selected, the unknown can be placed in a species using information contained in the generic and species descriptions. Tables and written descriptive material from both generic and species descriptions should be employed in the final identification because characteristics discussed in the generic descriptions are not normally repeated in the species descriptions.

Two examples of unknown identification using *Bergey's Manual* may be helpful in understanding this approach to unknown identification. First, suppose that you are trying to identify a gram-negative, facultatively anaerobic straight rod that is oxidase negative and motile by peritrichous flagella. Since the unknown is a regular gram-negative bacterium, you would use volume 1. Inspection of the table of contents will lead to section 5 because the unknown is a facultatively anaerobic, gram-negative rod. Table 5.1 on page 408 differentiates between the three families in section 5. The unknown is a straight rod that is oxidase negative and peritrichous; thus, it is a member of the family *Enterobacteriaceae*. Properties of the family are described on pages 408–420. Tentative identification of the genus can be made using the biochemical characteristics given in table 5.3, pages 414–417. Genus identity can be confirmed from the descriptions beginning on page 420 with particular emphasis on the properties given in boldface in the capsule descriptions of the genera.

Now imagine that the unknown is a gram-positive, facultatively anaerobic coccus that grows in chains and pairs, is catalase negative, and ferments glucose to lactate. Because the bacterium is gram positive and not an actinomycete, one would use volume 2. Inspection of the table of contents shows that only section 12 of the six sections contains gram-positive cocci. Table 12.1, pages 999–1,001, can be used to identify the genus. Since the unknown is a facultative anaerobe, one would turn to the part of the table that describes facultatively anaerobic genera (page 1,000). The genus *Streptococcus* best fits

the unknown because it forms chains and pairs and produces lactate. This can be confirmed by testing for the characteristics given in the capsule description of the genus on page 1,043 and the section entitled "Differentiation of the genus *Streptococcus* from other taxa" on page 1,046. Identification of the species is made from the descriptive material in the list of species that begins on page 1,047. In particular, the summary tables on page 1,048 and elsewhere are especially useful in species identification.

One must very carefully use the table of contents as a key to the sections. If there are many discrepancies between the unknown's properties and the descriptions of the leading candidates within a section, an error may have been made in Gram staining or some other basic laboratory procedure so that the unknown has been placed in the wrong section. If faced with this situation, it is best to recheck the results used to select the volume and the section.

It should also be kept in mind that there is often a great deal of variation between different strains of a species. A bacterium might differ from the species description in one test and match the description in all others. Sometimes all one can do is select the species or genus with the best overall match to the data. If possible, extra tests should be run to eliminate alternate candidates.

As mentioned earlier, *Bergey's Manual of Systematic Bacteriology* also serves as an excellent source of information about the biology of bacteria. The table of contents of each volume may be used to find the bacterial group of interest. However, it is usually easier to simply look for the desired family, genus, or species in the index of scientific names of bacteria at the end of each volume. All references to each bacterial taxon are noted in the index. The page numbers in boldface indicate the page on which the actual description of the taxon begins. Each volume also contains an extensive bibliography that lists in alphabetical order all the papers cited in the volume. These articles may be consulted for more information about the genus or species of interest.

Laboratory Report 37

Name: _____

Date: _____

Lab Section: _____

Using the First Edition of *Bergey's Manual of Systematic Bacteriology* to Identify Bacteria

1. Find the pages in volume 1 or 2 on which the following topics are discussed:

 a. How to use the manual. _____

 b. The nature of taxonomic ranks. _____

 c. Major developments in bacterial classification. _____

 d. The use of nucleic acids in bacterial classification. _____

2. Give the page in volume 1 or 2 on which R.G.E. Murray proposes a new arrangement of higher bacterial taxa (divisions and classes). _____

 Where does he describe the kingdom *Procaryotae* and its four divisions? _____

3. Using the indexes in volumes 1 and 2, find the pages on which descriptions of the following taxa begin:

 a. *Enterobacteriaceae* _____ e. *Rickettsiales* _____

 b. *Pseudomonas aeruginosa* _____ f. *Treponema* _____

 c. *Salmonella* _____ g. *Lactobacillus* _____

 d. *Mycobacterium tuberculosis* _____ h. *Staphylococcus* _____

4. Use *Bergey's Manual* to find the genus and species of an unknown with the following characteristics: gram-negative straight rod, facultatively anaerobic, motile by peritrichous flagella, oxidase negative, hydrolyzes urea and gelatin, produces hydrogen sulfide, oxidatively deaminates the amino acids phenylalanine and tryptophan, indole positive, and ornithine-decarboxylase negative.

 Genus: _____

 Species: _____

5. Use *Bergey's Manual* to find the genus and species of an unknown with the following characteristics: gram-positive straight rod, cell diameter 1.0 to 1.2 μm, motile by peritrichous flagella, grows in chains, facultative and can grow anaerobically, catalase positive, reduces nitrate to nitrite, elliptical endospores formed centrally in an unswollen cell (sporangium), colonies dull or frosted in appearance with an undulated margin, no parasporal body, Voges-Proskauer test positive, gelatin and starch hydrolyzed, and citrate catabolized.

 Genus: _____

 Species: _____

Review Questions

1. What is found in each section of the first edition of *Bergey's Manual of Systematic Bacteriology?*

2. In articles on the bacterial genera, what types of tables are used to summarize data?

3. How would you use *Bergey's Manual* to identify an unknown?

4. When using the table of contents in *Bergey's Manual,* why must you be very careful?

5. Of what value is the bibliography in *Bergey's Manual?*

6. How many current volumes are there of *Bergey's Manual?*

7. What are the three types of tables found in *Bergey's Manual?*

General Unknown

Be careful with the Bunsen burner flame. No mouth pipetting. Barritt's reagent contains naphthol, which is toxic and may cause peeling of the skin; thus, wear gloves when using this reagent. Kovacs' reagent is also caustic to the skin and mucous membranes due to the concentrated HCl and p-dimethylaminobenzaldehyde. Handle all cultures with care. Some may be pathogens or potential pathogens.

MATERIALS PER STUDENT

number-coded, 24-hour tryptic soy broth culture containing a mixture of any two bacteria (one gram positive and one gram negative) listed in the table on page 234
1 mannitol salt agar plate
1 EMB plate
1 blood agar plate
4 TSA slants—2 screw cap and 2 regular

MATERIALS TO BE REQUESTED

2 tryptic soy agar plates
2 nutrient gelatin deep tubes or gelatin test strips
2 starch agar plates
2 tributyrin agar plates
2 SIM agar deeps
2 Simmons citrate slants
2 urea broth tubes or urease test disks/strips
2 phenol red dextrose broth tubes or dextrose tablets
2 phenol red lactose broth tubes or lactose tablets
2 phenol red sucrose broth tubes or sucrose tablets
2 tryptic nitrate broth tubes
2 litmus milk tubes
2 MR-VP broth tubes

GENERAL LABORATORY MATERIAL

Gram-stain and acid-fast materials
3% hydrogen peroxide
Barritt's reagents
Kovacs' reagent
zinc powder
tetramethyl-p-phenylenediamine dihydrochloride or oxidase test disk/strip
glass slides
wax pencil
35°C incubator
Bunsen burner
inoculating loop and needle
disposable gloves
Pasteur pipettes with pipettor

Learning Objectives

Each student should be able to

1. Perform an isolation streak to separate two unknown bacteria
2. Identify unknown gram-positive and gram-negative bacteria

PRINCIPLES

The identification of unknown bacteria is one of the major responsibilities of the microbiologist. At this point in the course, students should have developed sufficient knowledge in staining methods, isolation techniques, microbial nutrition, biochemical activities, and characteristics of microorganisms to be able to work independently in identifying unknown bacteria.

This exercise is called a general unknown because unknown gram-positive and gram-negative bacteria will be used. Any of the following bacteria may be present in the unknown.

Gram Positive	Gram Negative
Bacillus cereus	*Acinetobacter calcouceticus*
B. megaterium	*Citrobacter freundii*
B. subtilis	*Enterobacter aerogenes*
Enterococcus faecalis	*Escherichia coli*
Listeria monocytogenes	*Klebsiella pneumonide*
Micrococcus luteus	*Proteus vulgaris*
Staphylococcus aureus	*Pseudomonas aeruginosa*
Others that the instructor may wish to add:	*Salmonella arizonae*
	Shigella sonnei
	Others that the instructor may wish to add:

From the first edition of *Bergey's Manual of Systematic Bacteriology,* which will be available in the laboratory, and from material in the textbook, each student is to work out a separation scheme (dichotomous key) in outline or flow-sheet form showing the important tests to use in distinguishing among the various genera. This should be done before the first period in order to use time and the necessary media most expediently. One example of a separation scheme (dichotomous key) is shown in figure 38.1. From *Bergey's Manual,* each student is to list the significant features of the possible bacteria in the above gram-positive and gram-negative list. After culturing and Gram staining, many bacterial groups can be eliminated. General morphological properties are most often used in placing a bacterium in the proper order and family. Biochemical and physiological characteristics are most useful at the genus and species level.

Success in the unknown identification will depend on (1) adequate planning of the approach, (2) careful selection of the biochemical tests, (3) weighing the value of one test over another, and (4) execution and interpretation of results.

Procedure

First Period

1. After obtaining the unknown mixture (containing one gram-positive and one gram-negative bacterium) from your instructor, immediately record the code number in the report for exercise 38.

2. Perform an isolation streak (*see exercise 16*) on the mannitol salt, EMB, and blood agar plates.

3. Label the plates with your name, date, and medium.

4. Incubate inverted for 24 to 48 hours at 35°C. During incubation, note the growth characteristics and describe in the report for exercise 38. Record bacterial colony morphology (*see exercise 15*).

5. Perform a Gram stain on well-isolated colonies. Make sure to mark the bottom of the plates indicating which colony is gram positive and which is gram negative.

Second Period

1. Using aseptic technique, inoculate two TSA slants with the gram-positive unknown and two TSA slants with the gram-negative unknown by means of a streak inoculation. (One of the two TSA slants for each unknown should be kept in a screw-cap tube.) Incubate at 35°C for 24 to 48 hours. One slant of the gram-positive and one of the gram-negative will be used as a stock culture. The other two slants will be used to determine the cultural characteristics of the unknowns. After 24 hours, the stock cultures can be kept in a refrigerator for later use if necessary. When the working culture is contaminated or not viable, subculture onto another slant from the screw-cap stock culture.

2. Perform a Gram stain on well-isolated, young (18- to 24-hour) colonies and record the reaction, morphology, and arrangement of the cells in the report for exercise 38. Mark the location of the colonies you Gram stained. If the unknown appears gram negative, stain a younger culture (about 18 hours old) to be sure that your unknown is a gram-positive bacterium that has become gram negative with age.

Third Period

1. Examine the tryptic soy agar slants for general culture characteristics. Record your observations in the report for exercise 38.

2. From your separation scheme, using aseptic technique, inoculate only the necessary media with the gram-positive and gram-negative unknowns.

3. Use a hanging drop slide (*see exercise 2*) to confirm motility, shape, and cell arrangement. Carefully measure your unknowns using a hanging drop slide or wet-mount (*see exercise 1*).

Figure 38.1 Separation Scheme for Some Common Bacterial Genera.

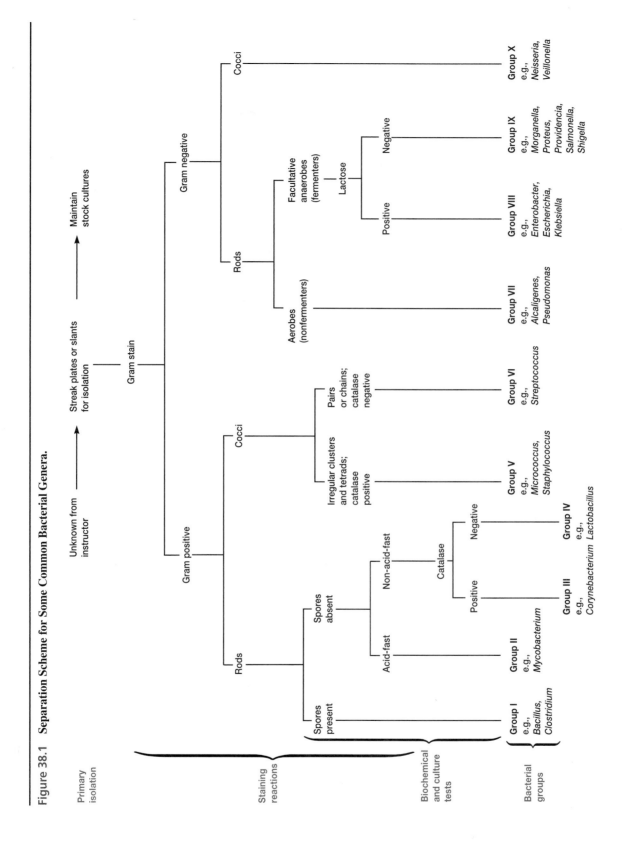

4. If your unknown is a gram-positive rod, stain for endospores (*see exercise 10*) after incubating a nutrient agar slant culture for 2 to 4 days.

Medium	Test
Phenol red broth tubes or tryptic agar base tubes for dextrose, lactose, and sucrose	Carbohydrate fermentation
Starch agar plate	Starch hydrolysis
Tryptic soy agar plate	Oxidase test
Nutrient gelatin deep tube, gelatin test strip	Gelatin liquefaction
SIM medium	H₂S, indole, motility
Tributyrin agar plate	Lipid hydrolysis
Tryptic nitrate broth	Nitrate reduction
MR-VP broth	Methyl red test, Voges-Proskauer test
Tryptic soy agar slant	Catalase activity
Simmons citrate agar slant	Citrate utilization
Urea broth, urease test disk/strip	Urease activity

Fourth Period

1. Based on your results, identify the unknown bacteria as to their genus and species. Use figure 38.1, your own outline or flow sheet (figure 38.2 is an example), and *Bergey's Manual* as guides.

Figure38.2 An Example of a Dichotomous Key Leading to the Identification of *Escherichia coli*.

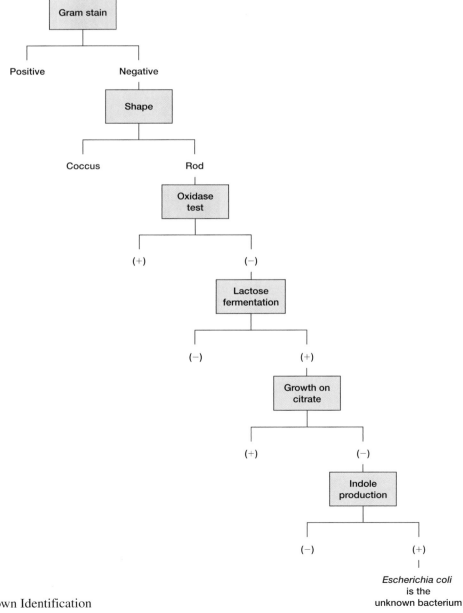

General Unknown

Characteristic*	Bacterium 1	Bacterium 2
TSA slant characteristics		
Amount of growth	_____	_____
Consistency	_____	_____
Form	_____	_____
Pigmentation	_____	_____
Macroscopic appearance		
Blood agar	_____	_____
EMB	_____	_____
Mannitol salt agar	_____	_____
Morphology		
Cell arrangement	_____	_____
Cell shape	_____	_____
Endospores' present position	_____	_____
Flagella present arrangement	_____	_____
Gram reaction	_____	_____
Biochemistry		
Gelatin liquefaction	_____	_____
Lipid hydrolysis	_____	_____
Starch hydrolysis	_____	_____
Fermentation	_____	_____
dextrose	_____	_____
lactose	_____	_____
sucrose	_____	_____
Catalase activity	_____	_____
Citrate utilization	_____	_____
H_2S production	_____	_____
Indole production	_____	_____
Methyl red test	_____	_____
Motility	_____	_____
Nitrate reduction	_____	_____
Oxidase activity	_____	_____
Urease activity	_____	_____

*Write "not tested or done" next to the characteristics or tests that were not done.

1. Answer the following questions regarding your unknown.

The code number for my unknown is _____

The genus of my unknown is _____

The species of my unknown is _____

Sketch of unknowns—label and give magnification.

Bacterium 1 **Bacterium 2**

Additional Tests Performed

	Characteristic	
Test	Bacterium 1	Bacterium 2
_____	_____	_____
_____	_____	_____
_____	_____	_____
_____	_____	_____
_____	_____	_____
_____	_____	_____

Be prepared to hand in on time a neat copy of your flowcharts.

2. In summary, what is the major purpose of this experiment?

Review Questions

1. Will the length of time each bacterium has remained in stock culture affect the biochemical results? Explain your answer.

2. Were there any tests that you carried out in this exercise for which you are not absolutely sure of the outcome for your unknowns? If so, list the tests and explain your reservations.

3. Outline the steps that you used to identify your unknown.

4. Is there another scheme that you might have used to shorten your identification of the unknowns?

5. Why is it necessary to complete the identification of a bacterium based on its physiology rather than just its morphology?

Environmental Factors Affecting Growth of Microorganisms

The paramount evolutionary accomplishment of bacteria as a group is rapid, efficient cell growth in many environments. Bacteria grow and divide as rapidly as the environment permits.
John Lyman Ingraham (Professor Emeritus of Bacteriology, University of California, Davis, 1924–)

The growth of microorganisms is greatly affected by the chemical and physical nature of their environment. An understanding of the environmental factors that promote microbial growth aids in understanding the ecological distribution of microorganisms. Therefore, the nature of some of these influences will be surveyed in this part of the manual. Some of these factors include temperature, pH, and osmotic pressure.

These same environmental factors that maximize microbial growth can also be manipulated to inhibit or retard the growth of unwanted microorganisms. Microbial control using antibiotics and disinfectants (chemicals) has become an important aspect of microbiology. Several exercises that cover control of microbial growth are included in this part of the manual.

After completing some (choice is up to the instructor) of the exercises in part seven, you will, at the minimum, be able to demonstrate an increased level in analysis skills, including

(a) collecting and organizing data in a systematic fashion;

(b) presenting data in an appropriate form (graphs, tables, figures, or descriptive paragraphs);

(c) assessing the validity of data (including integrity and significance); and

(d) drawing appropriate conclusions based on the results. This will meet the American Society for Microbiology Core Curriculum Laboratory Thinking Skill number 2 (*see pp. vi–viii*).

Barron Joseph Lister (1827–1912)

Dr. Lister was an English surgeon who, following Pasteur's theory that bacteria cause infection, introduced to surgery the practice of antisepsis. (*©Bettman/Corbis*)

In 1865, Dr. Joseph Lister, using carbolic acid as his antiseptic agent together with heat-sterilized instruments, greatly reduced postoperative mortality. He wrote the following lines in the *Lancet* ("On a New Method of Treating Compound Fractures . . ." *Lancet,* pp. 364–73, 1867):

> Applying [Pasteur's] principles to the treatment of compound fracture . . . it appears that all that is requisite is to dress the wound with some material capable of killing these germs, provided that any substance can be found reliable for this purpose, yet not too potent as a caustic.
>
> In the course of the year 1864 I was struck with an account of the remarkable effects produced by carbolic acid upon the sewage of the town of Carlisle, the admixture of a very small proportion not only preventing all odor from the lands irrigated with the refuse material, but, as it was stated, destroying entozoa which usually infest cattle fed upon such pastures.
>
> Carbolic acid proved in various ways well adapted to the purpose of an antiseptic in surgery.

Temperature

MATERIALS PER GROUP OF STUDENTS

24- to 48-hour tryptic soy broth cultures of *Escherichia coli* (ATCC 11229), *Bacillus stearothermophilus* (ATCC 7953), *Bacillus globisporus* (ATCC 23301), *Pseudomonas aeruginosa* (ATCC 10145), *Staphylococcus aureus* (ATCC 25923), and spore suspension of *Bacillus subtilis* (ATCC 6051). (To produce endospores, grow *B. subtilis* for 48 hours at 35°C on endospore agar, nutrient agar plus 0.002% $MnCl_2 \cdot 4H_2O$. Resuspend the paste in at least 7 ml of sterile diluent.)

12 tryptic soy agar slants
Bunsen burner
inoculating loop
15 tryptic soy broth tubes (9.9 ml per tube)
test-tube rack
18 sterile 1-ml pipettes with pipettor
3 sterile test tubes
refrigerator set at 4°C
incubators or water baths set at 4°, 23° to 25° (room temperature), 60°, 85°, and 100°C (The instructor or students are not limited to these temperatures. Modifications can be instituted based on incubators or water baths available.)
wax pencil
sterile water

Learning Objectives

Each student should be able to

1. Understand how microorganisms are affected by the temperature of their environment
2. Carry out an experiment that differentiates between several bacteria based on temperature sensitivity
3. Classify these same bacteria based on their temperature preference for growth
4. Determine the effects of heat on bacteria

SUGGESTED READING IN TEXTBOOK

1. Temperature, section 6.4; see also figures 6.12 and 6.13, and table 6.5.
2. Heat, section 7.4, see also table 7.2.

Pronunciation Guide

Bacillus globisporus (bah-SIL-lus glob-EE-spor-us)
B. stearothermophilus (ste-row-ther-MAH-fil-us)
B. subtilis (sub-til-us)
Escherichia coli (esh-er-I-ke-a KOH-lee)
Pseudomonas aeruginosa (soo-do-MO-nas a-ruh-jin-OH-sah)
Staphylococcus aureus (staf-il-oh-KOK-kus ORE-ee-us)

Why Are the Above Bacteria Used in This Exercise?

In this exercise, the student will gain expertise in differentiating between bacteria based on temperature sensitivity and classifying bacteria based on their temperature preference for growth. The six bacteria the authors have chosen to accomplish the above are the following. *Bacillus globisporus* (L. *globus,* a sphere) is an endospore-forming rod that has an optimum growth temperature of 20° to 25°C. *Bacillus stearothermophilus* (Gr. *thermus,* heat + *philus,* loving) is an endospore-forming rod that has an optimum growth temperature of 60° to 65°C. Its spores are more

heat resistant than those of any mesophilic species in the genus. *Bacillus subtilis* (L. *subtilis,* slender) is an endospore-forming rod that has an optimum growth temperature of about 30° to 40°C. *Escherichia coli* is a gramnegative, facultatively anaerobic rod that does not form spores and has an optimum growth temperature of 37°C. *Pseudomonas aeruginosa* is a gram-negative rod with an optimum growth temperature of 37°C. *Staphylococcus aureus* is a gram-positive coccus with an optimum growth temperature of 30° to 37°C.

PRINCIPLES

Each microbial species requires a temperature growth range that is determined by the heat sensitivity of its particular enzymes, membranes, ribosomes, and other components. As a consequence, microbial growth has a fairly characteristic temperature dependence with distinct **cardinal temperatures**—minimum, maximum, and optimum. **Minimum growth temperature** is the lowest temperature at which growth will occur; **maximum growth temperature** is the highest temperature at which growth will occur; and **optimum growth temperature** is the temperature at which the rate of cellular reproduction is most rapid. The optimum temperature for the growth of a given microorganism is correlated with the temperature of the normal habitat of the microorganism. For example, the optimum temperature for the growth of bacteria pathogenic to humans is near that of the temperature of human blood (35° to 37°C).

Most bacteria can be classified into one of three major groups based on their temperature requirements. **Psychrophiles** can grow at 0°C and have an optimum growth temperature of 15°C or lower; the maximum is around 20°C. **Mesophiles** have growth optima between 20° and 45°C. The majority of bacteria fall into this category. **Thermophiles** can grow at temperatures of 55°C or higher.

Boiling is probably one of the easiest methods of ridding materials of harmful bacteria. However, not all bacteria are equally sensitive to this high temperature. Some bacteria may be able to survive boiling even though they are unable to grow. These bacteria are termed **thermoduric.** Many of the spore formers (such as *B. subtilis*) can withstand boiling for 15 minutes because of their resistant endospores. Thus, both temperature and the species of bacteria will affect the disinfection of certain specimens. This is important to know when trying to kill pathogenic bacteria with heat.

Procedure

First Period

1. Work in groups of three to four students. Each group of students will be assigned one temperature to study: 4°, 23° to 25°, 60°, 85°, or 100°C.
2. Label each of the tryptic soy agar slants with the name of the test bacterium to be inoculated (*E. coli, B. stearothermophilus,* and *B. globisporus*), your name, and date.
3. Using aseptic technique (*see figure 14.3*), streak the surface of each slant with the appropriate bacterium. Incubate the slants for 24 to 48 hours at the temperature assigned to your group.
4. Take three sterile test tubes and label one *S. aureus,* the second *B. subtilis spores,* and the third *P. aeruginosa.* Add your name and date.
5. With a sterile pipette, aseptically add 1 ml of bacterial culture or spore suspension to the respective tubes.
6. Subject your tubes to the temperature you are studying for 15 minutes (i.e., place them in the refrigerator or let them stand at room temperature or put them in one of the water baths).
7. After 15 minutes, let the samples cool or warm up to room temperature. For each bacterial sample, make a dilution series as follows (*see appendix A*):
 a. Pipette 0.1 ml of the incubated sample into the 9.9 ml tryptic soy broth (10^{-2} dilution). Mix the tube thoroughly. With a fresh pipette, transfer 0.1 ml of this 10^{-2} dilution into 9.9 ml of broth (10^{-4} dilution) and mix. In the same way, prepare a 10^{-6}, 10^{-8}, and 10^{-10} dilution.
8. Incubate all dilutions at 35°C for 24 to 48 hours.

Second Period

1. At the end of incubation, observe the slants for the presence of growth. Record your observations and those of your classmates; use a + for the presence of growth and a − for the absence of growth in part 1 of the report for exercise 39.
2. Observe your dilution series to see which tubes have bacterial growth as indicated by turbidity. The logic of this procedure is that reproduction will have occurred in each tube that received at least one living bacterium. The greater the number of bacteria present in the particular sample, the more such a sample can be diluted and still contain bacteria in the aliquot transferred. Thus, if bacteria **A** are less susceptible to heat than bacteria **B**, bacteria **A** will require more dilutions in order to obtain a sterile sample as indicated by no growth.

3. From your results and those of your classmates, complete part 2 of the report for exercise 39 by indicating the last dilution in which growth occurred.

Temperature

1. Based on your observations of bacterial growth and those of your classmates, complete the following table.

Bacterium	Growth (+) or (−) 4°C	23° to 25°C	60°C	85°C	100°C	Temperature Classification
E. coli						
B. globisporus						
B. stearothermophilus						

2. Based on your observations and those of your classmates, complete the following table, showing range of surviving bacteria.

Temperature for 15 Minutes	Last Dilution in Which Growth Occurred *S. aureus*	*P. aeruginosa*	*B. subtilis*
4°C			
Room temperature (23° to 25°C)			
60°C			
85°C			
100°C			

3. In summary, what is the major purpose of this experiment?

Review Questions

1. How can you be sure that the turbidity produced in the broth tubes was caused by the bacteria used for the inoculation?

2. How can you determine experimentally whether a bacterium is a psychrophile or a mesophile?

3. What limitations are there for using boiling water as a means of sterilizing materials?

4. Is *S. aureus* a mesophile? Explain your answer.

5. Describe the three cardinal temperatures.

6. Which bacterium had the widest range of temperature tolerance? The narrowest range of temperature tolerance?

7. What are thermoduric bacteria?

pH

MATERIALS PER GROUP OF STUDENTS

saline suspensions of 24-hour tryptic soy broth cultures of *Alcaligenes faecalis* (ATCC 8750), *Escherichia coli* (ATCC 11229), and a Sabouraud dextrose agar slant (48 hours at 30°C) of *Saccharomyces cerevisiae* (ATCC 2366, a yeast). Add 3 ml of sterile saline to each *S. cerevisiae* slant and resuspend the yeast with a sterile inoculating loop. Transfer the suspension to a sterile culture tube. Adjust the suspensions to an absorbance of 0.05 at a wavelength of 550 to 600 nm by adding either more saline or culture.

pH meter or pH paper
4 tryptic soy broth tubes, pH 3.0
4 tryptic soy broth tubes, pH 5.0
4 tryptic soy broth tubes, pH 7.0
4 tryptic soy broth tubes, pH 9.0
(the pH of the above tubes is adjusted with either 1 N sodium hydroxide or 1 N hydrochloric acid)
Bunsen burner
sterile 1-ml pipettes with pipettor
spectrophotometer
cuvettes
wax pencil
test-tube rack

Learning Objectives

Each student should be able to

1. Understand how pH affects the growth of bacteria

2. Perform an experiment that relates bacterial growth to pH

SUGGESTED READING IN TEXTBOOK

1. pH, section 6.4; see also figure 6.11.

Pronunciation Guide

Alcaligenes faecalis (al-kah-LIJ-e-neez fee-KAL-iss)
Escherichia coli (esh-er-I-ke-a KOH-lee)
Saccharomyces cerevisiae (sak-ah-ro-MI-seez ser-ah-VEES-ee-eye)

Why Are the Above Microorganisms Used in This Exercise?

This exercise demonstrates the effect of pH on microbial growth. Students will use two bacteria and one yeast that have different pH ranges for optimal growth. After performing an experiment that relates microbial growth to pH, the student should appreciate the fact that *Alcaligenes faecalis* has optimal growth at about pH 7.0 and *Escherichia coli* at pH 6.0 to 7.0. The yeast, *Saccharomyces cerevisiae* grows best at a pH of 5.6.

PRINCIPLES

It is not surprising that **pH** (acidity; log $1/[H^+]$; *see appendix D*) dramatically affects bacterial growth. The pH affects the activity of enzymes—especially those that are involved in biosynthesis and growth. Each microbial species possesses a definite pH **growth range** and a distinct pH **growth optimum. Acidophiles** have a growth optimum between pH 0.0 and 5.5; **neutrophiles** between 5.5 and 8.0; and **alkalophiles** 8.5 to

11.5. In general, different microbial groups have characteristic pH optima. The majority of bacteria and protozoa are neutrophiles. Most molds and yeasts occupy slightly acidic environments in the pH range of 4 to 6; algae also seem to favor acidity.

Many bacteria produce metabolic acids that may lower the pH and inhibit their growth (*see exercises 20 and 25*). To prevent this, **buffers** that produce a pH equilibrium are added to culture media to neutralize these acids. For example, the peptones in complex media act as buffers. Phosphate salts are often added as buffers in chemically defined media.

In this exercise, you will work in groups to see how the pH affects the growth of several microorganisms.

Procedure

First Period

1. Label each of the tryptic soy broth tubes with the pH of the medium, your name, date, and the microorganism to be inoculated.
2. Using a sterile pipette, add 0.1 ml of the *E. coli* saline culture to the tube that has a pH of 3.0. Do the same for the tubes that have pH values of 5.0, 7.0, and 9.0.
3. Repeat the above for *A. faecalis* and *S. cerevisiae.*

4. Incubate the *E. coli* and *A. faecalis* cultures for 24 to 48 hours at 35°C, and the *S. cerevisiae* culture for 48 to 72 hours at room temperature.

Second Period

1. Use the spectrophotometer as described in exercise 19. Set the wavelength at 550 to 600 nm. Calibrate the spectrophotometer, using a tryptic soy broth blank of each pH for each respective set of cultures.
2. Fill each cuvette two-thirds full of the respective pH culture and read the absorbance. Blank the spectrophotometer with tryptic soy broth.
3. If no spectrophotometer is available, record your visual results as − (no growth), +, ++, +++, and ++++ (for increasing growth, respectively).
4. Record your results and those of your classmates in the report for exercise 40.

HINTS AND PRECAUTIONS

(1) Be sure to thoroughly mix the microbial suspensions just before reading the absorbance.

(2) Wipe any fingerprints off the cuvette before making a reading.

Laboratory Report 40

Name: _____

Date: _____

Lab Section: _____

pH

1. Summarize your results with respect to the pH and growth of each microorganism in the following table. Indicate the optimal pH with an asterisk.

Microorganism	pH	Visual Results (−, + → ++++)	Absorbance	% Transmission	Optimal pH
A. faecalis	3	_____	_____	_____	_____
	5	_____	_____	_____	_____
	7	_____	_____	_____	_____
	9	_____	_____	_____	_____
E. coli	3	_____	_____	_____	_____
	5	_____	_____	_____	_____
	7	_____	_____	_____	_____
	9	_____	_____	_____	_____
S. cerevisiae	3	_____	_____	_____	_____
	5	_____	_____	_____	_____
	7	_____	_____	_____	_____
	9	_____	_____	_____	_____

2. Which microorganism grew best in the acid pH range? _____

3. Which microorganism grew best in the neutral pH range? _____

4. Which microorganism grew best in the alkaline pH range? _____

5. Which microorganism has the widest pH growth range? _____

6. Which microorganism has the narrowest pH growth range? _____

7. Using a different color line for each bacterium, graph the growth responses to pH variations.

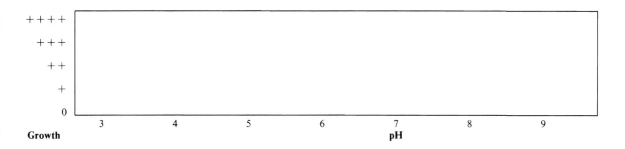

249

Review Questions

1. Why are buffers added to culture media?

2. Why do microorganisms differ in their pH requirements for growth?

3. What inhibits microbial growth at nonoptimal pHs?

4. What is the pH tolerance of bacteria compared to yeasts?

5. List and describe the chemistry of several common buffers used in microbiological media.

6. How would you define a buffer?

7. How do microorganisms change the pH of their own environment?

Osmotic Pressure

MATERIALS PER STUDENT

24- to 48-hour tryptic soy broth cultures of *Escherichia coli* (ATCC 11229) and *Staphylococcus aureus* (ATCC 25923). A 48-hour salt broth culture of *Halobacterium salinarium* (ATCC 19700)
1 petri plate nutrient agar with 0% NaCl
1 petri plate nutrient agar with 0.5% NaCl
1 petri plate nutrient agar with 5% NaCl
1 petri plate nutrient agar with 10% NaCl
1 petri plate nutrient agar with 20% NaCl
wax pencil
inoculation loop
Bunsen burner
Difco Manual or *BBL Manual* for the laboratory report

Learning Objectives

Each student should be able to

1. Define osmotic pressure and explain how it affects a bacterial cell
2. Explain how bacterial growth is related to osmotic pressure in the environment
3. Carry out an experiment that differentiates among three different bacteria based on their tolerance for different salt concentrations (osmotic pressure)

SUGGESTED READING IN TEXTBOOK

1. The Cell Wall and Osmotic Protection, section 3.5.
2. Solutes and Water Activity, section 6.4; see also table 6.4.

Pronunciation Guide

Escherichia coli (esh-er-I-ke-a KOH-lee)
Halobacterium salinarium (hal-o-bak-TE-re-um sal-i-nar-e-um)
Staphylococcus aureus (staf-il-oh-KOK-kus ORE-ee-us)

Why Are the Above Bacteria Used in This Exercise?

In this exercise, the student will differentiate three different bacteria based on their tolerance for different salt concentrations (osmotic pressure). The authors have thus chosen three bacteria that vary widely in osmotic tolerance. *Halobacterium salinarium* (Genus = Gr. *hals,* the sea + *bacterium,* small rod; species = L. *salinarium,* pertaining to salt works) is a gram-negative aerobic rod that requires a high salt concentration for growth. It is normally found in highly saline environments such as salt lakes and marine salterns and is associated with the spoilage of salted fish and hides. Mature cells retain their rod shape in 3.5 to about 5.2 M NaCl ($a_w \cong$ 0.75); at lower concentrations, pleomorphic forms appear, and at 1.5 M the cells are spherical because of loss of the cell wall. Suspensions are viscous at 1.5 M because of partial cell lysis; at 0.5 M, few, if any, cells can be detected. Sodium, chloride, and magnesium are required to maintain cell structure and rigidity. At low concentrations, the cell wall dissolves, and the cell membrane breaks up into tiny fragments. *Staphylococcus aureus* is a gram-positive coccus. *S. aureus* can grow in the presence of 15% (about 2.5 M) sodium chloride, or at an a_w of 0.86. *Escherichia coli* is a gram-negative rod that occurs as normal flora in the lower part of the intestine of warm-blooded animals. It is very sensitive to high salt concentrations—it lacks osmotic tolerance and can't grow below an a_w of 0.95.

PRINCIPLES

Since bacteria are separated from their environment by a selectively permeable plasma membrane, they can be affected by changes in the osmotic pressure or water availability of their surroundings. **Osmotic pressure** is the force developed when two solutions of different solute concentrations are separated by a membrane that is permeable only to the solvent. The **solvent** is the liquid, usually water, that dissolves a substance (the **solute**). Water availability is expressed quantitatively in terms of **water activity (a_w)**. Pure water has an a_w of 1.00, whereas cereals and other dried foods may have a_w values of 0.60 or lower. If a bacterium is placed in a **hypotonic solution** (low solute, high water content), water will enter the cell and cause it to burst (figure 41.1a) unless something is done to prevent the influx. Most bacteria have rigid cell walls that maintain the shape and integrity of the cell; thus, hypotonic solutions are not harmful to these bacteria. When bacteria are placed in a **hypertonic solution** (high solute, lower water content), water leaves, and the plasma membrane shrinks away from the wall (figure 41.1b), a process known as **plasmolysis.** This dehydrates the cell, and it ceases to grow. A few bacteria, called **halophiles,** are able to tolerate high (hyper-

tonic) salt concentrations. Bacteria that can live in very salty environments are called **extreme halophiles** to distinguish them from the **moderate halophiles** that live in the sea. In an **isotonic solution,** the concentration of solutes is the same (iso means equal) outside and inside the bacterium. The bacterium is in osmotic equilibrium with its environment and does not change volume (figure 41.1c).

In this exercise, you will examine the effect of various salt concentrations (osmotic pressure or water activity, a_w) on the growth of three species of bacteria. This tolerance (or lack of it) can be observed by the amount (or lack) of growth of the three bacterial species.

Procedure

First Period

1. With a wax pencil, divide the bottom of each of the five petri plates into thirds as indicated in figure 41.2. Place the name of the bacterium to be inoculated in each section. Add your name, salt concentration, and date.
2. Streak the respective bacteria onto the five different petri plates.
3. Incubate the plates, inverted, for 48 hours at 35°C.

Second Period

1. Observe the relative amount of growth in each section at each salt concentration. Record this growth as − (none), +, ++, +++, and ++++ (the most).
2. Record your results in the report for exercise 41.

Figure 41.1 **Effect of Osmotic Pressure on a Bacterial Cell.** The dots represent solute (NaCl) molecules. The shaded area represents water (solvent).

	Solution	Before	After
(a)	Hypotonic	H₂0 movement	Cell wall / Cell membrane
(b)	Hypertonic	H₂0 movement	Cell wall / Cell membrane
(c)	Isotonic	Cell wall / Cell membrane	H₂0 movement

Figure 41.2 **Streak Pattern.**

Osmotic Pressure

1. Record the amount of growth of the three bacteria at the different salt concentrations in the following table. Use $-$, $+$, $++$, $+++$, $++++$ to indicate the relative amount of growth.

Medium	E. coli	H. salinarium	S. aureus
0% NaCl	_____	_____	_____
0.5% NaCl	_____	_____	_____
5.0% NaCl	_____	_____	_____
10% NaCl	_____	_____	_____
20% NaCl	_____	_____	_____

2. Which of these bacteria tolerates the most salt? _____

3. Which of these bacteria tolerates the least salt? _____

4. Which of these bacteria tolerates a broad range of salt? _____

5. How would you classify *H. salinarium* as to its salt needs? _____

6. Using the *Difco Manual* or *BBL Manual,* list the ingredients of mannitol salt agar.

Ingredient	Amount
_____	_____
_____	_____
_____	_____
_____	_____
_____	_____
_____	_____

7. Why is mannitol salt agar used to isolate staphylococci?

8. In summary, what is the major purpose of this experiment?

Review Questions

1. Compare isotonic, hypotonic, and hypertonic solutions and their effects on bacterial cells.

2. Define:
 a. osmosis
 b. osmotic pressure
 c. plasmolysis
 d. halophilic

3. How is it possible for a bacterium to grow in a hypertonic environment?

4. What concentrations of NaCl are optimal for most bacteria?

5. What is unique about *H. salinarium?*

6. What foods can you think of that are protected from microbial destruction by salting?

7. Why don't bacteria lyse when placed in a hypotonic solution?

The Effects of Chemical Agents on Bacteria I: Disinfectants

SAFETY CONSIDERATIONS

Be careful with the Bunsen burner flame. No mouth pipetting. Always handle cultures with care since they may be potential pathogens. Keep all culture tubes upright in a test-tube rack or in a can. Phenol is poisonous and caustic. Do not handle with bare hands.

MATERIALS PER GROUP OF STUDENTS

20-hour tryptic soy broth cultures of *Staphylococcus aureus* (ATCC 25923) and *Pseudomonas aeruginosa* (ATCC 10145)
2 sterile screw-cap test tubes
1 sterile 5-ml pipette with pipettor
12 sterile 1-ml pipettes
48 tryptic soy broth tubes (10 ml per tube)
sterile water in Erlenmeyer flask
12 sterile tubes for making dilutions
Lysol
commercial disinfectants such as 3% hydrogen peroxide, 70% isopropyl alcohol, bleach, or Lysol cleaner. Many others may be used (see table 42.1), or students can bring in their own to test. If commercial disinfectants are used, note the use-dilution and active ingredients. Dilute with normal tap water. The tap water need not be sterilized for commercial disinfectants. Note if any of the disinfectants contain triclosan. Why is this important?
phenol (carbolic acid)
wax pencil
35°C incubator
test-tube rack
Bunsen burner
inoculating loop

Learning Objectives

Each student should be able to

1. Determine the effectiveness of some chemical disinfectants used in hospitals or homes as antimicrobial agents
2. Calculate a phenol coefficient

SUGGESTED READING IN TEXTBOOK

1. The Use of Chemical Agents in Control, section 7.5; see also tables 7.4 and 7.5.
2. Phenolics, section 7.5.
3. Alcohols, section 7.5.
4. Halogens, section 7.5.
5. Quaternary Ammonium Compounds, section 7.5.
6. Aldehydes, section 7.5.
7. Evaluation of Antimicrobial Agent Effectiveness, section 7.6; see also table 7.6.

Pronunciation Guide

Pseudomonas aeruginosa (soo-do-MO-nas a-ruh-jin-OH-sah)
Staphylococcus aureus (staf-il-oh-KOK-kus ORE-ee-us)

Why Are the Above Bacteria Used in This Exercise?

In this exercise, the student will determine the antimicrobial effectiveness of some common disinfectants and calculate a phenol coefficient. The authors have chosen two common bacteria (*Staphylococcus aureus* and *Pseudomonas aeruginosa*) for use. Both of these bacteria give excellent positive or negative results depending on the disinfectant used. *S. aureus* is often used by laboratories as a test microorganism to determine phenol coefficients.

PRINCIPLES

Many factors influence the effectiveness of chemical disinfectants and antiseptics. The **microbicidal** (to kill) or **microbiostatic** (to inhibit) **efficiency** of a chemical is often determined with respect to its ability to deter microbial growth. The first part of this exercise will examine this effect of several chemicals.

More specifically, the microbicidal efficiency of a chemical is often determined with respect to phenol and is known as the **phenol coefficient (PC).** The phenol coefficient is calculated by dividing the highest dilution of the antimicrobial of interest, which kills all organisms after incubation for 10 minutes but not after 5 minutes, by the highest dilution of phenol that has the same characteristics. Chemicals that have a phenol coefficient greater than 1 are more effective than phenol, and those that have a phenol coefficient less than 1 are less effective than phenol. However, this comparison should only be used for phenol-like compounds that do not exert bacteriostatic effects and are not neutralized by the subculture media used. The second part of this experiment will enable you to calculate a phenol coefficient for a select chemical. A list of commonly used antiseptics and disinfectants and their area of application is shown in table 42.1

Procedure

First Period
Growth Inhibition

1. Each group of students should select one of the disinfectants and, if necessary, dilute it according to the specifications on the label (the **use-dilution**).
2. Place 5 ml of disinfectant into two sterile tubes. Add 0.05 ml of *P. aeruginosa* to one tube and 0.05 ml of *S. aureus* to the other.
3. Using the wax pencil, label the tubes with your name and those of the respective bacteria. Mix each of the tubes in order to obtain a homogeneous suspension.
4. At intervals of 1, 2, 5, 10, and 15 minutes, transfer 0.1 ml of the mixture containing the bacteria and disinfectant to separate tubes of tryptic soy broth. Do this for both bacteria. Also inoculate two tubes of broth with 0.1 ml of both bacteria and mark these "controls."
5. Incubate all tubes for 48 hours at 35°C.

Phenol Coefficient (*See Safety Considerations*)

1. Dilute phenol in sterile distilled water 1/80, 1/90, and 1/100; dilute the Lysol 1/400, 1/450, and 1/500 so that the final volume in each tube is 5 ml.

Table 42.1		
Some Chemical Compounds Commonly Used for Controlling the Growth of Microorganisms		
Sodium hypochlorite (5%)	Disinfectant	External surfaces, such as tables
Iodine (1% in 70% alcohol)	Disinfectant	External surfaces, such as tables
Iodophors (70 ppm avail. I_2)	Disinfectant	External surfaces, such as tables
Lysol (5%), a solution of phenolics and/or quaternary ammonium compounds with soap	Disinfectant	External surfaces, such as tables
Phenol (5%), carbolic acid, source coal tar	Disinfectant	External surfaces, such as tables
Hexachlorophene (pHisoHex. Dial soap)	Disinfectant	Presurgical hand washing
Formaldehyde (4%)	Disinfectant	Oral and rectal thermometers
Iodophors (70 ppm avail. I_2)	Disinfectant	Oral and rectal thermometers
Zephrin (0.001%)	Disinfectant	Oral and rectal thermometers
Alcohol, ethanol (70%)	Antiseptic	Skin
Iodine (tincture in alcohol with Kl)	Antiseptic	Skin
Iodophors	Antiseptic	Skin
Organic mercury compounds (merthiolate, mercurochrome)	Antiseptic	Skin
Hydrogen peroxide (3%)	Antiseptic	Superficial skin infections
Potassium permanganate	Antiseptic	Urethral, superficial skin fungus infections
Silver nitrate (1%)(Argyrol)	Antiseptic	Prevention of eye infections in newborn babies
Zinc oxide paste	Antiseptic	Diaper rash
Zinc salts of fatty acids (Desenex)	Antiseptic	Treatment of athlete's foot
Glycerol (50%)	Antiseptic	Prevent bacterial growth in stool and surgical specimens
Ethylene oxide gas (12%)	Sterilization	Linens, syringes, etc.
Formaldehyde (20% in 70% alcohol)	Sterilization	Metal instruments
Glutaraldehyde (pH 7.5 of more)	Sterilization	Metal instruments

Source: Adapted from *Microbiology Experiments: A Health Science Perspective.* Kleyn, Bicknell, Gilstrap. 1999. McGraw-Hill. By permission

2. Label 18 tryptic soy broth tubes with the name and dilution of disinfectant, the time interval of the subculture (e.g., 5 minutes, phenol 1/80), and your name. Each dilution should be tested after 5-, 10-, and 15-minute incubations.
3. Place in order in a test-tube rack, one test tube of each of the different Lysol and phenol dilutions for each time interval.
4. Add 0.5 ml of *S. aureus* to each tube of disinfectant and note the time. Mix each of the tubes in order to obtain a homogeneous suspension and allow the disinfectant to come into contact with the bacteria.
5. Using aseptic technique, at intervals of 5, 10, and 15 minutes, transfer one loopful from each disinfectant tube into the appropriately labeled tryptic soy broth tube.
6. Incubate all tubes for 48 hours at 35°C.
7. The experiment can be repeated with *P. aeruginosa*.

Second Period
Growth Inhibition
1. Shake and observe each of the tubes for growth. Record the presence of growth as + and the absence of growth as −. Tabulate your results as well as the results of the class in part 1 of the report for exercise 42.

Phenol Coefficient
1. Shake and observe all tryptic soy broth cultures for the presence (+) or absence (−) of growth.
2. Record your observations in part 2 of the report for exercise 42.
3. From your data, calculate the phenol coefficient for Lysol. For example, assume a 1/20 dilution of phenol (1 part phenol in a total of 20 parts liquid) kills *S. aureus* within 10 minutes. A 1/300 (1 to 300) dilution of Lysol also kills *S. aureus* within 10 minutes.

$$PC = \frac{300}{20} \text{ or } \frac{1/20}{1/300}$$

$$PC = 15$$

Thus, Lysol is 15 times more effective than phenol in killing *S. aureus*.

Name: _____

Date: _____

Lab Section: _____

The Effects of Chemical Agents on Bacteria I: Disinfectants

1. Based on your observations of bacterial growth and those of your classmates, complete the following table.

| Bacterium | Name of Disinfectant plus Active Ingredients* | Use Dilution | Control | Degree of Growth Time of Exposure in Minutes | | | | |
				1	2	5	10	15

*Is triclosan present? Why is this important? _____

2. Based on your observations of bacterial growth and the observations of your classmates, complete the following table.

| Disinfectant | Dilution | Growth in Subculture (Minutes) | | |
		5	10	15
Phenol	1/80			
	1/90			
	1/100			
Lysol	1/400			
	1/450			
	1/500			

3. From the above data, calculate the phenol coefficient of Lysol. For example, if the Lysol dilution of 1/450 showed no growth at 10 minutes but growth at 5 minutes, and the phenol dilution of 1/90 showed no growth at 10 minutes but growth at 5 minutes, then:

$$\text{Phenol coefficient of Lysol} = \frac{1/450}{1/90} = 5$$

Review Questions

1. What are some limitations of a test such as you performed on the evaluation of a disinfectant?

2. List some criteria of a good disinfectant.

3. What is the phenol coefficient technique?

4. A disinfectant diluted 1/500 with water kills a bacterium after 10 minutes but not after 5 minutes. A 1/100 dilution of phenol kills the same bacterium after 10 minutes but not after 5 minutes. What is the phenol coefficient of the disinfectant?

5. What is the difference between microbicidal and microbiostatic?

6. What physical factors can influence the activity of a disinfectant?

7. Why do microorganisms differ in their response to disinfectants?

The Effects of Chemical Agents on Bacteria II: Antimicrobial Agents (Kirby-Bauer Method)

MATERIALS PER GROUP OF STUDENTS

4 150 × 15 mm Mueller-Hinton agar plates
antibiotic disk dispensers (BBL or Difco) or assorted individual vials containing antibiotic disks
4 sterile swabs
4- to 6-hour tryptic soy broth cultures of *Staphylococcus aureus* (ATCC 25903), *Escherichia coli* (ATCC 11229), *Pseudomonas aeruginosa* (ATCC 10145), and *Klebsiella pneumoniae* (ATCC e13883)
35°C incubator
forceps
metric rulers
wax pencil
70% ethyl alcohol and beakers
Bunsen burner

Learning Objectives

Each student should be able to

1. Appreciate the scope of antimicrobial activity of selected antibiotics
2. Perform the Kirby-Bauer method for determination of antibiotic sensitivity
3. Correctly interpret a Kirby-Bauer plate

SUGGESTED READING IN TEXTBOOK

1. Disk Diffusion Tests, section 35.3; also see table 35.3 and figures 35.1 and 35.2.
2. Susceptibility Testing, section 36.3; see also figure 36.5o.

Pronunciation Guide

Escherichia coli (esh-er-I-ke-a KOH-lee)
Klebsiella pneumoniae (kleb-se-EL-lah nu-MO-ne-ah)
Pseudomonas aeruginosa (soo-do-MO-nas a-ruh-jin-OH-sah)
Staphylococcus aureus (staf-il-oh-KOK-kus ORE-ee-us)

Why Are the Above Bacteria Used in This Exercise?

In this exercise, the student will evaluate the antimicrobial activity of selected antibiotics using the Kirby-Bauer method. To accomplish this objective, the authors have chosen four bacteria that the student has worked with in previous exercises and that often infect humans. The four bacteria are *Staphylococcus aureus, Escherichia coli, Pseudomonas aeruginosa,* and *Klebsiella pneumoniae.*

Medical Application

The number of antibiotics (and other antimicrobics) available today is larger than ever before. New antibiotics are continuously being developed and discovered; thus there is an increasing demand on the clinical laboratory to determine the antibiotic susceptibility or resistance of various pathogenic bacteria. In most clinical laboratories, the antibiogram has been replaced with molecular techniques.

PRINCIPLES

One method that is used to determine antibiotic susceptibility is the sensitivity disk method of **Kirby-Bauer** (named after W. Kirby and A. W. Bauer in 1966). In this method, antibiotics are impregnated onto paper disks and

Figure 43.1 A Kirby-Bauer Plate. A Mueller-Hinton agar plate inoculated with *S. aureus* and various antibiotics. Notice the diameter of the various zones of inhibition.

then placed on a seeded Mueller-Hinton agar plate using a mechanical dispenser or sterile forceps. The plate is then incubated for 16 to 18 hours, and the diameter of the **zone of inhibition** around the disk is measured to the nearest millimeter. The inhibition zone diameter that is produced will indicate the susceptibility or resistance of a bacterium to the antibiotic (figure 43.1). Antibiotic susceptibility patterns are called **antibiograms.** Antibiograms can be determined by comparing the zone diameter obtained with the known zone diameter size for susceptibility (table 43.1). For example, a zone of a certain size indicates susceptibility, zones of a smaller diameter or no zone at all show that the bacterium is resistant to the antibiotic. Frequently one will see colonies within the zone of inhibition when the strain is antibiotic resistant.

Many factors are involved in sensitivity disk testing and must be carefully controlled. These include size of the inoculum, distribution of the inoculum, incubation period, depth of the agar, diffusion rate of the antibiotic, concentration of antibiotic in the disk, and growth rate of the bacterium. If all of these factors are carefully controlled, this type of testing is highly satisfactory for determining the degree of susceptibility of a bacterium to a certain antibiotic.

The Kirby-Bauer method is not restricted to antibiotics. It may also be used to measure the sensitivity of any microorganism to a variety of antimicrobial agents such as sulfonamides and synthetic chemotherapeutics. Figures 43.1 and 43.2 illustrate the Kirby-Bauer method.

Procedure

First Period

1. With a wax pencil, mark the lid of each Mueller-Hinton agar plate with your name, date, and the name of the bacterium to be inoculated. Each group of students will inoculate the surface of four Mueller-Hinton plates with *S. aureus, E. coli, P. aeruginosa,* and *K. pneumoniae,* respectively. Use a separate, sterile cotton swab for each bacterium. The swab is immersed in the culture tube, and the excess culture is squeezed on the inner side of the test tube. If there are sufficient supplies you may wish to analyze the antimicrobial sensitivity of microorganisms from your throat.

2. The swab is then taken and streaked on the surface of the Mueller-Hinton plate three times, rotating the plate 60° after each streaking. Finally, run the swab around the edge of the agar. This procedure ensures that the whole surface has been seeded. Allow the culture to dry on the plate for 5 to 10 minutes at room temperature with the top in place.

3. Dispense the antibiotics onto the plate either with the multiple dispenser or individually with the single unit dispenser. Make sure that contact is made between the antibiotic disk and the culture by gently pressing the disk with alcohol-flamed forceps. DO NOT PRESS THE DISK INTO THE AGAR, AND DO NOT MOVE THE DISK ONCE IT IS PLACED ON THE AGAR.

4. Incubate the plates for 16 to 18 hours at 35°C. DO NOT INVERT THE PLATES.

Second Period

1. Measure the zones of inhibition to the nearest mm for each of the antibiotics tested. Record the results in the report for exercise 43. Use table 43.1 as an aid. For each antibiotic, determine whether the bacteria are resistant or susceptible.

HINTS AND PRECAUTIONS

(1) If the plate is satisfactorily inoculated, and the inoculum is sufficient, the zones of inhibition will be uniformly circular and confluent growth should be seen over the entire plate. If you see isolated colonies on your plate, then the technique performed is less than adequate, and the procedure should be repeated.

(2) Colonies growing within the zone of inhibition usually result in considering the bacteria drug resistant.

Table 43.1

Interpretation of Inhibition Zones of Test Cultures

Disk Symbol	Antibiotic	Disk Content	Diameter of Zones of Inhibition (mm)		
			Resistant	Intermediate	Susceptible
AM	Ampicillin[a] when testing gram-negative microorganisms and enterococci	10 μg	16 or less		17 or more
AM	Ampicillin[a] when testing staphylococci and penicillin G-susceptible microorganisms	10 μg	28 or less		29 or more
B	Bacitracin	10 units	8 or less	9–12	13 or more
CB	Carbenicillin when testing *Proteus* species and *E. coli*	50 μg	19 or less	18–22	23 or more
CB	Carbenicillin when testing *P. aeruginosa*	50 μg	13 or less	14–16	17 or more
C	Chloramphenicol (Chloromycetic®)	30 μg	12 or less	13–17	18 or more
CC	Clindamycin[c] when reporting susceptibility to clindamycin	2 μg	14 or less	15–20	21 or more
CC	Clindamycin[c] when reporting susceptibility to lincomycin	2 μg	16 or less	17–20	21 or more
CL	Colistin[d] (Coly-mycin®)	10 μg	8 or less	9–10	11 or more
E	Erythromycin	15 μg	13 or less	14–22	23 or more
GM	Gentamicin	10 μg	12 or less	13–14	15 or more
K	Kanamycin	30 μg	13 or less	14–17	18 or more
ME	Methicillin[c]	5 μg	9 or less	10–13	14 or more
N	Neomycin	30 μg	12 or less	13–16	17 or more
NB	Novobiocin[f]	30 μg	17 or less	18–21	22 or more
OL	Oleandomycin[g]	15 μg	11 or less	12–16	17 or more
P	Penicillin G. when testing staphylococci[h]	10 units	28 or less		29 or more
P	Penicillin G. when testing other microorganisms[h,i]	10 units	14 or less		22 or more
PB	Polymyxin B[d]	300 units	8 or less	9–11	15 or more
R	Rifampin when testing *N. meningitidis* susceptibility only	5 μg	16 or less	17–19	20 or more
S	Streptomycin	10 μg	6 or less	7–9	10 or more
S	Sulfonamides	300 μg	12 or less	13–16	17 or more
T(TE)	Tetracycline[j]	30 μg	14 or less	15–18	19 or more
VA	Vancomycin	30 μg	14 or less	15–16	17 or more

Source: Based on data from the National Committee for Clinical Laboratory Standards (NCCLS).

[a]The ampicillin disk is used for testing susceptibility of both ampicillin and betacillin.

[b]Staphylococci exhibiting resistance to the penicillinase-resistant penicillin class disks should be reported as resistant to cephalosporin class antibiotics. The 30 mcg cephalothin disk cannot be relied upon to detect resistance of methicillin-resistant staphylococci to cephalosporin class antibiotics.

[c]The clindamycin disk is used for testing susceptibility to both clindamycin and lincomycin.

[d]Colistin and polymyxin B diffuse poorly in agar, and the accuracy of the diffusion method is thus less than with other antibiotics. Resistance is always significant, but when treatment of systemic infections due to susceptible strains is considered, it is wise to confirm the results of diffusion test with a dilution method.

[e]The methicillin disk is used for testing susceptibility of all penicillinase-resistant penicillins; that is, methicillin, cloxacillin dicloxacillin, oxacillin, and nafcillin.

[f]Not applicable to medium that contains blood.

[g]The oleandomycin disk is used for testing susceptibility to oleandomycin and trioleandomycin.

[h]The penicillin G. disk is used for testing susceptibility to all penicillinase-susceptible penicillins except ampicillin and carbenicillin: that is, penicillin G., phenoxymethyl penicillin, and phenethicillin.

[i]This category includes some organisms such as enterococci and gram-negative bacilli that may cause systemic infections treatable with high doses of penicillin G. Such organisms should only be reported susceptible to penicillin G. and not to phenoxymethyl penicillin or phenethicillin.

[j]The tetracycline disk is used for testing susceptibility to all tetracyclines; that is chlorotetracycline, demeclocycline, doxycycline, methacycline, oxytetracycline, minocycline, and tetracycline.

Figure 43.2 **Antimicrobic Sensitivity Testing.**

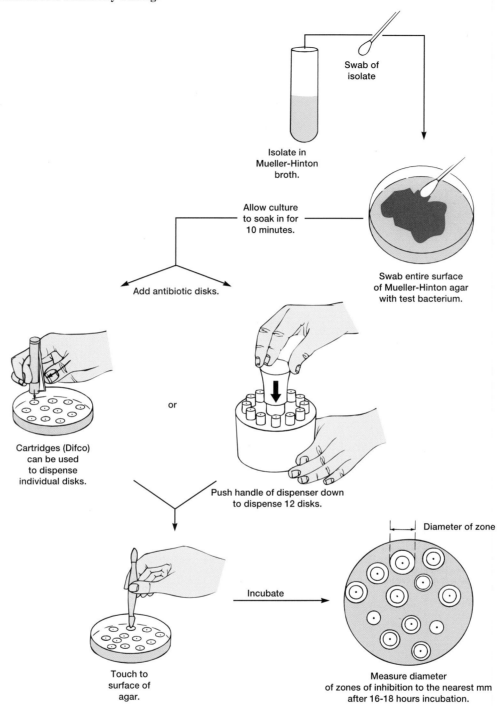

Swab of isolate

Isolate in Mueller-Hinton broth.

Allow culture to soak in for 10 minutes.

Swab entire surface of Mueller-Hinton agar with test bacterium.

Add antibiotic disks.

Cartridges (Difco) can be used to dispense individual disks.

or

Push handle of dispenser down to dispense 12 disks.

Touch to surface of agar.

Incubate

Diameter of zone

Measure diameter of zones of inhibition to the nearest mm after 16-18 hours incubation.

Name: _____

Date: _____

Lab Section: _____

The Effects of Chemical Agents on Bacteria II: Antimicrobial Agents (Kirby–Bauer Method)

1. Based on your measurements, complete the following table on the susceptibility of each test bacterium to the antibiotic by using an R (resistant), I (intermediate), or S (sensitive).

	Disk Code	*S. aureus*		*E. coli*		*P. aeruginosa*		*K. pneumoniae*	
Antibiotic or Antimicrobic		Zone Size	S, I, R	Zone Size	S, I, R	Zone Size	S, I, R	Zone Size	S, I, R
1. _____									
2. _____									
3. _____									
4. _____									
5. _____									
6. _____									
7. _____									
8. _____									
9. _____									
10. _____									
11. _____									
12. _____									

2. From the above table, which antibiotic (antimicrobic) would you use against each of the following?

S. aureus _____

E. coli _____

P. aeruginosa _____

K. pneumoniae _____

Review Questions

1. How can you determine whether the zone of inhibition is due to death or to inhibition of a bacterium?

2. What factors must be carefully controlled in the Kirby-Bauer method?

3. In which growth phase is a bacterium most sensitive to an antibiotic?

4. If the clinical laboratory reports bacterial susceptibility to an antibiotic but the patient is not responding to it, what could have gone wrong?

5. What are the similarities and differences in response to plates with gram-positive and gram-negative bacteria? Between enterics and nonenterics?

6. What is the difference between an antibiotic and an antimicrobic?

7. What are some reasons bacteria are becoming more resistant to antibiotics?

Hand Washing, Environmental Sampling, and Microbiological Monitoring

MATERIALS PER STUDENT

2 petri plates of TSA containing lecithin and polysorbate 80

2 RODAC plates containing TSA with lecithin and polysorbate 80 for environmental sampling

HYcheck System for Disinfection Control (Difco, 9039-36-4)

Millipore Membrane Sampler and Swab Test Kits (Coli-Count Sampler [MCOO] or Swab [MCSK], Yeast and Mold Sampler [MYOO] or Swab [MYSK], Total Count Sampler [MTOO] or Swab [MTSK])

variety of chemical disinfectants such as Staphene (Vestal Labs), Lysol, Zephiran (Winthrop Labs), alcohol (70%), and others

variety of hand soaps such as Ivory, Dial, Dove, Zest, Palmolive, Lifebuoy, Phisohex (Winthrop Labs), Betadine (Purdue Federick Co.), Septisoft (Vestal Labs), and others. Note if any of these products contain triclosan!

1-ml and 10-ml pipettes with pipettor for preparing dilutions

paper towels

wax pencil

35°C incubator

sterile single-use surgical scrub brush

receptacle for used brushes

Learning Objectives

Each student should be able to

1. Understand the value of proper hand washing
2. Understand the importance of efficient decontamination procedures
3. Evaluate the effectiveness of hand washing and decontamination
4. Perform routine microbiological monitoring

SUGGESTED READING IN TEXTBOOK

1. Definition of Frequently Used Terms, section 7.1.
2. The Use of Chemical Agents in Control, section 7.5; see also tables 7.4 and 7.5.

PRINCIPLES

Most commercial hand-washing products contain anti-bacterial chemicals. These products are widely used in the home and workplace. There are two types available: hand soaps containing the phenolic antiseptic called triclosan and hand gels containing ethyl alcohol. Hand gels can be used without water whereas soaps require the addition of water.

Proper hand-washing technique performed by clinical personnel is the most effective method of controlling infections, especially **nosocomial** (hospital-acquired) **infections.** A layer of oil and the structure of the skin prevent the removal of microorganisms by simple hand washing. Using a soap or gel will help remove the oil, and scrubbing with a brush for 7 to 8 minutes will maximize the removal of both **transient** (contaminated) and **resident microorganisms.** In this exercise, each student will be given a specific soap or gel to evaluate. The proper hand-washing technique will be demonstrated by the instructor. The effectiveness of the hand-washing technique and the soap or gel will then be shown by the presence or absence of growth on a culture plate.

Also in this exercise, various environmental surfaces will be **disinfected** with chemicals generally employed in a clinical setting. **Disinfection** is the killing of vegetative forms of pathogenic microorganisms or viruses. **Disinfectants** are agents, usually chemicals, used to carry out disinfection and are normally used only on inanimate objects. The purpose for this part of

the exercise is to show the student the importance of both chemical and physical removal of potentially harmful microorganisms by disinfectants and, at the same time, to provide experience in routine biological monitoring. **RODAC** (**r**eplicate **o**rganism **d**etection **a**nd **c**ounting) plates will be used for this part of the exercise. The design of the plate permits the pouring of a raised surface of the culture medium for total surface contact of the area being sampled. The 10-mm grid on the bottom of the plate facilitates counting and colony location. These plates are used for the detection and enumeration of microorganisms present on surfaces that must be kept sanitary.

Alternatives for environmental sampling are either Difco's HYcheck System for Disinfection Control or Millipore's Sampler and Swab Test Kits for microbiological monitoring.

Procedure

First Period

Hand Washing

1. Before doing any hand washing, using one of your hands, gently make a five-finger impression on one of the TSA plates by rolling each finger and your thumb on the agar. With the wax pencil, label this plate with your name, date, and "before" hand washing.

2. Take one of the soaps supplied and wash your hands according to the directions of the manufacturer or your instructor. Wash your hands for the length of time assigned by the instructor. Some appropriate intervals are 10 seconds, 30 seconds, and 3 minutes, but other times may be chosen. In this way, the effect of both the soap selected and the length of scrubbing can be studied. The hand-washing technique involves:

 a. Using continuously running hot water
 b. Using plenty of soap
 c. Applying the soap with vigorous contact on all surfaces of the hands
 d. Washing hands with a surgical scrub brush
 e. Keeping hands down at all times, so any runoff will go into the sink and not down the arms
 f. Avoiding splashing; rinsing thoroughly
 g. Drying well with paper towels
 h. Discarding the towels into a bag provided for that purpose
 i. Using a paper towel to turn off the faucet

3. After washing, make another five-finger impression (using the same hand and fingers) on a different TSA plate. Mark this plate with your name, date, and the duration of hand washing.

4. Incubate both plates at 35°C for 24 to 48 hours.

First Period

Environmental Sampling

1. Choose any environmental surface such as a floor, wall, table, or drinking fountain, and place a RODAC plate on its surface (figure 44.1). Make sure that a perfect contact has occurred. Mark this plate with your name, date, and "before" disinfection.

2. Take one of the disinfectants supplied and make the proper use-dilution as suggested by the manufacturer. Pour some of the chemical (5 to 10 ml) on the area that was previously sampled with the RODAC plate. Allow sufficient time for contact (your instructor will suggest time intervals). Wipe the area with a paper towel to absorb the excess fluid. Scrubbing with the towel can also be done.

3. Make another impression of the area with another RODAC plate. Mark this plate with your name, date, and "after" disinfection.

4. Place both RODAC plates in a 35°C incubator for 24 to 48 hours.

5. The HYcheck contact slide for Disinfection Control (Difco) is a convenient alternative for testing disinfectant effectiveness. The hinged slide paddle has tryptic soy agar on one side and D/E neutralizing agar for disinfectant testing on the other. The paddle's surface is marked off in one-square-centimeter units for ease in colony counting. Surfaces can be quickly tested in the following way (your instructor will demonstrate the procedure).

 a. Press the paddle's terminal spike against the surface to be tested and bend the paddle hinge to a convenient angle.
 b. Holding the slide by its cap, gently press the TSA plate firmly against the surface to be sampled and hold it there for a few seconds. Then bend the paddle back straight and return it to its container.
 c. Disinfect the surface as described in step 2.
 d. Remove the slide and press the purple D/E neutralizing agar against the disinfected surface; hold it there for a few seconds.
 e. Return the slide to its container, seal it tightly, and label the container. Incubate at 35° to 37°C and count any colonies on the two sides after about 24 and 48 hours incubation.

6. Sampler and Swab Test Kits from Millipore can also be used for routine microbiological sampling and monitoring of a specific environment. Advantages of these kits include no media to mix, no agar plates to prepare, and nothing to sterilize and clean. Testing is as simple as sam-

Figure 44.1 RODAC Plates. (a) A RODAC plate. **(b)** Applying a RODAC plate to an environmental surface.

(a)

(b)

ple, incubate, and count (figures 44.2 and 44.3). Specific temperatures and incubation times are given in table 44.1.

Second Period
Hand Washing

1. Examine the TSA plates around the area where the finger impressions were made. Compare the "before" and "after" plates. Complete the data asked for in part 1 of the report for exercise 44.
2. It will be necessary to exchange data with other students in order to answer some of the review questions.

Second Period
Environmental Sampling

1. Examine the RODAC plates, HYcheck slide, and/or Millipore test kit/sampler. Count the number of colonies on each plate, slide, and/or

test kit/sampler. Record your results and complete the data asked for in part 2 of the report for exercise 44.
2. It will be necessary to exchange data with other students in order to answer some of the review questions.

HINTS AND PRECAUTIONS

(1) If you have to dilute concentrated disinfectants, wear gloves and carry out the dilutions in the chemical (fume) hood.

(2) Millipore samplers should be used only when counts >10 colonies/ml are anticipated.

(3) Samplers are not recommended for testing drinking water or when a 100-ml sample is required due to a low number of microorganisms or for marine water where salt concentrations might alter nutrients.

(4) Triclosan is primarily effective against gram-positive bacteria.

Figure 44.2 The Self-Contained Membrane Filter Test (Millipore). (**a**) To sample surfaces, wipe the surface with the swab; return the swab to the plastic case containing sterile buffer; shake 30 times; discard the swab. (**b**) To sample liquids, fill the case with the fluid to be tested; insert the sampler into the case and lay it horizontally with the membrane side down for 30 seconds; pour out the test fluid. (**c**) Reinsert the sampler into the case and incubate with the membrane side down. (**d**) Count the colonies or make a quick comparison using the comparison chart provided in the package (figure 44.3). (*Courtesy of Millipore Corporation*)

(a)

(b)

(c)

(d)

Figure 44.3 Small and Large Colony Counts. Simply align the sampler with photo showing same density of colonies and record that number. (*Courtesy of Millipore Corporation*)

Environmental Factors Affecting Growth of Microorganisms

Table 44.1

Temperature and Incubation Times for Various Millipore Test Kits

	Incubation Organism	Test Kit	Temperature	Sampler Time	Color
SWAB TEST KIT					
Food Applications					
Raw Materials	Coliform	Coli-Count	35°C±1°C	22–24 hrs	Blue
Saws	Bacteria	HPC	25°C–35°C	48–72 hrs	Red
Grinders	Bacteria	HPC	25°C–35°C	48–72 hrs	Red
Beverage Applications					
Equipment Surfaces	Bacteria	HPC	25°C–35°C	48–72 hrs	Red
	Yeast, Mold	Yeast & Mold	28°C–32°C	48–72 hrs	Yellow
Public Health Inspections	Coliform	Coli-Count	35°C±1°C	22–24 hrs	Blue
	Bacteria	HPC	25°C–35°C	48–72 hrs	Red
SAMPLER					
Food Applications					
Treated Waters	Bacteria	HPC	25°C–35°C	48–72 hrs	Red
Raw Materials	Bacteria	HPC	25°C–35°C	48–72 hrs	Red
Liquid Raw Ingredients with fine particulate suspensions	Yeast, Mold	Yeast & Mold	28°C–32°C	22–24 hrs	Yellow
	Coliform	Coli-Count	35°C±1°C	22–24 hrs	Blue
Finished Liquid Product	Bacteria	HPC	25°C–35°C	48–72 hrs	Red
Beverage Applications					
Treated Waters	Bacteria	HPC	25°C–35°C	48–72 hrs	Red
	Coliform	Coli-Count	35°C±1°C	22–24 hrs	Blue
Finished Syrup	Bacteria	HPC	25°C–35°C	48–72 hrs	Red
Packaged Product	Bacteria	HPC	25°C–35°C	48–72 hrs	Red
	Yeast, Mold	Yeast & Mold	28°C–32°C	48–72 hrs	Yellow
Electronics High Purity Water	Bacteria	HPC	28°C±2°C	48–96 hrs	Red
Cooling Tower Waters	Bacteria	HPC	28°C–35°C	48–72 hrs	Red
	Bacteria	Total Count	28°C–35°C	48–72 hrs	White
	Yeast, Mold	Yeast & Mold	28°C–32°C	48–72 hrs	Yellow
Process Water	Bacteria	HPC	25°C–35°C	48–72 hrs	Red
	Coliform	Coli-Count	35°C±1°C	22–24 hrs	Blue
	Bacteria	Total Count	25°C–35°C	48–72 hrs	White
Dialysis Water					
Make-Up Water	Bacteria	HPC	35°C±2°C	48–72 hrs	Red
Finished Dialysate	Bacteria	Total Count	35°C±2°C	48–72 hrs	White
Public Health Inspections	Coliform	Coli-Count	35°C±1°C	22–24 hrs	Blue
	Bacteria	HPC	25°C–35°C	48–72 hrs	Red
Environmental Waters	Coliform	Coli-Count	35°C±1°C	22–24 hrs	Blue
	Bacteria	HPC	28°C–35°C	48–72 hrs	Red
	Bacteria	Total Count	25°C–35°C	48–72 hrs	White
Laboratory Grade Water	Bacteria	HPC	28°C–35°C	48–72 hrs	Red
Cosmetic and Pharmaceutical Manufacturing	Bacteria	HPC	25°C–35°C	48–72 hrs	Red

Note: Except for Coli-Count Sampler and Swab Test Kits, the incubation times and temperatures listed above are accepted ranges and can vary depending on specific situations. The same incubation time and temperature should be used for routine testing to establish a standard for comparison. Incubation time and temperature for coliform testing is specific and follows accepted standards.

Source: Copyright © Millipore Corporation, Bedford, MA. Reprinted by permission.

Laboratory Report 44

Name: _____

Date: _____

Lab Section: _____

Hand Washing, Environmental Sampling, and Microbiological Monitoring

1. Data from hand-washing experiment.

 a. Type of hand-washing material used: _____

 b. Length of time of hand washing: _____

 c. Type of culture medium used: _____

 d. Hours of incubation: _____

 e. Temperature of incubation: _____

 f. Colony count prior to hand washing: _____

 g. Colony count after hand washing: _____

 h. Colony shape (*see figure 15.1*): _____

 i. Interpretation: _____

2. Data from environmental sampling.

 a. Chemical used: _____

 b. Use-dilution of chemical: _____

 c. Contact time used: _____

 d. Hours of incubation: _____

 e. Temperature of incubation: _____

 f. Colony count prior to disinfection: _____

 g. Colony count after disinfection: _____

 h. Colony shape (*see figure 15.1*): _____

 i. Interpretation: _____

3. Additional data from microbiological monitoring.

4. In summary, what is the major purpose of this experiment?

Review Questions

1. Why is contact so important in the disinfection process?

2. What disinfectant proved to be the most effective? The least effective? Suggest reasons why.

3. What soap proved to be the most effective? The least effective? Did any contain triclosan? Suggest reasons why and the significance of triclosan.

4. Since most normal flora are not harmful, why must they be removed in a surgical scrub?

5. List some of the reasons why hand washing fails to remove all microorganisms. In your discussion, analyze the effect of hand-scrubbing time (as well as the soap).

6. Why is liquid soap preferred to bar soap during a surgical scrub?

7. What microbial advantage is there to paper towels in a rest room over a continuous-feed cloth towel?

Determination of a Bacterial Growth Curve: Classical and Two-Hour Methods

MATERIALS PER GROUP OF STUDENTS (CLASSICAL METHOD)

- 10- to 12-hour (log phase) tryptic soy broth cultures of *Escherichia coli* (ATCC 11229). Cultures can be maintained in log phase by immersion in an ice-water bath
- 100 ml of brain-heart infusion in a 250-ml Erlenmeyer flask
- 21 99-ml saline saline blanks
- 3 100-ml bottles of tryptic soy agar or 20 individual 15-ml tubes of TSA
- 37°C water bath with shaker or temperature controlled shaker incubator
- spectrophotometer
- 13 × 100 mm cuvettes
- Nephelo culture flasks (Bellco Glass Co., 340 Edrudo Rd., Vineland, NJ 08360; 800-257-7043; *www.bellcoglass.com*)
- colony counter
- 28 petri plates
- 1-ml and 10-ml sterile pipettes with pipettor
- Bunsen burner
- wax pencil
- 1,000-ml beaker
- ruler

MATERIALS PER GROUP OF STUDENTS (TWO-HOUR METHOD)

- 6-hour starter culture of *Vibrio natriegens* (ATCC 14048; Preceptrol)
- 500-ml flask containing approximately 300 ml of *V. natriegens* medium (3.7% [37 g/liter]

brain-heart infusion + 2% [20 g/liter] NaCl, final pH 7.4)
- 1- and 10-ml pipettes with pipettor
- water bath or incubator at 37°C

Learning Objectives

Each student should be able to

1. Understand the growth dynamics of a bacterial culture
2. Identify the typical phases of a bacterial growth curve
3. Use a spectrophotometer
4. Measure bacterial growth and turbidity
5. Plot a growth curve and determine the generation time of a culture of *E. coli* and/or *Vibrio natriegens*

SUGGESTED READING IN TEXTBOOK

1. The Growth Curve, section 6.1; see also figures 6.1 to 6.3.
2. Measurement of Microbial Growth, section 6.2.

Pronunciation Guide

Escherichia coli (esh-er-I-ke-a KOH-lee)
Vibrio natriegens (VIB-ree-o nat-re-gens)

Why Are the Above Bacteria Used in This Exercise?

In this experiment, the student will learn how to measure bacterial growth, plot a bacterial growth curve, and determine the generation time. To accomplish this, the authors have chosen the common bacterium *Escherichia coli* that

has a generation time of approximately 21 minutes at 40°C and *Vibrio natriegens* that has a generation time of less than 10 minutes at 37°C.

PRINCIPLES

Classical Growth Curve

The four phases **(lag, logarithmic, stationary,** and **death or decline)** of growth of a bacterial population can be determined by measuring the **turbidity** of the population in a broth culture. Turbidity is not a direct measure of bacterial numbers but an indirect measure of biomass, which can be correlated with cell density during the log growth phase. Since about 10^7 bacterial cells per milliliter must be present to detect turbidity with the unaided eye, a spectrophotometer can be used to achieve increased sensitivity and obtain quantitative data.

The construction of a complete **bacterial growth curve** (increase and decrease in cell numbers versus time) requires that aliquots of a shake-flask culture be measured for population size at intervals over an extended period. Because this may take many hours, such a procedure does not lend itself to a regular laboratory session. Therefore, the first part of this exercise has been designed to demonstrate only the lag and log phases of a bacterial growth curve. The bacterial population will be plotted on graph paper by using both an indirect and direct method for the measurement of growth. The resulting growth curve can be used to delineate stages of the growth cycle. It also makes possible the determination of the growth rate of a particular bacterium under standardized conditions in terms of its **generation time**—the time required for a bacterial population to double.

The indirect method uses spectrophotometric measurements of the developing turbidity in a bacterial culture taken at regular intervals. These samples serve as an index of increasing cellular mass. The graphical determination of generation time is made by extrapolation from the log phase, as illustrated in figure 45.1. For example, select two points (0.2 and 0.4) on the absorbance (*A*) scale that represent a doubling of turbidity. Using a ruler, extrapolate by drawing a line between each absorbance on the ordinate, and the plotted log or exponential phase of the growth curve. From these two points, draw perpendicular lines to the time intervals on the abscissa. From these data, the generation time can be calculated as follows:

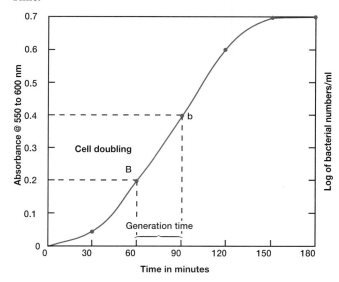

Figure 45.1 Indirect Method of Determining Generation Time.

Generation time = t (A of 0.4) − t (A of 0.2)

Generation time = 90 minutes − 60 minutes
= 30 minutes.

The same graphical generation time determination can be done with a plot of population counts.

The growth rate constant can also be determined from the data. When the \log_{10} of the cell numbers or absorbance is plotted versus time, a straight line is obtained, the slope of which can be used to determine the value of g and k. The dimensions of k are reciprocal hours or per hour. The growth rate constant will be the same during exponential growth regardless of the component measured (e.g., cell biomass, numbers). The growth rate constant provides the microbiologist with a valuable tool for comparison between different microbial species when standard growth and environmental conditions are maintained.

Once the growth rate constant is known, the **mean generation time (doubling time)** can be calculated from the following equation:

$$g = \frac{1}{k}$$

This equation also allows one to calculate the growth rate constant from the generation time.

As mentioned previously, the generation time can be read directly from the bacterial growth curve plot, and the growth rate constant then determined. To calculate the generation time (g) from these data with an equation, use the following formula:

$$\text{Generation time} = \frac{0.301t}{\log_{10}N_t - \log_{10}N_0}$$

where N_0 = bacterial population at point B or any other point at the beginning of the log phase

N_t = bacterial population at point b or any other point at or near the end of the log phase

t = time in minutes between b and B (figure 45.1).

From the previous equation, one can also determine the specific **mean growth rate constant (k)** for any culture during unrestricted growth. During this time, the rate of increase of cells is proportional to the number of cells present at any particular time. In mathematical terms, the growth rate is expressed as

$$k = \frac{n}{t}$$

where n is the number of generations per unit time. The symbol k represents the mean growth rate constant. Converting the equation to logarithms:

$$k = \frac{\log N_t - \log N_0}{0.301t}$$

Two-Hour Method

Vibrio natriegens is a facultative anaerobic rod with a single polar flagellum. It was first isolated in 1958 from a salt marsh on Sapelo Island, Georgia. The rapid growth of *V. natriegens* (a generation time of less than 10 minutes) makes this bacterium useful for determining a complete bacterial growth curve. The growth cycle, from lag through log and stationary phases, can be measured in approximately 2 hours.

Procedure (Classical Method)

First Period

1. Separate the 21 sterile saline blanks (99 ml each) into seven sets of three each. Using the wax pencil, label each set as to the time of inoculation ($t = 0$, $t = 30$, $t = 60$, $t = 90$, $t = 120$, $t = 150$, and $t = 180$) and the dilution in each blank (10^{-2}, 10^{-4}, and 10^{-6}) (figure 45.2).
2. Using the wax pencil, label seven sets of four plates each with your name, the time of inoculation (use the same times as in step 1), and the dilution (10^{-4}, 10^{-5}, 10^{-6}, 10^{-7}) to be plated.

3. Melt three 100-ml bottles or 20 15-ml tubes of tryptic soy agar in a water bath, cool to 45°C, and leave in the 45°C water bath until needed (see step 7).
4. Using a sterile pipette, transfer 5 ml of the log phase *E. coli* culture to the flask containing 100 ml of brain-heart infusion broth. Label with your name, time, and date. The approximate absorbance (*A*) of this broth should be about 0.1 at 550 to 600 nm (*see exercise 19 for proper use of the spectrophotometer*).
5. After the initial *A* has been determined, shake the culture and aseptically transfer 1 ml to the 99-ml water blank labeled 10^{-2} and continue to serially dilute to 10^{-6} (*see appendix F and figure 45.2*).
6. Place the culture flask in the shaker water bath or incubator, set at 37°C and 120 rpm. If a shaker bath is not available, the flask should be shaken periodically.
7. Plate the 0 time dilutions into the appropriately labeled petri plates, using the amounts indicated in figure 45.2. Pour 15 ml of the melted agar into each plate and mix by gentle rotation on a flat surface.
8. Thereafter, at 30-minute intervals, transfer 5 ml of the broth culture to a cuvette and determine the *A* of the culture at 550 to 600 nm. Be sure to suspend the bacteria thoroughly each time before taking a sample.
9. At the same time interval, transfer 1 ml of the culture into the 10^{-2} water blank of the set labeled with the appropriate time (see step 1). Complete the serial dilution once again as indicated in figure 45.2 and plate into the labeled (see step 2) petri plates. Add melted agar as per step 7.
10. When the media in the petri plates hardens, incubate them in an inverted position for 24 hours at 35°C.

Second Period

1. Perform colony counts on all plates as described in exercise 19.
2. Record all measurements and corresponding bacterial counts in the table in the report for exercise 45.
3. On the paper provided, plot the following:
 a. Log absorbances on the ordinate, and incubation times on the abscissa. Use figure 45.1 as an example.
 b. \log_{10} values of the bacterial counts on the ordinate, and incubation times on the abscissa. Connect the points with a ruler.
 c. Also construct a graph of the data using semilog graph paper. Calculate generation time and mean growth rate constant. Employ both the graphical method and growth equations.

Procedure (Two-Hour Method)

1. Zero the spectrophotometer at 550 to 600 nm with the *V. natriegens* medium from the flask.
2. Place the flask containing the brain-heart infusion medium in the 37°C water bath or incubator for 15 minutes.
3. While slowly agitating the flask in the water bath, inoculate it with 10 ml of 6-hour *V. natriegens* culture.
4. Read and record the %*T* of this initial culture (0 time) and every 10 minutes thereafter for about 2 hours. Be sure to suspend the bacteria thoroughly each time before taking a sample. Use appendix B to accurately convert %*T* (transmittance) to *A* (absorbancy). If your spectrophotometer has a digital readout, measure the absorbance directly rather than calculating it from the %*T*.

5. Construct a growth curve by plotting *A* against time on semilog graph paper. Calculate the mean generation time and growth rate constant using both equations and the graphical method.

HINTS AND PRECAUTIONS

(1) Be sure to maintain good aseptic technique when making transfers and report any spills to your instructor. Carefully clean and decontaminate your work area at the end of the experiment.

(2) The bacterial growth curve study can be very easily run using Nephlo culture flasks. These flasks have side arms that fit into the spectrophotometer's cuvette compartment so that growth can be followed without removing samples.

Laboratory Report 45

Name: _____

Date: _____

Lab Section: _____

Determination of a Bacterial Growth Curve: Classical and Two-Hour Methods

Classical Method

1. Based on your data on absorbance and plate counts (bacterial cells per milliliter), complete the following table.

Incubation Time in Minutes	Absorbance @550 to 600 nm	Plate Counts, Bacteria/ml	Dilution Factor	Log of Bacteria/ml
0				
30				
60				
90				
120				
150				
180				

2. Calculate the generation time for this *E. coli* culture by the indirect method, using the formula given in the Principles section and by the indirect method using your growth curve and extrapolations from the absorbances for doubling. Show all calculations in the space provided.

 a. From formula

 b. From growth curve

3. What is the *k* value for your *E. coli* culture?

Two-Hour Method

1. Based on your data on absorbance, complete the following table.

Incubation Time in Minutes	%T	A
0	_____	_____
20	_____	_____
40	_____	_____
60	_____	_____
80	_____	_____
100	_____	_____
120	_____	_____

2. In summary, what is the major purpose of this experiment?

Review Questions

1. Define generation time.

2. When following bacterial growth, why is absorbance plotted instead of percent transmission?

3. Can generation time be calculated from any phase of the growth curve? Explain your answer.

4. What is occurring in a bacterial culture during the lag phase? During the growth phase?

5. What is the significance of a k value?

6. What is meant by the turbidity of a culture?

7. How can the mean generation time be determined for a bacterial culture?

Absorbance

Log cell number

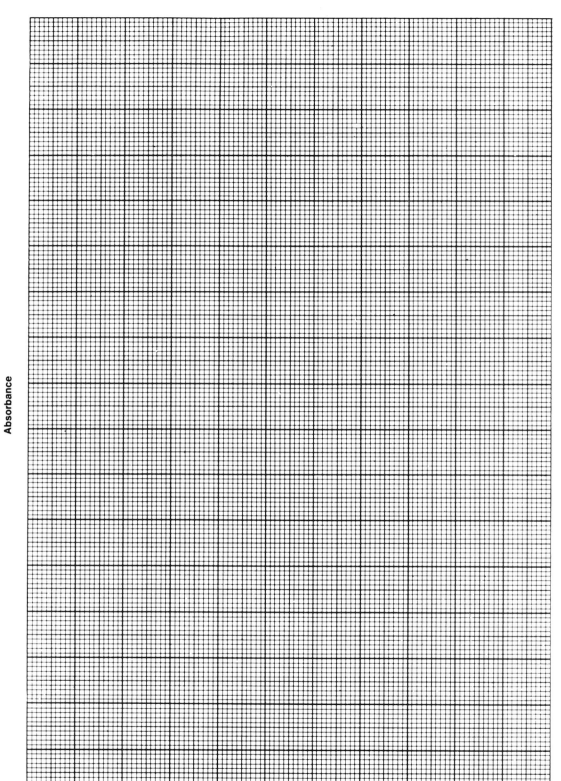

Time

Environmental Factors Affecting Growth of Microorganisms

Absorbance

Log cell number

Time

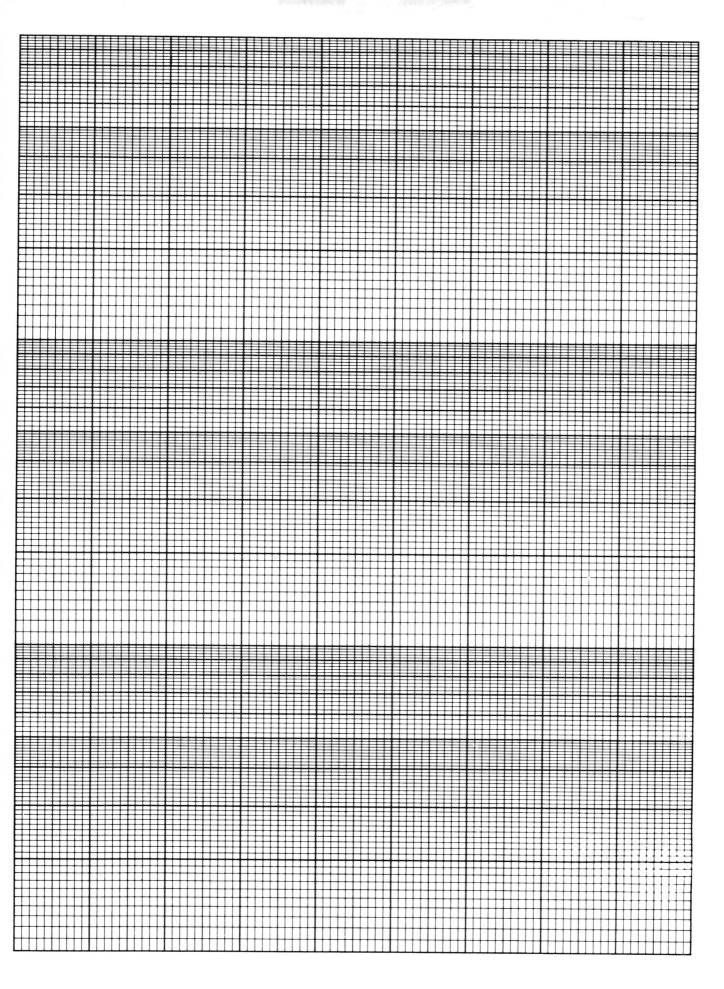

Environmental and Food Microbiology

Like a bridge over troubled water I will lay me down.
Paul Simon songwriter and singer, 1942–

This part of the manual contains exercises dealing with environmental and food microbiology.

The quality of **potable** (drinking) **water** available for public use is of major concern to everyone. Municipal and rural water supplies can transmit human diseases such as cholera, typhoid fever, shigellosis, salmonellosis, and gastroenteritis. It is generally time consuming and difficult to directly isolate and identify the relevant human pathogens because of their low numbers. Therefore, the amount of fecal contamination is monitored by counting the number of nonpathogenic, or **indicator,** bacteria, almost completely fecal in origin, present in human sewage in large numbers, and that survive in water long enough to be satisfactorily counted. The bacterium that satisfies these criteria best is *Escherichia coli.* It is a member of the **coliform group** of bacteria. Even though all coliforms do not come from human feces, the **total coliform count** is used as an index of sewage pollution. Coliform bacteria are counted during water quality monitoring in four ways:

(1) The most probable number (MPN) of coliforms is determined by the use of lactose or lauryl tryptose broth fermentation tubes.
(2) The presence-absence test (P-A) for coliforms and fecal coliforms is a modification of the MPN procedure in which a large water sample (100 ml) is incubated in a single culture bottle.
(3) The membrane filter technique (MF) employs a membrane filter. Use of the proper culture medium to grow the bacteria trapped on the filter allows the rapid detection of total coliforms, fecal coliforms, or fecal streptococci by the appearance of their characteristic colonies.
(4) To test for both coliforms and *E. coli*, the related KONFIRM Test can be used. All four of these procedures follow the American Public Health Association's recommendations in its *Standard Methods for the Examination of Water and Wastewater* (19th ed) and are the basis for the first two exercises in this part of the lab manual.

Louis Pasteur (1822–1895)

Louis Pasteur was the developer of pasteurization.

In 1866, the first edition of *Studies on Wine* was published. Pasteur showed that sour wines were caused by contamination of the wine yeast with acid-producing organisms. He also showed that other contaminating organisms were responsible for undesirable secondary fermentations, producing tasteless wines. Having shown that wine was the result of the action of the right yeast and proper method of maturation, Pasteur pointed out that the application of heat was a practical method for killing undesirable ferments. He stated:

> One should not confuse . . . the slow heating of wine, which is used only to help oxygen change the color of wine, with my process of heating which aims essentially at destroying parasites.

Viruses are extremely small infective agents. A complete virus particle, or **virion,** has a much simpler structure than a cell. It essentially consists of a block of genetic material (DNA or RNA) surrounded by a proteinaceous coat that protects it from the environment and aids in its transmission from host to host. The protein coat of a virus is called the **capsid.** Capsids normally have one of three shapes: (1) icosahedral (as in the poliovirus and adenovirus); (2) helical (as in the tobacco mosaic virus); or (3) complex (as in the vaccinia virus). An excellent example of a complex virus is the bacteriophage, or phage, a virus that reproduces within bacterial cells and destroys them in the process. It is called complex because additional structures are attached to the portion of the capsid that surrounds its nucleic acid. In the third exercise in this part of the manual, we will work with bacteriophages that reproduce within the common bacterium, *Escherichia coli.* One of the most thoroughly studied of these phages is the **T4 bacteriophage.** Its capsid consists of a head containing the double-stranded DNA genome and a complex tail. The virus attaches to its host cell by the base plate on its tail. The sheath then contracts and injects the viral DNA into the host cell.

Soils are complex environments in which microorganisms play a major role. For example, the soil microbial community is a source of nutrients, contributing to nutrient cycling and decomposition. Complex microbial interactions with other microorganisms, macroorganisms, and nutrients influence degradation processes. These organisms interact with each other and with plants in the formation and maintenance of soils. Unfortunately, most microorganisms in soils have not been cultured and studied. Nevertheless, life on earth could not be sustained in the absence of the soil microbiota and their many important interactions. This section of the manual contains one exercise that is intended to acquaint students with the characteristics and activities of soil microorganisms and to familiarize the student with the quantification of soil microbiota.

Most of the foods we consume contain microorganisms. In many instances, these microorganisms have been introduced into the food from the environment or during food preparation. In some cases, specific microorganisms are purposely added to a food product as part of the production process (e.g., cheese, pickles, buttermilk, sauerkraut, yogurt, and sausage). Once in the food product and when a suitable temperature exists, the microorganisms will use the food product as an energy source, metabolizing it, and excreting waste products. Some of these waste products may make the food product unpalatable, and, as a result, the food source is said to be spoiled. At other times, the waste products from the microorganisms may cause **food poisonings** or **food intoxications** in the consumer. In yet other instances, the waste products of the microorganisms may be desirable (this is called "beneficial spoilage") since their products aid in the production of a specific food or food product (e.g., cheese, yogurt). Finally, there are microorganisms whose presence indicates the sanitary quality of the food product. These **indi-**cator microorganisms are primarily bacteria. The exercise included in this part of the manual is intended to introduce some of the microorganisms involved in food spoilage, food poisonings, food production, and the sanitary analysis of food products. The methods for the latter can be found in the *Compendium of Methods for the Microbiological Examination of Foods,* prepared by the American Public Health Association, 800 I St. NW, Washington, DC 20001.

When suitable temperatures exist, milk provides an excellent medium for the growth of many bacteria that cause diseases such as salmonellosis (*Salmonella* spp.), brucellosis (*Brucella* spp.), listeriosis (*Listeria* spp.), and tuberculosis (*Mycobacterium*). Although these diseases are still transmitted by contaminated milk (and milk products), their incidence has been greatly reduced since the introduction of **pasteurization.** Pasteurization of milk is designed to expose the milk to a high enough temperature for a long enough period to destroy all pathogens. In this process, the nutritional quality and taste of milk is not altered. As a result, most disease outbreaks today are the result of consuming raw, or unpasteurized, milk. It is also vital that milk be routinely subjected to bacterial counts in order to prevent any disease outbreaks. If the bacterial population of milk is high, this may indicate that the milk has been improperly collected, handled, or stored or that the dairy cows from which the milk was obtained were diseased. The dairy industry (just as is done for water) uses coliform and total bacterial counts as indicators of the sanitary quality of milk. Various methods are used in the analysis of milk and dairy products. These methods can be found in *Standard Methods for the Examination of Dairy Products,* published by the American Public Health Association, 800 I St., NW, Washington, DC 20001. The last exercise in this part of the manual demonstrates several of these methods.

After completing the exercises in part eight, you will, at the minimum, be able to demonstrate an increased level in analysis skills, including: (a) collecting and organizing data in a systematic fashion; (b) presenting data in an appropriate form (graphs, tables, figures, or descriptive paragraphs); (c) obtaining microbial samples; (d) assessing the validity of the data (integrity and significance); and (e) drawing appropriate conclusions based on the results. This will meet the American Society for Microbiology Core Curriculum Laboratory Thinking Skills number 2. Laboratory skills numbers 3, 4, and 5 will also be improved. These include:

(a) using biochemical test media and accurately recording macroscopic observations;

(b) performing aseptic transfers;

(c) isolating colonies and/or plaques;

(d) correctly spreading appropriate dilutions;

(e) estimating the number of microorganisms in a sample, using serial dilution techniques; and

(f) extrapolating plate counts to obtain correct CCU or PAU in the starting sample (*see pp. vi–viii*).

Standard Coliform Most Probable Number (MPN) Test and Presence-Absence Coliform Test

MATERIALS PER GROUP OF STUDENTS

10 10-ml single-strength lactose broth (SSLB) in Durham fermentation tubes (lauryl tryptose broth or presence/absence broth can also be used)

5 10-ml double-strength lactose broth (DSLB) in Durham fermentation tubes

125-ml water sample (each group of students should bring in their own from a possible contaminated water system) at room temperature. (If the water samples are collected early, they should be refrigerated until analyzed.)

Gram-staining reagents

petri plate containing Levine's EMB agar (or LES Endo agar)

1 tryptic agar slant

3 tubes brilliant green lactose bile broth (Difco's Bacto brilliant green bile broth 2%) or 2 tubes lauryl tryptose broth containing Durham tubes

1 sterile 10-ml pipette with pipettor

2 sterile 1-ml pipettes

wax pencil

test-tube rack

35°C incubator

inoculating loop and needle

Bunsen burner

1 P-A culture bottle (250-ml milk dilution bottle) containing 50 ml of triple-strength Presence-Absence or P-A broth

Learning Objectives

Each student should be able to

1. Determine the presence of coliform bacteria in a water sample
2. Obtain some index as to the possible number of coliform bacteria present in the water sample being tested
3. List and explain each step (presumptive, confirmed, completed) in the multiple-tube technique for determining coliforms in the water sample
4. Perform the presence-absence coliform test

SUGGESTED READING IN TEXTBOOK

1. Sanitary Analysis of Waters, section 29.5.
2. Waters and Disease Transmission, section 29.5.

Pronunciation Guide

Citrobacter (SIT-ro-bac-ter)
Escherichia (esh-er-I-ke-a)
Enterobacter (en-ter-oh-BAK-ter)
Klebsiella (kleb-se-EL-lah)

PRINCIPLES

The number of total coliforms (*Enterobacter, Klebsiella, Citrobacter, Escherichia*) in a water sample can be determined by a statistical estimation called the **most probable number (MPN) test** (figure 46.1). This test involves a multiple series of Durham fermentation tubes and is divided into three parts: the **presumptive, confirmed,** and **completed tests.**

In the presumptive test, dilutions from the water sample are added to lactose or lauryl tryptose broth fermentation tubes. After 24 to 48 hours of incubation at 35°C, one looks for bacteria capable of fermenting lactose with gas production, presumably coliforms. (The lauryl tryptose broth is selective for gram-negative bacteria due to the presence of lauryl sulfate.)

Figure 46.1 The Most Probable Number (MPN) Procedure for Water Examination for the Presence of Coliforms by the Presumptive, Confirmed, and Completed Tests.

Environmental and Food Microbiology

Table 46.1

Most Probable Number (MPN) Index for Various Combinations of Positive and Negative Results When Five 10-ml Portions, Five 1-ml Portions, and Five 0.1-ml Portions Are Used

No. of Tubes Giving Positive Reaction Out of				No. of Tubes Giving Positive Reaction Out of			
5 of 10 ml Each	5 of 1 ml Each	5 of 0.1 ml Each	MPN Index per 100 ml	5 of 10 ml Each	5 of 1 ml Each	5 of 0.1 ml Each	MPN Index per 100 ml
0	0	0	<2	4	2	1	26
0	0	1	2	4	3	0	27
0	1	0	2	4	3	1	33
0	2	0	4	4	4	0	34
1	0	0	2	5	0	0	23
1	0	1	4	5	0	1	31
1	1	0	4	5	0	2	43
1	1	1	6	5	1	0	33
1	2	0	6	5	1	1	46
2	0	0	5	5	1	2	63
2	0	1	7	5	2	0	49
2	1	0	7	5	2	1	70
2	1	1	9	5	2	2	94
2	2	0	9	5	3	0	79
2	3	0	12	5	3	1	110
3	0	0	8	5	3	2	140
3	0	1	11	5	3	3	180
3	1	0	11	5	4	0	130
3	1	1	14	5	4	1	170
3	2	0	14	5	4	2	220
3	2	1	17	5	4	3	280
3	3	0	17	5	4	4	350
4	0	0	13	5	5	0	240
4	0	1	17	5	5	1	350
4	1	0	17	5	5	2	540
4	1	1	21	5	5	3	920
4	1	2	26	5	5	4	1,600
4	2	0	22	5	5	5	≥2,400

In the confirmed test, one transfers material from the highest dilution of those lactose broth tubes that showed growth and gas production into brilliant green lactose bile broth, which is selective and differential for coliforms. The tube is incubated for 48 ± 3 hours at 35°C. Gas formation in the Durham tube is a confirmed test for total coliforms.

In the completed test, a sample from the positive green lactose bile broth is streaked onto Levine's EMB or LES Endo agar and incubated for 18 to 24 hours at 35°C. On EMB agar, coliforms produce small colonies with dark centers. On LES Endo agar, coliforms produce reddish colonies. Samples are then inoculated into brilliant green lactose bile broth and onto a nutrient agar slant. These tubes are incubated for 24 hours at 35°C. If gas is produced in the lactose broth (*see figure 20.2*), and the isolated bacterium is a gram-negative (based on a Gram stain) nonsporing rod, the **completed test** is **positive.**

An estimate of the number of coliforms (most probable number) can also be done in the presumptive test. In this procedure, 15 lactose broth tubes are inoculated with the water sample. Five tubes receive 10 ml of water, 5 tubes receive 1 ml of water, and 5 tubes receive 0.1 ml of water. A count of the number of tubes showing gas production is then made, and the figure is compared to a table (table 46.1) developed by the American Public Health Association. The number is the most probable number (MPN) of coliforms per 100 ml of the water sample. (It should be noted that the MPN index usually comes from the presumptive test if raw sewage is being tested and comes from confirmed or completed tests for other types of samples.)

More recently, a simple and very sensitive alternative to the classical MPN procedure has been developed: the **presence-absence (P-A) coliform test.** The P-A test is a modification of the MPN procedure in which a large water sample (100 ml) is incubated in a single culture bottle with triple-strength broth containing lactose, sodium lauryl sulfate, and bromcresol purple indicator. The P-A test is based on the assumption that no coliforms should be present in 100 ml of drinking water. Sodium lauryl sulfate inhibits many bacteria, but not coliforms. A positive test results in the production of acid from lactose fermentation (bromcresol purple changes

from purple to yellow) and constitutes a positive presumptive test. As with the MPN test, it requires confirmation. If there is no color change, the results are negative for coliforms in the 100-ml water sample.

Procedure for the MPN Test

First Period
Presumptive Test

1. Mix the bottle of water to be tested 25 times. Inoculate five of the double-strength lactose (or lauryl tryptose) broth tubes with 10 ml of the water sample; five single-strength tubes with 1 ml of the water sample; and five single-strength tubes with 0.1 ml of the water sample. Carefully mix the contents of each tube without spilling any of the broth by rolling the tubes between the palms of your hands. Using the wax pencil, label all tubes with your name, date, and the amount of water added.

2. Incubate the three sets of tubes for 24 to 48 hours at 35°C.

3. Observe after 24 ± 2 and 48 ± 3 hours. The presence of gas in any tube after 24 hours is a positive presumptive test. The formation of gas during the next 24 hours is a doubtful test. The absence of gas after 48 hours is a negative test.

4. Determine the number of coliforms per 100 ml of water sample (see Principles section and use table 46.1). For example, if gas were present in all five of the 10-ml tubes, in only one of the 1-ml series, and in none of the 0.1-ml series, your test results would read 5-1-0. Table 46.1 indicates that the MPN for this reading would be 33 coliforms per 100 ml of water sample.

Second Period
Confirmed Test

1. Record your results of the presumptive test in the report for exercise 46.

2. Using an inoculating loop, from the tube that has the highest dilution of water sample and shows gas production transfer one loopful of culture to the brilliant green lactose bile broth tube. Incubate for 48 ± 3 hours at 35°C. The formation of gas at any time within 48 hours constitutes a **positive confirmed test.**

Third Period
Completed Test

1. Record your results of the confirmed test in the report for exercise 46.

2. From the positive brilliant green lactose bile broth tube, streak a LES Endo or Levine's EMB plate.

3. Incubate the plate inverted for 24 hours at 35°C.

4. If coliforms are present, select a well-isolated colony and inoculate a single-strength, brilliant green lactose bile broth tube and streak a nutrient agar slant.

5. Gram stain any bacteria found on the slant.

6. The formation of gas in the lactose broth and the demonstration of gram-negative, nonsporing rods in the agar culture is a satisfactorily completed test revealing the presence of coliforms and indicating that the water sample was polluted. This is a **positive completed test.**

Procedure for the P-A Coliform Test

First Period

1. Inoculate 100 ml of the water sample into a 250-ml P-A culture bottle containing 50 ml of triple-strength P-A broth. Mix thoroughly by inverting the bottle five times to achieve even distribution of the triple-strength medium throughout the sample.

2. Incubate at 35°C.

Second Period

1. Inspect the P-A culture bottle after 24 and 48 hours for acid production. A distinct yellow color forms in the medium when acid conditions exist following lactose fermentation. If gas also is being produced, gently shaking the bottle will result in a foaming reaction.

2. With an inoculating loop, transfer any culture that shows acid production or acid and gas (a positive presumptive test) to a tube of brilliant green lactose bile (BGLB) broth containing a Durham tube for incubation at 35°C.

Third Period

1. Turbidity in the BGLB broth and gas in the Durham tube within 48 hours confirm the presence of coliform bacteria (e.g., *Escherichia coli*).

2. Record results as presence-absence test positive or negative for coliforms in 100 ml of water sample in the report for exercise 46.

> ### HINTS AND PRECAUTIONS
>
> **(1)** Do not confuse the appearance of an air bubble in a clear Durham tube with actual gas production. If gas is formed as a result of fermentation, the broth medium will become cloudy. Active fermentation may be shown by the continued appearance of small bubbles of gas throughout the medium outside the inner vial when the fermentation tube is gently shaken.
>
> **(2)** When taking your water sample, the upper 38 cm of most waters usually contains the greatest numbers of live bacteria.
>
> **(3)** Use sterile containers to collect water samples.

Name: _____

Date: _____

Lab Section: _____

Standard Coliform Most Probable Number (MPN) Test and Presence-Absence Coliform Test

1. Results of presumptive test.

Number of Positive Tubes			
5 tubes DSLB	5 tubes SSLB	5 tubes SSLB	MPN
10 ml H_2O added	1.0 ml H_2O added	0.1 ml H_2O added	
24 hours 48 hours	24 hours 48 hours	24 hours 48 hours	
_____ _____	_____ _____	_____ _____	_____

2. Results of confirmed test.

Positive		Negative	
24 Hours	48 Hours	24 Hours	48 Hours
_____	_____	_____	_____

EMB or LES Endo Agar Results _____

3. Results of completed test.

Lactos Fermentation Results	Morphology
_____	_____

Presence-Absence (P-A) Coliform Test

1. Results in P-A culture bottle. _____

2. Results from brilliant green lactose bile (BGLB) broth.

 Acid: _____ Gas: _____

Review Questions

1. Why are coliforms selected as the indicator of water potability?

2. Does a positive presumptive test indicate that water is potable?

3. Why is the MPN test qualitative rather than quantitative?

4. What is the function of the following in the MPN test?
 a. lactose broth
 b. Levine's EMB or LES Endo agar
 c. nutrient agar slant
 d. Gram stain

5. What does a metallic green sheen indicate on an EMB plate? Pink to dark red colonies with a metallic surface sheen on LES Endo agar?

6. What bacterial diseases can be transmitted by polluted water?

7. What does a positive presence-absence test indicate? A negative presence-absence test?

Environmental and Food Microbiology

Membrane Filter Technique for Coliforms and Fecal Streptococci; KONFIRM Test for Fecal Coliforms

SAFETY CONSIDERATIONS

In this experiment, students will be taking unknown samples and growing them to large concentrations. Any of these samples could contain human pathogens; thus, extreme caution should be taken when working with and disposing of the final products. Be careful with the Bunsen burner flame. The 95% ethyl alcohol is flammable—do not use around an open flame. No mouth pipetting. Dispose of all water samples properly. Do not look directly at the ultraviolet light.

MATERIALS PER GROUP OF STUDENTS

vacuum pump or water faucet aspirator
sterile membrane filter apparatus
9 sterile plastic 50-mm diameter petri plates,
 Millipore (No. PD10 047 00 or Gelman plates)
sterile membrane filter disks, Millipore
 (No. HAWG 047 AO) or Gelman
sterile absorbent disk pads (packed with filters)
5-ml pipettes with pipettor
M-Endo Broth MF
M-FC broth
KF streptococcus agar
LES Endo agar
HACH m-ColiBlue 24 broth
waterproof tape
Whirlpak bag
44.5 ±2°C water bath
300-ml water sample that each group of students
 brings in from different sources
95% ethyl alcohol for sterilizing forceps
sterile forceps
Bunsen burner
sterile distilled water
35°C incubator
wax pencil
KONFIRM test tablets (for 100-ml samples)
1 comparator, 2 mls of developer (KEY Scientific
 Products)
1 sterile cotton swab
1 nonfluorescent container (150 ml)
long wavelength ultraviolet lamp (Wood's lamp),
 KS 1699 or equal

Learning Objectives

Each student should be able to

1. Understand the principles of the membrane filter technique
2. Determine the quality of a water sample using the membrane filter technique and/or the KONFIRM test
3. Quantitate the number of coliforms in a water sample using the membrane filter technique
4. Perform a KONFIRM water test for fecal coliforms

SUGGESTED READING IN TEXTBOOK

1. Microbial growth in natural environments, section 6.5.
2. Sanitary Analysis of Waters, section 29.5.

 Medical Application

Historically, contaminated water has always been a serious public health concern and remains so today. In the United States, public drinking water is tested and treated daily to maintain potability. This assures the safety of the public water consumers. The membrane filter technique is one test commonly used in combination with other tests to test for the presence of fecal coliforms. Fecal contamination of drinking water can lead to the bacterial transmission of *Salmonella typhi* (typhoid fever), *S. paratyphi* (paratyphoid fever), *Shigella dysenteriae* (bacillary dysentery), and *Vibrio cholerae* (cholera) as well as viruses (poliomyelitis and hepatitis A) and parasites (*Cryptosporidium*).

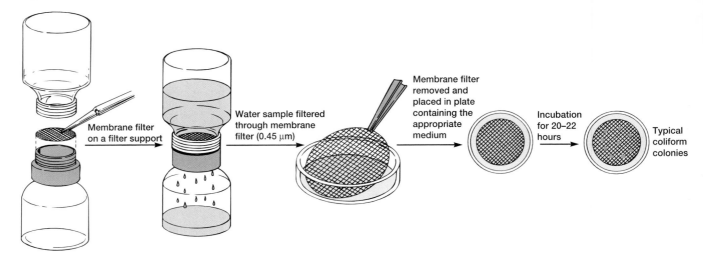

PRINCIPLES FOR MEMBRANE FILTER TECHNIQUE

The second standard test for measuring coliform numbers (quantity) in water is the **membrane filter technique** (figure 47.1).[a] This technique involves filtering a known volume (100 ml for drinking water samples) of water through a special sterile filter. These filters are made of nitrocellulose acetate or polycarbonate, are 150 μm thick, and have 0.45 μm diameter pores. A grid pattern is typically printed on these filter disks in order to facilitate colony counting. When the water sample is filtered, bacteria (larger than 0.45 μm) in the sample are trapped on the surface of the filter. The filter is then carefully removed, placed in a sterile petri plate on a pad saturated with a liquid or agar-based medium, and incubated for 20 to 22 hours at 35°C. One assumes that each bacterium trapped on the filter will then grow into a separate colony. By counting the colonies, one can directly determine the number of bacteria in the water sample that was filtered.

The broth medium usually employed in detecting total coliforms is M-Endo broth MF. Other media, such as M-FC broth and KF streptococcus agar, are available for the detection of both fecal coliforms and fecal streptococci, respectively. Fecal streptococci are the Lancefield Group D streptococci that occur in the feces of humans and other warm-blooded mammals. Total coliform colonies will be pink to dark red in color and will appear to have a golden green metallic sheen or luster (figure 47.2*a,d*). Fecal coliform colonies will appear

blue (figure 47.2*b*), and fecal streptococci (figure 47.2*c*) colonies will appear light pink and flat, or dark red.

In determining total coliforms, the amount of water filtered should be enough to result in the growth of about 20 to 80 colonies and no more than a total of 200 bacterial colonies of all types. About 50 to 200 ml of unpolluted water are often adequate for such bacterial counts. Polluted water may contain so many coliforms that it will be necessary to dilute 1 ml or less of sample with about 50 ml of sterile water. This is done in order to provide enough volume for uniform bacterial dispersion across the filter surface, in addition to providing an appropriately low coliform count.

Coliform density is expressed in terms of the number of coliforms per 100 ml of water and is calculated according to the following formula:

$$\text{Coliform colonies/100 ml} = \frac{\text{coliform colonies counted} \times 100}{\text{ml of water sample filtered}}$$

The number of coliforms should be given to two significant figures per 100 ml.

The standard set for potable (drinking) water is a limit of 1 coliform per 100 ml and an action limit of 4 coliforms per 100 ml. An **action limit** means that the water company or other provider must take immediate action to remedy the problem(s) that is/are responsible for the presence of coliforms.

From positive fecal coliform and fecal streptococci test results, one can be fairly certain that the water pollution is from a fecal source. However, in order to determine whether the fecal source is from human or animal feces, public health authorities rely on a ratio expressed as

[a]As applied to the membrane filter technique, the coliform group may be defined as comprising all facultatively anaerobic, gram-negative, nonspore-forming, rod-shaped bacteria that produce a red colony with a metallic sheen within 24 hours at 35°C on an Endo-type medium containing lactose.

Figure 47.2 **Membrane Filter Technique.** (**a**) Total coliform test: Total coliforms on a membrane filter (M-Endo MF Broth medium). Notice the dark red to purple colonies with a metallic sheen. (**b**) Fecal coliform test: Fecal coliforms on a membrane filter (M-FC broth medium). Notice the blue-colored colonies. (**c**) Fecal streptococcus test: Fecal streptococci on KF streptococcus agar. Notice the light pink colonies. (**d**) HACH m-ColiBlue 24 broth: The sum of red and blue colonies indicates total coliforms, while the blue specifies *E. coli*.

(a) (b) (c) (d)

$$\text{FC/FS Ratio} = \frac{\text{Number of fecal coliforms}}{\text{Number of fecal streptococci}}$$

Those waters showing a higher fecal coliform count than fecal strep count are likely to contain wastes from humans. In most cases, the **FC/FS ratio** will be greater than two. When the ratio is equal to or greater than four, one can be assured that the pollution is from human fecal material.

When water shows a higher fecal strep count than fecal coliform count, it is most likely that the pollution is from animal origin. If ratios fall in between two and four, estimates must be made as to how close the ratio is to either the human or animal values. The following table shows some typical FC/FS ratios:

Animal	Fecal coliform (millions)	Fecal streptococci (millions)	Ratio FC/FS
Human	13.00	3.00	4.40
Cow	0.23	1.30	0.20
Sheep	16.00	38.00	0.40
Pig	3.30	84.00	0.04
Turkey	0.29	2.80	0.10

PRINCIPLES FOR KONFIRM TEST FOR FECAL COLIFORMS

KONFIRM is an ONPG-MUG test that also can be used to detect and confirm the presence of total coliforms or *E. coli* in a 100-ml water sample (figure 47.3). The active ingredients in KONFIRM are incorporated into a tablet containing growth factors and indicators that will reveal and confirm the presence of fecal contamination. Noncoliform bacteria are inhibited and do not interfere with the test.

ONPG and MUG are enzyme substrates bound to specific indicators. When the enzyme galactosidase hy-

Figure 47.3 KONFIRM Water Test (KEY Scientific Products; 149C Texas Avenue, Round Rock, Texas 78664). Here you see all the materials needed to perform the water test, with the exception of the ultraviolet lamp. (*Courtesy of Key Scientific Products.*)

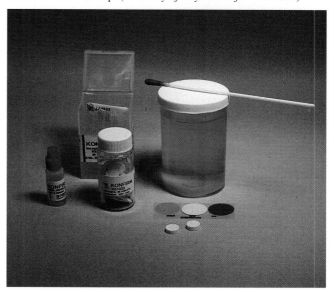

drolyses the colorless ONPG (O-nitrophenyl-β-D-galactopyranoside), it produces galactopyranose and O-nitrophenyl, which is yellow (figure 47.4*a*).

Tryptophan is an amino acid that is degraded to indole by some coliforms. The presence of indole is demonstrated by the addition of a developer. Development of a green color is a positive test for indole (figure 47.4*a*).

The enzyme glucuronidase cleaves the MUG compound (4-methylumbelliferyl-β-D-glucuronide) into its components; the product 4-methylumbelliferone is fluorescent (figure 47.4*b*).

Therefore, if coliforms are present in the water sample, the medium turns yellow within 24 hours at 35°C due to the cleavage of ONPG. To check for *E. coli*, the medium is observed under long-wavelength UV light

Figure 47.4 Use of the KONFIRM Comparator. (a) The comparator shows the approximate color of the positive yellow ONPG, green indole, and white MUG. ONPG and MUG comparisons are more accurate by holding the comparator behind the bottle, half above and half below the water line and observing through the bottle. To observe the desired color of a positive ONPG test, the yellow dot is used. Water that is yellow or brown at the onset should develop to this compared color or darker. **(b)** To observe fluorescence, hold the Wood's lamp over the white dot. A positive MUG (left) will be only slightly duller than this color. A positive indole will progress to the color of the green dot in 1 to 5 minutes. (*b: Courtesy of Key Scientific Products.*)

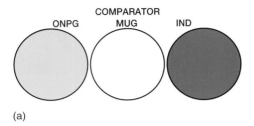

COMPARATOR

ONPG MUG IND

(a)

(b)

for fluorescence. When *E. coli* is present, the MUG is hydrolyzed to yield a fluorescent product. After developer is added to the water, the development of a green color is positive for indole. A positive indole along with a positive ONPG and MUG is confirmatory for *E. coli;* only *E. coli* is positive for all three substrates.

If the above tests are negative for fecal coliforms, the water is considered acceptable for human consumption; if it is positive for coliforms or *E. coli,* it is not acceptable for human consumption.

Procedure

General Procedure for the Membrane Filter Apparatus (figure 47.1)

1. Using alcohol-sterilized forceps, place a sterile absorbent pad in a sterile petri plate.
2. Transfer 2.0 ml of sterile M-Endo broth to the absorbent pad using a 5-ml pipette. Place the lid on the dish.
3. Aseptically place a sterile membrane filter holder on a filter flask.
4. Using sterile forceps, place a sterile membrane filter on the filter holder with its grid side up.
5. Place a sterile filter funnel on top of the membrane filter disk and secure it to the holder.
6. Attach the assembled membrane filter outfit to a vacuum source.
7. Pour the premeasured water sample into the funnel. If the sample is less than 20 ml, add about 25 to 50 ml of sterile water to the funnel first. Mix after adding water sample. Apply the vacuum to filter the sample through the membrane.
8. Rinse the walls of the funnel with 20 ml of sterile distilled water and pull the rinse water through the filter.
9. Remove the funnel and use sterile forceps to carefully transfer the filter to the absorbent pad soaked with M-Endo media or onto a plate of LES Endo agar. Make sure that the grid side is up. The membrane should be rolled onto the agar or filter pad so that no air pockets form between the membrane and medium. Incubate the plate at 35°C. Don't invert.
10. After 24 hours, remove the filter from the dish and dry it on a piece of absorbent paper at room temperature for 1/2 to 1 hour.
11. Count the colonies and record your results in the report for exercise 47.

Total Coliform Test

1. Using the sterile forceps, place a sterile absorbent pad into each of three petri plates. With the wax pencil, label these plates with your name, date, and TCT (total coliform test).
2. Add 2.0 ml of M-Endo Broth MF (or m-ColiBlue 24 broth) to the surface of each pad.
3. Filter 1, 5, and 15 ml of the water sample, and add the membranes respectively to each plate.
4. Incubate the plates at 35°C for 22 to 24 hours.
5. Count only those colonies that are pink to dark red with a metallic sheen (figure 47.2*a*). Use a plate containing 20 to 80 colonies and no more than 200 of all types of colonies. If the m-ColiBlue 24 broth is used, blue to purple colonies indicate *E. coli.* The total coliform count is given by the sum of red and blue colonies (figure 47.2*d*).
6. Record your results in the report for exercise 47.

Fecal Coliform Test

1. Using the sterile forceps, aseptically insert sterile pads into three snap-lid petri plates. Using the wax pencil, label these plates with your name, date, and FCT (fecal coliform test).
2. Add 2.0 ml of M-FC broth to the surface of each pad.
3. Filter 1, 5, and 15 ml of the water sample, and add the membranes respectively to each plate.
4. Snap the lids of the petri plates, seal them with waterproof tape, and place them in a Whirlpak bag.
5. Incubate the plates in a 44.5° ±0.2°C water bath for 22 to 26 hours. Make sure the bags are beneath the surface.
6. Count only blue-colored colonies on a plate containing 20 to 60 fecal coliform colonies.
7. Record your results in the report for exercise 47.

Fecal Streptococcus Test

1. Aseptically insert a sterile absorbent pad into each of three petri plates. Using the wax pencil, label these plates with your name, date, and FST (fecal streptococcus test).
2. Add 2 ml of KF streptococcus agar; allow the agar to cool.
3. Filter water sample volumes of 1, 5, and 15 ml as per previous test. Place the membranes in the labeled petri plates.
4. Incubate these plates for 48 hours at 35°C.
5. Count only those colonies that are light pink and flat, or smooth, dark, and red with or without pink margins (figure 47.2c). Use the plate containing 20 to 100 colonies.
6. Record the number of colonies in the report for exercise 47.

GENERAL PROCEDURE FOR THE KONFIRM WATER TEST FOR FECAL COLIFORMS

First Period

1. Using sterile forceps, remove one KONFIRM tablet and place it into a sterile, nonfluorescent container (≥150 ml).
2. Add 100 ml of water sample.
3. Cap the bottle and shake briefly. Small bubbles should appear on the surface of the tablet as it is activated. (Normally, the tablet will not dissolve completely; residue on the bottom does not affect the test.) NOTE: The water should be colorless after the tablet dissolves. The appearance of a yellow or brown color as the tablet dissolves is an indication of a high chlorine concentration in the sample

and negates the test. Water samples that are already brown or yellow will show a deeper color of brown or yellow if there is too much chlorine.
4. Incubate for 24 hours at 35°C.

Second Period

1. After 24 hours of incubation, look for a yellow color (figure 47.4). A lack of color with a dissolved tablet or around the undissolved tablet is a negative test. Any yellow color development is a positive test for total coliforms (a positive ONPG).
2. If the water sample is yellow, test for fluorescence by holding it 2 to 4 inches from a UV lamp in a darkened room. Look for a bright blue fluorescence, especially on the surface of the water. Blue fluorescence along with a yellow color is presumptive for *E. coli*.
3. If the ONPG is positive, do the indole test. Dip a sterile swab into the test container, remove, and add 1 to 2 drops of developer to the tip of the swab. (Alternately shake the bottle, remove the cap, and add the developer reagent to the substrate that clings to the inside of the lid.) Development of a green color is positive. A positive indole along with a positive ONPG and MUG is **confirmatory** for *E. coli;* only *E. coli* is positive for all three substrates.

A positive ONPG, negative MUG, and positive indole should be further evaluated for the possible **presence** of *E. coli*. A sample that is indole negative and MUG negative is unlikely to contain *E. coli*.

HINTS AND PRECAUTIONS

(1) The broth media should be freshly prepared on the day of the exercise.

(2) Water should not be used that is high in turbidity or contains a lot of algae. Coliform density is always expressed in terms of a 100-ml water sample. If the water sample is diluted, the number of colonies must still be calculated for a 100-ml sample. Similarly, if less than 100 ml of water is filtered, the coliform density must still be expressed in terms of 100 ml.

(3) If the water sample will not be tested immediately, store it in the refrigerator to prevent extra microbial growth.

(4) When using the membrane filter technique, the 20th edition of *Standard Methods for the Examination of Water and Wastewater* suggests that the following sample volumes be used for total coliform tests:

Drinking water	100 ml
Swimming pools	100 ml
Wells	10–100 ml
Lakes	10–100 ml
Bathing beaches	0.1–10 ml
River water	0.001–1.0 ml
Raw sewage	0.0001–0.1 ml

Name: _____

Date: _____

Lab Section: _____

Membrane Filter Technique for Coliforms and Fecal Streptococci; KONFIRM Test for Fecal Coliforms

1. What was the total coliform count of the water sample? Show your calculations in the space below.

2. What was the fecal coliform count of the water sample? Show your calculations in the space below.

3. What was the fecal streptococci count of the water sample? Show your calculations in the space below.

4. What are some other applications of the membrane filter technique?

5. Results for the KONFIRM water test.
 a. ONPG _____
 b. MUG _____
 c. Indole _____

6. In summary, what is the major purpose of this experiment?

Review Questions

1. What are some advantages of the membrane filter technique in the analysis of a water sample? Some disadvantages?

2. What color are coliform colonies?

3. Would you consider your water sample safe to drink? Why or why not?

4. What problems would be encountered if you used the membrane filter technique on a water sample with a lot of suspended solids in it?

5. What might your results be if a 0.10 μm filter is used to filter the water sample?

6. Other than for water examination, what are some additional applications of the membrane filter technique in microbiology?

7. What is the bacteriological standard for potable water?

Isolation of *Escherichia coli* Bacteriophages from Sewage and Determining Bacteriophage Titers

MATERIALS PER GROUP OF STUDENTS

flask of raw sewage
centrifuge
0.22 and 0.45 μm membrane filters
filter apparatus
aluminum foil
500-ml flask
graduated cylinder
1 tube T-broth
concentrated T-broth (2×)
Escherichia coli strain B (ATCC 11229) 24-hour tryptic soy broth culture
1-ml cotton-plugged serological pipettes with pipettors
cotton-plugged Pasteur pipettes
35°C incubator
Bunsen burner
1 ml chloroform
5 sterile screw-cap tubes
2 9.9-ml sterile saline (0.85% NaCl) blanks
4 4.5-ml sterile saline (0.85% NaCl) blanks
5 petri plates
5 tubes with 12 ml sterile bottom agar (nutrient agar pours)
5 tubes containing 4.5 ml top agar (nutrient broth plus 0.75% agar)

48° to 50°C water bath
thermometer
wax pencil

Learning Objectives

Each student should be able to

1. Describe a bacteriophage
2. Develop techniques for cultivating bacteriophages
3. Determine a bacteriophage titer

SUGGESTED READING IN TEXTBOOK

1. Cultivation of Viruses, section 16.3.
2. Reproduction of Double-Stranded DNA Phages: The Lytic Cycle, section 17.2.

Pronunciation Guide

Escherichia coli (esh-er-I-ke-a KOH-lee)

Why Is the Above Bacterium and Sewage Used in This Exercise?

The major objectives of this exercise are to have the student isolate *Escherichia coli* bacteriophages from sewage and determine bacteriophage titers. Raw sewage is a good source for the bacteriophages that prey upon the *E. coli.*

PRINCIPLES

Bacteriophages can be isolated from many different environments. Since they grow and reproduce within bacteria, one would expect to find them wherever a large

population of bacteria is present. For example, large numbers of *E. coli* grow in the intestinal tract of warm-blooded animals. Therefore, animal manure and untreated sewage are excellent sources of coliphages.

Because the concentration of bacteriophages specific for a particular host may be relatively low, the first step in the isolation procedure is an enrichment step. When the sample is incubated with a population of the proper host, phages specific for that bacterium will greatly multiply in number. The phages then are much easier to isolate since they are present in greater numbers.

Membrane filtration serves to remove cell debris and most bacteria. Any bacteria that may remain in the enriched culture can be killed and lysed by treatment with chloroform. The filtrate containing bacterial viruses may now be stored in the refrigerator for months.

Individual viruses in the filtrate can be isolated by mixing a small amount of filtrate with a young culture of the host bacterium and then spreading the mixture out on the surface of a petri plate containing nutrient agar. This is called the **double-layered culture technique.** In practice, the viruses and bacteria are mixed with a dilute agar medium (the **top agar**) and then poured in a thin layer on the surface of harder **bottom, or base, agar.** When the top agar jells, the viruses and bacteria are immobilized. Whenever a virus particle is present in the agar, it will infect an adjacent bacterial cell, reproduce, and **lyse** its host cell. The virus particle will then give rise to millions of virions, and a clear area of lysed bacteria will develop in the bacterial lawn. This clear area is called a **plaque** (figure 48.1). Ideally, each virus will produce one plaque containing enormous numbers of its progeny. Samples of each plaque can then be removed and used to culture large quantities of a single type of virus for further study. Because each plaque arises from a single virus particle, a count of the number of plaques will enable one to calculate the concentration of viruses in the original undiluted sample.

Figure 48.1 Plaques of Viruses. Plaques of a bacteriophage on a lawn of *Escherichia coli.*

Procedure

First Period

1. Obtain about 25 ml of raw sewage. Centrifuge it for 10 minutes at just below top speed in a table model clinical centrifuge (figure 48.2*a*).
2. Pass the supernatant through a 0.22 μm membrane filter to remove bacterial contaminants (figure 48.2*b*).
3. Add 20 ml of the filtered supernatant to an aluminum-foil-capped, 500-ml flask containing 20 ml of sterile, twofold concentrated T-broth (the final broth concentration will be normal).
4. Inoculate the mixture with 1 ml of an overnight (10^9 cells per milliliter) *E. coli* tryptic soy broth culture (figure 48.2*c*). Incubate for 24 hours at 35°C (figure 48.2*d*).
5. Melt five tubes with 12 ml sterile bottom agar, cool them in a water bath to 48° to 50°C, and pour into five sterile petri plates. This forms the base agar (figure 48.2*i*). After the plates have hardened, dry them overnight in a 35°C incubator.
6. Inoculate a T-broth culture tube with 1 ml of *E. coli* and incubate it at 35°C overnight or for 24 hours (figure 48.2*f*). This will serve as a stock culture.

Second Period

1. Filter the stock culture through another 0.45 μm membrane filter (figure 48.2*e*). Transfer 8.0-ml portions of the filtrate into sterile, screw-cap tubes (figure 48.2*h*).
2. Add 0.2 ml of chloroform to each screw-cap tube with a sterile pipette (figure 48.2*g*). Mix thoroughly.
3. The virus stock can be used immediately or kept in the refrigerator until it can be analyzed.
4. Prepare a serial dilution of the enriched bacteriophage sample as shown in figure 48.2*i*. Withdraw a sample after lowering the pipette tip into the medium as short a distance as possible so as not to transfer extra culture. Transfer an aliquot to the next tube and mix by drawing the culture in and out of the pipette several times. Discard the pipette. With a new pipette, transfer an aliquot to the next tube in the series. Use a new, sterile, cotton-plugged pipette for each transfer.
5. After all dilutions have been prepared, add 2 drops of the overnight *E. coli* culture to the melted, cooled 10^{-4} to 10^{-8} dilution tubes of top agar (figure 48.2*i*). Use a sterile, cotton-plugged Pasteur pipette for this. Keep the tubes in the 48° to 50°C water bath as much of the time as possible during this process.
6. As soon as possible after the top agar tubes have been inoculated, each dilution should be plated out in the following way. Use a new, sterile, 1-ml

Figure 48.2 The Isolation of *E. coli* Bacteriophages from Sewage.

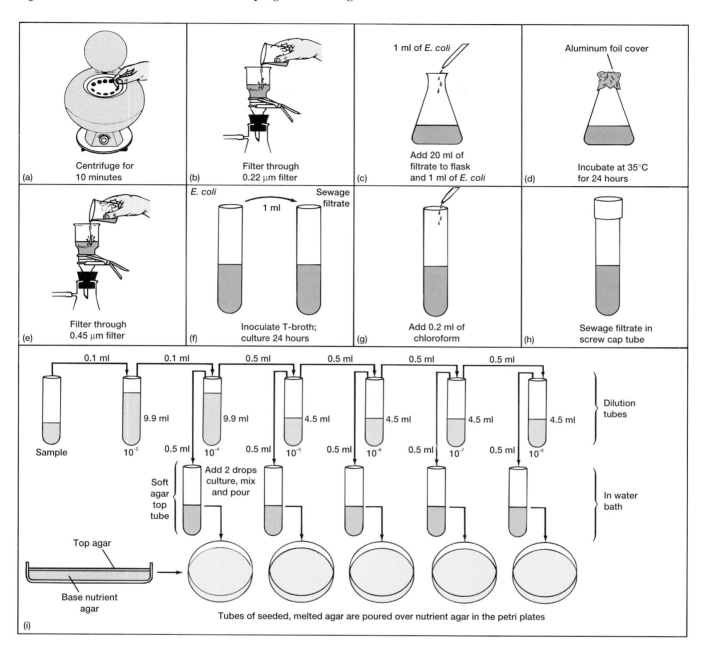

(a) Centrifuge for 10 minutes

(b) Filter through 0.22 μm filter

(c) 1 ml of *E. coli*. Add 20 ml of filtrate to flask and 1 ml of *E. coli*

(d) Aluminum foil cover. Incubate at 35°C for 24 hours

(e) Filter through 0.45 μm filter

(f) *E. coli* — 1 ml → Sewage filtrate. Inoculate T-broth; culture 24 hours

(g) Add 0.2 ml of chloroform

(h) Sewage filtrate in screw cap tube

(i) Dilution tubes: Sample — 0.1 ml — 10^{-2} (9.9 ml) — 0.1 ml — 10^{-4} (9.9 ml) — 0.5 ml — 10^{-5} (4.5 ml) — 0.5 ml — 10^{-6} (4.5 ml) — 0.5 ml — 10^{-7} (4.5 ml) — 0.5 ml — 10^{-8} (4.5 ml). In water bath. Soft agar top tube — Add 2 drops culture, mix and pour — 0.5 ml each. Top agar / Base nutrient agar. Tubes of seeded, melted agar are poured over nutrient agar in the petri plates

pipette and transfer 0.5 ml of 10^{-4} dilution to a 45°C soft agar top tube. Mix quickly and thoroughly, and pour the top agar aseptically into a petri plate containing sterile bottom agar (first period, step 5). Immediately spread the top agar over the surface of the base nutrient agar by tilting the plate. Set the plate aside to harden. Repeat this procedure with each of the other four dilutions. Use the wax pencil to label the plates with your name, date, and the dilution factor.

7. Incubate the plates in an inverted position at 35°C for 8 to 24 hours. The plates may be stored in the refrigerator after incubation and analyzed at a later time.

Third Period

1. Examine the plates carefully and, using a Quebec colony counter, count the number of plaques on each plate. Complete part 1 of the report for exercise 48. Also, carefully note the plaque morphology.

2. Use the plate(s) with the most favorable number of plaques (25 to 250) to determine the number of coliphages in 1 ml of the original enriched sample. This is done by dividing the plaque number by the dilution factor as shown by the following example:

$$\text{Plaque count} = \frac{250/\text{ml}}{0.5 \times 10^{-6}}$$
$$= 500 \times 10^{6}$$
$$= 5 \times 10^{8}$$

3. Complete part 2 of the report for exercise 48.

Laboratory Report *48*

Name: _____

Date: _____

Lab Section: _____

Isolation of *Escherichia coli* Bacteriophages from Sewage and Determining Bacteriophage Titers

1. Data: PFU _____ Dilution _____

 PFU _____ Dilution _____

2. a. How many coliphages are present in a milliliter of the enriched sewage isolate? Show your calculations in the space provided.

 b. Are there obvious variations in plaque size and morphology? Explain your answer.

 c. What does (*b*) indicate about the nature of the coliphage sample that you have isolated?

3. In summary, what is the major purpose of this experiment?

Review Questions

1. What is meant by a plaque-forming unit?

2. Determine the number of plaque-forming units per ml in a sample that has been diluted 10^{-8} and has given 200 plaque-forming units when 0.5 ml is plated.

3. What are coliphages?

4. Why is enrichment of the sewage necessary for the isolation of coliphages?

5. Describe the similarities between the plaque technique in this exercise and the standard plate count for bacteria.

6. What effect does chloroform have on viruses? On bacteria?

7. Why are two different size filters used in this experiment?

Enumeration of Soil Microorganisms

MATERIALS PER GROUP OF STUDENTS

- 1 g of rich garden soil in 99 ml sterile water
 Students may want to bring in their own soil
 and weigh out 1 g on a balance.
- 1 50-ml flask of melted (50°C) tryptic soy agar
- 1 50-ml flask of melted glycerol yeast extract agar
 that contains cyclohexamide (actidione 0.05 g/50
 ml) to inhibit fungal growth. Actidione® Agar
 (Cycloheximide Agar) can be purchased from
 Oxoid Unipath, 800 Proctor Ave., Ogdensburg,
 NY 13669, or by calling 1-800-567-8378.
- 1 50-ml flask of melted Sabouraud dextrose agar
- 2 99-ml bottles of sterile water
- 48° to 50°C water bath
- wax pencil
- Bunsen burner
- 1-ml pipettes with pipettor
- 12 petri plates
- *Difco Manual* or *BBL Manual*

Learning Objectives

Each student should be able to

1. Become acquainted with some of the
 microorganisms present in garden soil
2. Determine the number of bacteria,
 actinomycetes, and fungi in the garden soil
 using the plate count method

SUGGESTED READING IN TEXTBOOK

1. Microorganisms and the Formation of Different
 Soils, section 30.3; see also table 30.2.

Pronunciation Guide

Aspergillus (as-per-JIL-us)
Mucor (MU-kor)
Penicillium (pen-a-SIL-ee-um)
Rhizopus (rye-ZOH-pus)

PRINCIPLES

Actinomycetes (including actinoplanetes, nocardio-forms, and streptomycetes), other bacteria, and filamentous fungi (*Rhizopus, Mucor, Penicillium,* and *Aspergillus*) are all important members of the soil microbial community. Protozoa, algae, cyanobacteria, nematodes, insects and other invertebrates, and viruses are also important members but will not be studied in this exercise. Each gram of rich garden soil may contain millions of these micro- and macroorganisms.

Since soils vary greatly with respect to their physical features (e.g., pH, general type, temperature, and other related factors), the microorganisms present will also vary. For example, acid soils will have a higher number of fungi compared to alkaline soils, and rich garden soil will contain more actinomycetes than either the other bacteria or fungi. Not surprisingly, no single technique is available to count the microbial diversity found in average garden soil. Thus, in this exercise, each group of students will try to determine only the relative number of fungi, actinomycetes, and other bacteria in a sample of garden soil using the serial dilution agar plating procedure covered in exercise 17.

To support the three different groups of microorganisms, you will use three types of media: (1) Sabouraud dextrose agar for the isolation of fungi, (2) glycerol yeast agar for the isolation of actinomycetes, and (3) tryptic soy agar for the isolation of other bacteria.

Figure 49.1 **Plate Count Procedure for Enumerating Soil Microorganisms.**

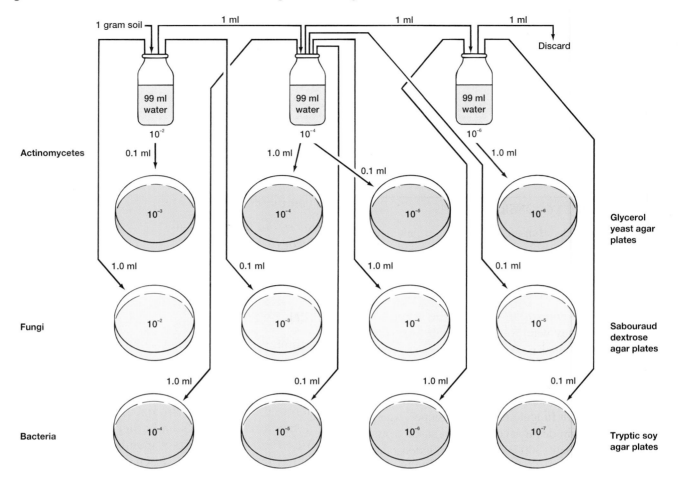

Procedure

1. The procedure for enumeration of soil microorganisms is illustrated in figure 49.1.

2. Place 1 g of garden soil in a 99-ml sterile water blank. Mix the soil and water thoroughly by shaking the water-soil mixture vigorously for 3 minutes, keeping your elbow on the lab table. Transfer 1 ml of this mixture to the second water blank and mix as above. Transfer 1 ml of this mixture to the third water blank and mix as above.

3. Using a wax pencil, label three sets of four petri plates each as follows: actinomycetes (10^{-3}, 10^{-4}, 10^{-5}, and 10^{-6}), fungi (10^{-2}, 10^{-3}, 10^{-4}, and 10^{-5}), and other bacteria (10^{-4}, 10^{-5}, 10^{-6}, and 10^{-7}). Be sure to use the correct medium for each type of microorganism.

4. Using a 1-ml pipette and aseptic technique, distribute the proper amount of each soil dilution to the respective petri plates as indicated in figure 49.1.

5. Remove the melted glycerol yeast agar from the water bath and pour 15 ml into each of the actinomycetes petri plates. Mix on a flat surface, using a circular motion, and allow to harden. Do the same for the Sabouraud dextrose and tryptic soy agars.

6. Invert the petri plates and incubate for 3 to 7 days at room temperature. Observe daily for the appearance of colonies. Count the plates with fewer than 250 colonies but more than 25. Designate plates with over 250 colonies as too numerous to count (TNTC) and those with less than 25 colonies as too few to count (TFTC). Record your data in the report for exercise 49.

7. Determine the number of respective microorganisms per milliliter of original culture (gram of soil) as follows:

$$\text{Microorganisms per gram of soil} = \frac{\text{count/plate}}{\text{dilution used}}$$

For example, if 200 colonies were present on the 10^{-7} plate, the calculation would be

Microorganisms per gram of soil

$$= \frac{200}{10^{-7}}$$

$$= 2.0 \times 10^9 \text{ colonies/gram}$$

Laboratory Report 49

Name: _____

Date: _____

Lab Section: _____

Enumeration of Soil Microorganisms

1. Number of microorganisms per gram of soil.

Microorganism	Dilution Factor	Number of Colonies	Microorganisms/Gram of Garden Soil
Actinomycetes	10^{-3}	_____	_____
	10^{-4}	_____	_____
	10^{-5}	_____	_____
	10^{-6}	_____	_____
Other bacteria	10^{-4}	_____	_____
	10^{-5}	_____	_____
	10^{-6}	_____	_____
	10^{-7}	_____	_____
Fungi	10^{-2}	_____	_____
	10^{-3}	_____	_____
	10^{-4}	_____	_____
	10^{-5}	_____	_____

2. Some normal population sizes for soil microorganisms are as follows:
 a. Actinomycetes = 7×10^6 per gram
 b. Other bacteria = 53×10^6 per gram
 c. Fungi = 1×10^5 per gram

3. In summary, what is the major purpose of this experiment?

Review Questions

1. Why were three different media used in this experiment?

2. Would the same results be obtained in this experiment if the soil sample were collected during a different time of the year? Explain your answer.

3. What general group of soil bacteria cannot be determined using the media and procedures in this exercise?

4. What generalizations can you make from this exercise with respect to your garden soil?

5. What are some physical features of soil that influence microbial populations?

6. Using a *Difco* or *BBL Manual,* what is the composition of glycerol yeast extract agar?

7. Why are different dilutions used for bacteria, fungi, and actinomycetes?

Bacterial Count of a Food Product

MATERIALS PER GROUP OF STUDENTS

20 g of hamburger or chicken
180 ml of sterile distilled water
weighing paper
scale or balance
Bunsen burner
colony counter
plate count agar (Standard Methods Agar)
2 99-ml sterile saline dilution blanks
sterile 1-ml pipettes with pipettor
5 sterile petri plates
35°C incubator
wax pencil
boiling water bath
48° to 50°C water bath for cooling tubes

Learning Objectives

Each student should be able to

1. Explain why the standard plate count is used in food quality control
2. Determine the number of bacteria in a food sample by performing a standard plate count

SUGGESTED READING IN TEXTBOOK

1. Food Spoilage Processes, section 41.2.

PRINCIPLES

The sanitary control of food quality is primarily concerned with testing food products for the presence of specific microorganisms. Food products are the primary vehicle responsible for the transmission of microbial diseases of the gastrointestinal system. For this reason, food products are routinely examined for the presence of bacteria.

The **heterotrophic plate count** can be used to determine the number of viable bacteria in a food sample. The larger the count, the greater the likelihood that specific pathogens capable of causing disease will be present and also that the food will spoil. Normally, raw hamburger should not contain over 10^6 bacteria per gram.

One of the limitations of the heterotrophic plate count is that only bacteria capable of growing in the culture medium under the environmental conditions provided will be counted. As a result, a medium that supports the growth of most heterotrophic (requiring organic carbon) bacteria is commonly used.

Figure 50.1 illustrates the procedure for performing a heterotrophic plate count on a food product.

Procedure

First Period

1. Weigh out 20 g of raw hamburger or chicken.
2. Blend the 20 g of the meat in a blender with 180 ml of sterile distilled water for 5 minutes. This gives a 1/10 dilution factor.
3. Make dilutions from 10^{-2} to 10^{-6} as indicated in figure 50.1.
4. Using the wax pencil, label the petri plates with your name, date, and dilution.
5. Using aseptic technique, pipette aliquots from the dilution blanks to the petri plates.

*Discard pipette

6. Melt the plate count agar (PCA) pours in a water bath and cool to about 48° to 50°C. Add the cooled PCA agar to the plates. Gently swirl on a flat surface in a figure-eight motion and allow to harden.

7. Incubate the plates in an inverted position for 24 to 48 hours at 35°C.

Second Period

1. Arrange the plates in order of lowest to highest dilution.

2. Count the number of colonies on the plates that have between 25 to 250 colonies. Designate plates with fewer than 25 colonies as too few to count (TFTC) and plates with more than 250 colonies as too numerous to count (TNTC).

3. Calculate the average number of bacteria per gram of hamburger as follows:

Number of bacteria/gram
= number of colonies

$$\times \frac{1}{\text{dilution factor}}$$

$$\times \frac{1}{\text{volume plated}}$$

$$\times \text{ sample volume} \times \frac{1}{\text{weight of sample}}$$

For example, if 200 colonies were counted from a 10^{-3} dilution of a 20-gram sample and 1 ml was plated:

$$\text{Number bacteria/gram} = 200 \times \frac{1}{10^{-3}} \times \frac{1}{1 \text{ ml}}$$

$$\times 20 \times \frac{1}{20 \text{ g}} = 200,000$$

4. Record your results in the report for exercise 50.

Laboratory Report 50

Name: _____

Date: _____

Lab Section: _____

Bacterial Count of a Food Product

1. Record in the table the number of colonies per plate.

Dilution	Number of Colonies/Plate	
10^{-3}	_____	
10^{-4}	_____	
10^{-5}	_____	
10^{-6}	_____	

2. Number of bacteria per gram of hamburger _____

 In the space provided, do your calculations from the plate that had between 25 and 250 colonies.

3. In summary, what is the major purpose of this experiment?

Review Questions

1. Why is a standard plate count performed on food products?

2. Does the number of bacteria recorded by the standard plate count method accurately reflect the total bacterial count from that sample? Explain.

3. Why are 25 to 250 colonies used for calculations?

4. Why is a 20-gram sample of hamburger blended with 180 ml of diluent to yield a 1/10 dilution before making further dilutions?

5. What is the difference between food poisoning and food intoxication?

6. Why is plate count agar (Standard Methods Agar) used in this experiment?

7. Why should hamburger not be repeatedly frozen and thawed?

Examination of Milk for Bacteria

MATERIALS PER GROUP OF STUDENTS

 pasteurized milk sample
 unpasteurized milk may be obtained from any dairy farm. If a sample is not available, pasteurized milk incubated at 35°C may be used or the instructor may wish to contaminate the sample with coliforms such as *Escherichia coli, Klebsiella pneumoniae,* or *Enterobacter aerogenes*
 32°C incubator
 6 15-ml and 6 5-ml violet red bile agar tubes
 2 9-ml and 2 99-ml saline blanks
 6 petri plates
 wax pencil
 boiling water bath
 methylene blue solution (1/25,000)
 2 screw-cap test tubes
 sterile 10-ml and 1-ml pipettes with pipettor
 37°C water bath
 Bunsen burner
 test-tube rack
 colony counter
 48° to 50°C water bath for cooling tubes

Learning Objectives

Each student should be able to

1. Explain the purpose of pasteurization
2. Determine the sanitary quality of milk by performing a coliform analysis and methylene blue reductase test

SUGGESTED READING IN TEXTBOOK

1. The Use of Physical Methods in Control, section 7.4.
2. Microbiology of Fermented Foods, section 41.6.
3. Diseases Transmitted by Foods, section 41.4.

Pronunciation Guide

Enterobacter aerogenes (en-ter-oh-BAK-ter a-RAH-jen-eez)
Klebsiella pneumoniae (kleb-se-EL-lah nu-MO-ne-ah)

PRINCIPLES

Pasteurization is a means of processing raw milk, before it is distributed, to assure that it is relatively free of bacteria and safe for human consumption. It is a heat process gentle enough to preserve the physical and nutrient properties of milk, but sufficient to destroy pathogenic microorganisms. The two methods most commonly used for pasteurization of milk are (1) heating at 62.9°C (145°F) for 30 minutes or (2) heating to 71.6°C (161°F) for a minimum of 15 seconds.

The presence of coliforms in milk (and milk products) is a major indicator of the sanitary quality of milk. Their presence can be determined by a coliform plate count. A high count means that there is the possibility of the presence of disease-causing bacteria. A low count decreases this possibility but does not completely rule out the absence of disease-causing bacteria. In the first part of this exercise, each group of students will do a plate count on a pasteurized and unpasteurized milk sample.

Milk that contains a large number of growing bacteria will have a lower concentration of O_2 (a lower oxidation-reduction potential) compared to milk with few bacteria. This is because growing aerobic and facultatively anaerobic bacteria (*see exercise 18*) use oxygen as a final electron acceptor in cellular respiration. The dye, **methylene blue,** is a redox indicator. It

319

Figure 51.1 **Figure 51.1 Methylene Blue Reduction Test for Bacteria. (a)** No reduction. **(b)** Complete reduction.

loses its blue color in an anaerobic environment and is reduced to leuco-methylene blue (figure 51.1). As a result, the **methylene blue reductase test** can be used to rapidly screen the quality of milk for the load of coliforms and *Lactococcus* (*Streptococcus*) *lactis*, strong reducers of methylene blue and indicators of contamination. The larger the bacterial load, the more quickly the milk will spoil. The speed at which the reduction occurs and the blue color disappears indicates the quality of milk as follows:

a. Reduction within 30 minutes—very poor milk quality (class 4 milk)
b. Reduction between 30 minutes and 2 hours—poor milk quality (class 3 milk)
c. Reduction between 2 to 6 hours—fair quality (class 2 milk)
d. Reduction between 6 to 8 hours—good quality (class 1 milk)

In the second part of this exercise, each group of students will perform a methylene blue reductase test on a pasteurized and unpasteurized milk sample.

Procedure

Coliform Analysis

1. Shake the milk sample 25 times. Make dilutions of the pasteurized and unpasteurized milk samples as indicated in figure 51.2a.
2. Use the wax pencil to label the petri plates with your name, date, the respective dilution, and either pasteurized or unpasteurized milk.
3. Pipette 1-ml milk aliquots of each dilution into the appropriate plates.
4. Melt and cool the violet red bile agar (VRBA) tubes and add 15 ml of it to each of the plates. Swirl

Figure 51.2 **Examination of Milk for Bacteria.** (a) Dilution series for coliform analysis of pasteurized and unpasteurized milk samples. (b) Methylene blue reductase test.

*Discard pipette.

gently on a flat surface and allow the agar to solidify. Afterward, add 5 ml of VRBA to each plate, swirl gently, and allow to solidify.

5. Incubate all plates at 32°C for 24 hours.

Second Period

1. For accuracy, select the plate that has between 25 to 250 colonies, which are located below the surface, are lens shaped, deep red, and surrounded by a pink halo. Record these as the coliform count per ml of milk.

Methylene Blue Reductase Test

1. Label the two screw-cap tubes with your name and pasteurized and unpasteurized, respectively.

2. Using the 10-ml pipette, transfer 10 ml of unpasteurized milk to one screw-cap tube and, with another pipette, 10 ml of pasteurized milk to the other tube (figure 51.2*b*).

3. Add 1 drop of methylene blue to each tube.

4. Cap tightly and invert the tubes several times.

5. Place the tubes in a test-tube rack and place the rack in the 37°C water bath. After a 5-minute incubation, remove the tubes from the water bath and invert several times to mix again.

6. Observe the tubes at 30-minute intervals for 8 hours. Reduction is demonstrated by a change in color of the milk sample from blue to white. When at least four-fifths of the tube has turned white, the end point of reduction has been reached, and the time should be recorded.

7. Record your results and the class of milk in the report for exercise 51.

Examination of Milk for Bacteria

1. Data from coliform analysis.

Milk Sample	Number of Colonies		
	Dilution		
	10^0	10^{-1}	10^{-2}
Pasteurized	_____	_____	_____
Unpasteurized	_____	_____	_____

Number of colonies per milliliter of pasteurized milk _____

Number of colonies per milliliter of unpasteurized milk _____

Calculations:

2. Data from methylene blue reductase analysis.

	Unpasteurized Milk	Pasteurized Milk
Reduction time	_____	_____
Milk class	_____	_____

3. In summary, what is the major purpose of this experiment?

Review Questions

1. What is the function of the methylene blue in the reductase test for milk quality?

2. Why does milk sour when it is not refrigerated?

3. How can a milk sample be contaminated by humans?

4. Why is milk pasteurized and not sterilized?

5. What are the differences in methylene blue reduction time between the different classes of milk? What do these differences signify?

6. As a bacteriological medium, how does milk differ from water?

7. What are some bacteria normally found in milk?

Medical Microbiology

The Greeks, whenever doubt arose with regard to any disease, always thought it best to rely on nature and her doings, which were sure, in the last resort, to banish the disease. They based their opinion on the following grounds: Nature being the servant and provider of all living things in healthy days, helps them also in disease.
Galen [Claudius Galenus] (Greek physician and teacher, 129–200 A.D.)

Medical, or **clinical, microbiology** deals with the diagnosis, treatment, and prevention of infectious diseases caused by all groups of microorganisms. As such, a complete coverage of these groups is not feasible in a few exercises. This part of the manual has, therefore, been constructed to enable the student to perform some of the routine techniques that are used in a **clinical microbiology laboratory.** For example, students will first do a routine blood-typing exercise. Then students will isolate some of the microorganisms that are part of their normal microbiota by using aseptic technique and selective and differential media. In subsequent exercises, students will use various techniques to examine, isolate, and identify some of the more prevalent and potentially pathogenic microorganisms such as staphylococci, pneumococci, *Neisseriae,* and aerobic and anaerobic spore-forming bacteria. As these exercises are performed, keep in mind that very similar if not identical procedures are routinely done for the many other microorganisms that are associated with the human body.

When carrying out the following exercises, view all microorganisms as **potential pathogens.** Take all necessary precautions to prevent infecting yourself or other members of your group (*See Use of Pathogens, p. xiv*).

Paul Ehrlich (1854–1915)

Paul Ehrlich was the cowinner in 1908 with Elie Metchnikoff of the Nobel Prize for Physiology or Medicine for developing the side chain theory.

Paul Ehrlich clearly enjoyed life. He smoked 25 cigars a day; he was fond of drinking a seidel of beer (publicly) with his old laboratory servant and many seidels of beer with colleagues. Although a modern man, there was still something medieval about him when he said: "We must learn to shoot microbes with magic bullets." Thus, chemotherapy was born.

Paul Ehrlich's concept of a "magic bullet" was a compound that would wipe out the microorganisms from the host's body following a single *dosa sterilisa magna.* Along these lines, he changed a drug that is the favorite poison of murderers (arsenic) into a saver of lives with respect to syphilis. With the cooperation of the German chemical industry, he produced and marketed the first chemotherapeutic agents. His theories of specific receptors and differential affinities produced such agents as mepercrine (the first synthetic malarial prophylactic) and Prontosil red (the parent compound of the sulphonamides).

In addition to his work in chemotherapy, Ehrlich invented the differential stain for white blood cells, initiated quantitative methods in immunology, and produced his famous side chain theory on the nature and origin of antibodies.

Overall, there have been few microbiologists who have made major contributions in so many different fields.

After completing either one or more of the exercises in part nine, you will, at the minimum, be able to demonstrate an increased level in analysis skills, including

(a) collecting and organizing data in a systematic fashion;

(b) presenting data in an appropriate form (graphs, tables, figures, or descriptive paragraphs);

(c) assessing the validity of the data (integrity and significance);

(d) drawing appropriate conclusions based on the results; and

(e) integrating knowledge and making informed judgments about microbiology in everyday life.

This will meet the American Society for Microbiology Core Curriculum Laboratory Thinking Skills numbers 2 and 4. Laboratory skills numbers 1–6 will also be improved. These include

(a) using a bright-field microscope to view and interpret slides;

(b) properly preparing slides for microbiological examination;

(c) proper use of aseptic technique for transferring and handling of microorganisms and instruments;

(d) use of appropriate microbiological media and test systems; and

(e) the correct use of standard laboratory equipment (*see pp. vi–viii*).

Agglutination Reactions: Blood Groups

MATERIALS PER STUDENT

sterile, unused blood lancets or an automatic lancet device
glass slide
toothpicks
70% ethyl alcohol and iodine
cotton applicators or cotton swabs
anti-A and anti-B sera
wax pencil
grease-cutting detergent (Bon Ami)
commercially prepared and tested blood (A, B, AB, and O) or WARD's artificial blood
disposable gloves
BIOHAZARD Bags and a SHARPS container for proper disposal of material

Learning Objectives

Each student should be able to

1. Appreciate and understand the potential hazards involved with using human blood
2. Understand the meaning of the terms agglutination and hemagglutination
3. Determine his or her own blood type (or the blood types of the commercial blood)

SUGGESTED READING IN TEXTBOOK

1. Agglutination, section 33.3.

PRINCIPLES

When agglutination reactions involve the clumping of red blood cells, they are termed **hemagglutinations.** Hemagglutination occurs because of the ability of antibodies to cross-link red blood cells by binding to surface antigens. Hemagglutination microscopic-slide tests are routinely used in **blood typing.**

The four major human blood types (A, B, AB, and O) are genetically determined. Individuals with blood types A, B, or AB have antigens on the surface of their red blood cells. These antigens, known as **agglutinogens,** correspond to the specific group (e.g., type A has agglutinogen A). An individual having blood type O lacks these agglutinogens. These surface antigens are also referred to as **isoantigens** (antigens that exist in alternative forms in a species and that are capable of causing an immune response in genetically different individuals of the same species, but not in individuals who carry them). The ABO blood grouping system is unique because individuals who have isoantigens on their red blood cells will have the opposite **isoantibodies (agglutinins)** in their serum. For example, persons with type A red blood cells have anti-B isoantibodies (table 52.1).

Table 52.1

Antigens and Antibodies in Human ABO Blood Groups

Blood Type	Antigen Present on Erythrocyte Membranes	Antibody in Plasma	Can Receive Blood from	Can Give Blood to
A	A	Anti-B	A and O	A and AB
B	B	Anti-A	B and O	B and AB
AB	A and B	Neither Anti-A nor Anti-B	AB; A; B; O	AB only
O	Neither	Anti-A and Anti-B	O	O; A; B; AB

Transfusion reactions, or incompatibility reactions, generally result when blood typing is done incorrectly. Such a reaction is caused by the transfusion of the wrong ABO group. This is due to the fact that the blood of a person receiving a different blood type from his/her own will agglutinate these cells. The ABO antibodies will cause rapid destruction of the transfused blood, and the clumped erythrocytes may block blood vessels.

In this exercise, students will use antibodies against type A and B antigens to determine their own (or commercially prepared and tested) blood type, using microscopic-slide agglutination reactions. In addition, data will be pooled for the whole class in order to determine blood group frequencies. (If a slide-warming, blood-typing box is available, your instructor may also want to demonstrate and include Rh typing data.)

Procedure

1. Obtain a microscope slide and wash it thoroughly in the grease-cutting detergent (Bon Ami). With the wax pencil, write the letters **A** and **B** on each side of the slide as illustrated in figure 52.1*a*. Draw two separate circles next to the letters, about 1.5 cm in diameter.

2. Obtain dropper bottles containing commercially prepared antibody solutions and place a drop of the antibody to a particular blood type within the circle near the corresponding letter.

3. To obtain some of your blood, clean the tip of your finger by rubbing it with an alcohol-saturated cotton swab or gauze pad with iodine and again with alcohol (figure 52.1*b*).

4. Obtain a new and clean sterile lancet from the package and jab the point into your finger (figure 52.1*c*). If an automatic lancet device is available, your instructor will demonstrate how to use it. **Never use a lancet that someone else has used because viral diseases such as AIDS and hepatitis can be transmitted in this way.**

5. Carefully massage your finger, and work out several drops of blood. Discard the first two drops since they

represent mostly tissue fluid. Let the third or fourth drop fall (figure 52.1*d*) into the anti-A antibody on the slide and the next two into the anti-B antibody circle. (Do not touch your finger to the slide.) Dispose of the lancet in the proper SHARPS container.

6. Stir the antibody and the blood in each circle immediately with clean toothpicks, using a rotary motion (figure 52.1*e*). Dispose of the toothpicks in the proper receptacle.

7. After donating blood, wipe your finger once more with an alcohol pad or gauze pad.

8. Observe the two mixtures against a light background with bright light, watching for the appearance of red granules, which indicate that the red blood cells are agglutinating or clumping (figure 52.2). If you cannot easily see differences between the two samples, study the slide with your microscope. Use the low-power objective with the substage condenser closed to increase contrast.

9. Determine your blood type according to the following chart and complete the report for exercise 52.

Antisera		Blood Group
A	B	
No clumping	No clumping	O
Clumping	No clumping	A
No clumping	Clumping	B
Clumping	Clumping	AB

HINTS AND PRECAUTIONS

(1) You should massage your finger after using the lancet and then release pressure to allow several drops of blood to flow freely and evenly. Discard the first few drops of blood as they may contain tissue fluid.

(2) If you are using an Autolet automatic lancet or a similar device to draw blood, carefully remove and discard both the lancet and the platform after every use. NEVER reuse the platform or you may transfer infected blood to another person.

(3) Usually it is less painful if you draw blood from the side of the fingertip rather than from the tip itself.

Figure 52.1 Procedure for Determining Blood Types.

(a) Place a drop of each solution on slide as indicated.

(b) Rub alcohol on tip of finger. Allow to dry.

(c) Prick finger with sterile lancet. Never use a lancet that someone else has used.

(d) Add a drop of blood to each circle. Do not touch finger to antiserum.

(e) Use seperate toothpicks to stir each drop.

Figure 52.2 Slides Illustrating Human ABO Blood Groups After Adding Antisera A to the Blood on the Left Part of the Slide and Antisera B to the Blood on the Right Part of the Slide.
The top slide shows no clumping; thus, it is blood group O. The second slide shows clumping on the right but not on the left; thus it is blood group B. The third slide shows clumping on the left but not on the right; thus it is blood group A. The bottom slide shows clumping on both left and right; it is blood group AB.

Agglutination Reactions: Blood Groups 329

Name: _____

Date: _____

Lab Section: _____

Agglutination Reactions: Blood Groups

1. Your blood type is _____.
2. Tabulate the class results and determine the distribution of blood types by completing the following table.

Blood Groups	Number of Students Tested	Number Positive	Percent
O	_____	_____	_____
A	_____	_____	_____
B	_____	_____	_____
AB	_____	_____	_____

3. How do the class results compare with table 52.1?

4. In summary, what is the major purpose of this experiment?

Review Questions

1. Could a type AB individual receive blood from a type B individual? Who would receive type O blood? Explain your answers.

2. Could a person with blood type AB receive blood from a primate with the same blood type? Explain your answer.

3. What is the blood type of the person to either your left or right? _____ Are your blood types compatible? Explain why or why not.

4. Why would a person with blood type A exhibit a transfusion reaction if given type B blood?

5. What causes clumping to occur?

6. What is an agglutination reaction?

7. What is serotyping?

Isolation of Normal Microbiota from the Human Body

MATERIALS PER STUDENT

1 blood agar plate
1 tryptic soy agar plate (TSA)
1 eosin methylene blue (EMB) plate
2 triple sugar iron (TSI) agar slants
2 Simmons citrate agar slants
2 lysine agar (LIA) slants
2 SIM tube agar deeps
2 methyl red–Voges-Proskauer (MR-VP) broth
3 sterile swabs
tube of 0.85% saline (NaCl)
3% hydrogen peroxide
sterile tongue depressors
Kovacs' solution
methyl red indicator
Voges-Proskauer reagent
wax pencil
Bunsen burner
inoculating loop
inoculating needle
35°C incubator
Gram-stain reagents
disinfectant for swabs and tongue depressors
test-tube rack or can

Learning Objectives

Each student should be able to

1. Isolate pure cultures of bacteria from his or her throat, skin, and rectal area
2. Become better acquainted with the normal microbiota of the human body
3. Learn several techniques for culturing microbiota from the human body

SUGGESTED READING IN TEXTBOOK

1. Normal Microbiota of the Human Body, section 31.2.
2. Skin, section 31.2.
3. Oropharynx, section 31.2.
4. Large Intestine (Colon), section 31.2.
5. Identification of Microorganisms from Specimens, section 36.2.

Pronunciation Guide

Enterobacter aerogenes (en-ter-oh-BAK-ter a-RAH-jen-eez)
Escherichia coli (esh-er-I-ke-a KOH-lee)

 Medical Application

EMB and MacConkey's agar allows a quick preliminary indication of whether a clinical specimen contains enteric bacteria. Enteric bacteria are facultatively anaerobic gram-negative rods. They can be divided into those bacteria that produce acid from lactose fermentation (the coliforms; e.g., the nonpathogenic *Escherichia coli* and *Enterobacter aerogenes*) and those that don't (nonlactose fermenters; e.g., the pathogenic *Salmonella typhi* [typhoid fever] and *Shigella dysenteriae* [bacillary dysentery]).

PRINCIPLES

The microorganisms that constitute the **normal microbiota** of the human body are usually harmless, although some are potential pathogens or opportunists. These latter microorganisms may cause disease under certain circumstances.

Four main reasons for learning about the normal human microbiota are (1) to gain an understanding of the different microorganisms at specific body locations, which provides greater insight into the possible infections that might result from injury to these body sites; (2) to develop a knowledge of the native microorganisms in any one part of the body, which helps the clinician put in perspective the possible source and significance of microorganisms isolated from clinical infections; (3) to gain a knowledge of indigenous microbiota aids in understanding the consequences of overgrowth of those microorganisms normally absent at a specific body site; and (4) to develop an increasing awareness of the role these indigenous microbiota play in stimulating the host immune response, which provides protection against microorganisms that might otherwise cause disease.

In this exercise, the student will isolate pure cultures of bacteria from his or her throat, skin, and rectal area in order to become better acquainted with the normal microbiota of these areas, and to become familiar with some of the techniques of culturing bacteria from the body.

Procedure

First Period

Throat Culture

1. Your instructor will demonstrate the proper procedure for doing a throat swab. Note the specific area that is being swabbed (figure 53.1).
2. Each student will then swab his or her partner's throat around the tonsillar area (not the roof of the mouth) using a sterile cotton swab. Use a sterile tongue depressor to keep the tongue from interfering with the swab. Roll the swab back and forth near one edge of a blood agar plate, then streak from this area with a sterile loop (*see figure 16.2*).
3. Using the wax pencil, label the plate with your name, body site, date, and type of medium.
4. Incubate the plate, inverted, at 35°C for 24 to 72 hours.

Figure 53.1 Taking a Throat Culture. Throat cultures are taken from the area at the very rear of the mouth, behind the uvula, by rotating a sterile swab over the area. Note that the swab should touch only the pharyngeal areas around the tonsils (if present), and not the roof of the mouth. Any inflamed area that might be observed should be cultured. The single most common error encountered by beginning students when taking throat cultures is that the culture is taken from the wrong part of the mouth.

Skin Culture

1. Each student is to take a swab, moisten it in 0.85% saline (squeeze out the excess fluid by pressing the swab against the side of the tube) and roll it over his or her arms. Streak a TSA plate in the same manner you used for the blood agar plate.
2. Label as in number 3 in the throat culture procedure.
3. Incubate, inverted, at room temperature for 24 to 72 hours.

Rectal Culture

1. Take a sterile swab in either a sealed tube or packet and an EMB plate to the rest room. Culture the rectum by inserting a swab 1/4 to 1/2 inch into the rectum. Streak one-third of the plate with the swab while you are in the rest room. Place the swab in the proper receptacle—**not in the commode or urinal.**
2. When you return to the laboratory, streak for isolation with a sterile loop. Label as in step 3 in the throat culture procedure.
3. Incubate, inverted, at 35°C for 24 to 72 hours.

Second Period

Throat Culture

1. Observe the blood agar plate for the various types of colonies (*see exercise 54 for hemolytic patterns*). Characterize the colonies and hemolytic types. Select a number of well-isolated colonies that fit the descriptions in table 53.1.
2. Stain each colony you select with the Gram stain (*see exercise 8*). Perform a catalase test on the same colonies you Gram stained (*see exercise 28 and figure 28.1*). Extreme care must be exercised if a colony is taken from a blood agar plate. Erythrocytes contain catalase, and a carryover of only a few blood cells can give a false-positive reaction.
3. Complete part 1 of the report for exercise 53.

Skin Culture

1. Observe the TSA plate for microorganisms that have been isolated from the skin. If no colonies grew after 24 hours, incubate for an additional 24 to 48 hours.
2. Select isolated colonies, describe their appearance (table 53.1), and do a Gram stain.
3. Complete part 2 of the report for exercise 53.

Rectal Culture

1. The most common rectal isolates will be *Escherichia coli* and *Enterobacter aerogenes*. These can be distinguished by biochemical tests. Select two different colonies from your EMB plate (*see exercise 16*) and inoculate the following media with each, respectively:
 a. TSI slant. With a needle, stab the butt and streak the surface (*see figure 14.3*).
 b. LIA slant. With a needle, stab the butt and streak the surface.
 c. SIM deep. Do a needle stab only.
 d. Simmons citrate slant. With a loop, streak the slant.
 e. MR-VP broth. Inoculate with a loop.
2. Incubate all tubes at 35°C for 18 to 24 hours.

Third Period

TSI

1. Identify your rectal bacteria from the following biochemical tests in the TSI tube: acid slant and acid butt with gas (A/Ag); acid slant and acid butt (A/A); acid slant and acid butt with H_2S (A/A with H_2S); alkaline slant and alkaline butt (K/K); alkaline slant and neutral butt (K/N).

 For acid production, the medium will turn yellow; for gas production, the medium will be broken up;

for an alkaline reaction, the medium will be pinkish; and for H_2S production, the medium in the butt will turn black. See figures 21.1 and 21.2 for examples of the above reactions.
2. Table 53.1 gives a list of some of the more frequently encountered *Enterobacteriaceae* found in the rectal area.

LIA

1. Observe the LIA tube for the following reactions: alkaline slant and alkaline or neutral butt (K/K or K/N); alkaline slant and acid butt (K/A); and red slant and acid butt (R/A). See figure 32.1 for an example of the above reactions.
2. Interpretation:

 K/K or K/N lysine-decarboxylase positive
 K/A lysine-decarboxylase negative
 R/A lysine-deaminase positive

SIM

1. First, observe the SIM tube for motility and H_2S production. Motile cultures will have growth diffusing from the stab line while nonmotile ones will not (*see figure 24.1c*).
2. H_2S production is indicated by blackening along the line of growth (*see figure 24.1c*).
3. Add 2 to 4 drops of Kovacs' reagent to the surface, shake gently, and allow to stand for a few minutes. Indole production is indicated by a red color (*see figure 25.1*).

Simmons Citrate

1. If the citrate slant turns blue, this indicates citrate utilization and is a positive test (*see figure 25.5*).

MR-VP Broth

1. Transfer half of the MR-VP broth culture to a clean tube. Add 2 drops of methyl red indicator. A pink color indicates a positive test. To the other portion of MR-VP broth add 3 to 4 drops of VP reagent. Shake gently and allow to stand at room temperature for 15 minutes (*see exercise 25*). If acetoin is produced, a red color will develop, and the test is positive.
2. Complete part 3 of the report for exercise 53.

HINTS AND PRECAUTIONS

When taking a throat culture, it helps if the patient is positioned in good light so that you can see the back of the throat and the position of the swab as you insert it.

Table 53.1

Some Common Microorganisms Associated with the Human Body and Their Characteristics

Throat Organisms	Gram's Stain; Morphological Features; Arrangement	Appearance on Blood Agar Plate
*Streptococcus viridans (alpha streptococcus)	Gram positive Cocci Chains	Small, dome-shaped colony surrounded by an area of greenish brown discoloration
Streptococcus pneumoniae (pneumococcus)	Gram positive Cocci Pairs and short chains	Similar to Streptococcus viridans except it has a depressed center; α-hemolysis
β-hemolytic streptococci, e.g., Streptococcus pyogenes	Gram positive Cocci Chains	Small, raised colonies surrounded by an area of clear β-hemolysis, with a sharp border
Staphylococcus aureus	Gram positive Cocci Clusters	Large, round, tan to yellowish colony surrounded with a clear zone of β-hemolysis
Staphylococcus epidermidis	Gram positive Cocci Clusters	Large, round and white, opaque colonies generally with no hemolysis
Neisseria spp.	Gram negative Cocci Pairs	Varies with species. Generally fairly small, whitish to yellowish colonies. Some are dry and crumbling.
Corynebacterium spp. (diphtheroids)	Gram positive Pleomorphic rods Club shaped, bipolar, or barred	Varies with species. Some are dry and wrinkled.

*Most common microorganism isolated.

Skin Organisms	Gram's Stain; Morphological Features; Arrangement	Appearance on TSA Plate
*Staphylococcus epidermidis	Gram positive Coccus Clusters	Large, round, white, glistening colony
Staphylococcus aureus	Gram positive Coccus Clusters	Large, round, glistening, generally yellow colony
*Corynebacterium spp. (diphtheroids)	Gram positive Pleomorphic rods Club shaped, bipolar, or barred	Generally very dry and wrinkled
Yeasts	Gram positive Large budding cells Single	Large, round, moist colony
*Bacillus spp.	Gram positive Rod Single and chains	Round or irregular; surface dull becoming thick and opaque
*Streptococcus viridans (alpha streptococcus)	Gram positive Cocci Chains	Small, dome-shaped colony surrounded by an area of discoloration
Micrococcus luteus	Gram positive Cocci Single, tetrads	Large, round, yellow, glistening colony

*Streptococcus viridans, Staphylococcus epidermidis, Corynebacterium spp., and Bacillus spp. are the most commonly isolated microorganisms from healthy skin.

Rectal Organisms	TSI	H₂S	LIA	Reactions SIM	Citrate	MR	VP	Pigmentation
Escherichia coli	K/A or A/Ag	±	K/K or K/N or K/A	− + ⊕ *	−	+	−	−
Citrobacter freundii	A/Ag	+	K/A	+ − +	+	+	−	−
Enterobacter aerogenes	A/Ag	−	K/K or K/N	− − +	+	−	+	−
Klebsiella pneumoniae	A/Ag	±	K/K or K/N	− ⊕ −	+	−	+	−
Proteus mirabilis	K/A	+	R/A	+ − +	+	+	−	−
Proteus vulgaris	K/A	+	R/A	+ + +	+	+	−	−
Alcaligenes faecalis	K/K or K/N	−	K/K or K/N	− − +	+	−	−	−
Pseudomonas aeruginosa	K/K or K/N	−	K/K or K/N	− − +	+	−	−	+

*− + ± Negative sulfur, positive indole, and variable motility.

Laboratory Report 53

Name: _____

Date: _____

Lab Section: _____

Isolation of Normal Microbiota from the Human Body

1. Bacteria isolated from the throat on the blood agar plate.

	Colony Type 1	Colony Type 2	Colony Type 3
Gram stain	_____	_____	_____
Colony morphology	_____	_____	_____
Catalase test	_____	_____	_____
Hemolysis	_____	_____	_____

2. Bacteria isolated from the skin on the TSA plate.

	Colony Type 1	Colony Type 2	Colony Type 3
Gram stain	_____	_____	_____
Morphological features	_____	_____	_____
Colony arrangement	_____	_____	_____

3. Bacteria isolated from the rectal area on the EMB plate and using the various biochemical reactions.

Bacteria	Reactions					
	TSI	H$_2$S	LIA	SIM	Citrate	MR-VP
_____	___	___	___	_____	_____	___
_____	___	___	___	_____	_____	___
_____	___	___	___	_____	_____	___

4. In summary, what is the major purpose of this experiment?

Review Questions

1. When doing a throat culture, what specific area of the throat is swabbed, and why?

2. When doing a skin culture, why is the swab first moistened with saline?

3. What are four reasons for knowing which microorganisms are associated with different parts of the body?

 a.

 b.

 c.

 d.

4. Why were the blood agar plates incubated at 35°C, whereas the TSA plates were at room temperature?

5. Why is EMB used for rectal cultures?

6. What are fastidious streptococci?

7. Would you expect to isolate β-hemolytic streptococci from the skin? Explain your answer.

Staphylococci

MATERIALS PER STUDENT

0.85% sterile saline (NaCl)
1 Vogel-Johnson plate (for coagulase positive [+] colonies)
1 Mueller-Hinton plate
if desired, m-staphylococcus broth tube and/or mannitol salt agar plate
4 sterile cotton swabs
1 tryptic soy broth tube (2 ml)
1 blood agar plate
1 citrated rabbit plasma tube (0.5 ml)
1 DNase test agar plate
1 nutrient gelatin deep or KEY gelatin test strip
antibiotic disks (including penicillin G, methicillin, and novobiocin; 5 μg)
β-lactamase detection disks (Difco or KEY)
forceps and 70% alcohol for disinfecting
1 N hydrochloric acid
37°C water bath
35°C incubator
STAPHAUREX rapid latex test kit (Wellcome Diagnostics)
disinfectant for swabs and mixing sticks
disposable gloves

Bunsen burner
inoculating loop
Gram-staining material

Learning Objectives

Each student should be able to

1. Recognize the medical significance of the staphylococci
2. Perform selected clinical procedures to isolate staphylococci from certain areas of the human body
3. Perform selected clinical procedures that distinguish the pathogenic species of staphylococci from the nonpathogenic species

SUGGESTED READING IN TEXTBOOK

1. The *Staphylococcaceae,* section 23.4; see also table 23.3.
2. Staphylococcal Diseases, section 39.3.

Pronunciation Guide

Micrococcus (my-kro-KOK-us)
Staphylococcus aureus (staf-il-oh-KOK-kus ORE-ee-us)
S. epidermidis (e-pee-DER-meh-diss)
S. saprophyticus (sa-pro-FIT-e-kus)
Streptococcus (strep-to-KOK-us)
S. salvarius (sal-vah-REE-uss)

 Medical Application

Mannitol salt and DNase test agars are used in the clinical laboratory to differentiate the pathogenic *Staphylococcus* spp., which ferment mannitol, from the nonpathogenic *Micrococcus* spp. The pathogen most often encountered is *Staphylococcus aureus,* which causes common food poisoning, osteomyelitis, bacterial pneumonia, and other diseases.

The coagulase test is used to distinguish between pathogenic and nonpathogenic members of the genus *Staphylococcus*. All pathogenic strains of *S. aureus* are coagulase positive, whereas the nonpathogenic species (*S. epidermidis*) are coagulase negative. The coagulase increases virulence by surrounding those bacteria that produce it with a clot that protects the bacteria from phagocytosis and antibodies.

PRINCIPLES

The genus *Staphylococcus* (Gr. *staphylo,* bunch of grapes) consists of gram-positive cocci, 0.5 to 1.5 μm in diameter, usually in irregular clusters, within which pairs and tetrads are commonly seen (figure 54.1). They are nonmotile and nonsporing. Members of this genus are facultatively anaerobic. Colonies are round, convex, mucoid, and adherent to the agar. They are chemoorganotrophic, requiring nutritionally rich media. Staphylococci have respiratory and fermentative metabolism, producing acid but no gas from carbohydrates (figure 54.2). They are able to grow on nutrient agar with 5% NaCl and are usually positive for catalase. Oxidase negative members contain cytochromes and are Voges-Proskauer positive. Most species reduce nitrate to nitrite. Optimum growth is at 37°C. Staphylococci are commensals on the skin and in the human mouth and upper respiratory tract. They can be human pathogens.

S. aureus (L. *aureus,* golden) is the most important clinical member of this genus. It may be isolated from the skin or mucous membranes of the body. It can cause various infections (e.g., carbuncles, abscesses, pneumonia, endocarditis, food poisoning, toxic shock syndrome) throughout the body. In addition, some of these staphylococci are resistant to penicillin. This resistance comes about when the bacteria produce penicillinase (β-lactamase), which hydrolyzes the β-lactam ring of penicillin.

S. epidermidis (Gr. *epidermidis,* the outer skin) is a nonpathogenic member of this genus (figure 54.3). It is part of the normal microbiota of the skin. Additionally, *S. saprophyticus* (Gr. *sapros,* putrid) can be isolated and may be responsible for urinary tract infections—especially in females.

Clinical tests for staphylococci are directed toward (1) separation of the genus *Staphylococcus* from other gram-positive cocci (such as *Micrococcus* or *Streptococcus*) and (2) separation of the species within the

Figure 54.2 Cultures of *Staphylococcus aureus* on Vogel-Johnson Agar. VJ agar is a modified tellurite-glycine agar containing phenol-red indicator and mannitol that is selective for recovery and identification of coagulase positive *S. aureus*. *S. aureus* grows on this medium and produces gray to black colonies from the production of free tellurium. A yellow halo is typically seen around the colonies from the fermentation of mannitol and the production of an acid pH. Tellurite, lithium chloride, and high glycine content inhibit the growth of many gram-positive and gram-negative bacteria.

Figure 54.1 Morphology of Staphylococci. Gram's stain smear prepared from a staphylococcal culture illustrating the characteristic grapelike clusters of gram-positive cocci (×1,000).

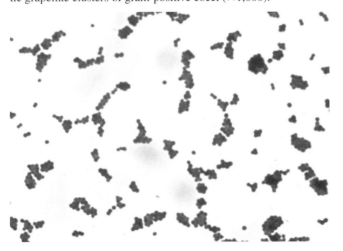

Figure 54.3 Culture of *Staphylococcus epidermidis* on Blood Agar. Notice the white, opaque, nonhemolytic, smooth colonies characteristic of *S. epidermidis*.

Table 54.1

Some Characteristics Used to Distinguish Gram-Positive Cocci

Characteristic	S. aureus	Micrococcus spp.	Streptococcus spp.
Ferments glucose	+	–	+
Mannitol salt agar			
Ferments	+	–	+&–
Growth	+	–	+&–
Nitrate reduction	+	Variable	–
Catalase production	+	+	–
Oxidase reaction	–	+&–	–
Hemolysis	β	None	α, β none
Bile solubility	–	–	–or+

Table 54.2

Some Characteristics of *Staphylococcus* Species

Characteristic	Species		
	S. aureus	S. epidermidis	S. saprophyticus
Coagulase	+	–	–
Blood agar lysis	β	–/weak	–
Nitrate reduction	+	+	–
Acid produced aerobically			
Mannitol	+	–	d
D-trehalose	+	–	+
Sucrose	+	+	+
D-xylose	–	–	–
Novobiocin resistance at an MIC of 5 μg/ml	– (S)	– (S)	+ (R)
DNase activity	+	–/weak	–
Gelatinase activity	+	–	–
Pigmentation	Yellow	White	White, yellow

d = 11–89% strains positive

genus. Table 54.1 lists some of the characteristics used to separate members of this genus from the other genera. Table 54.2 lists characteristics used to differentiate between the three most commonly isolated *Staphylococcus* species.

One of the most popular ways to isolate and identify *S. aureus* and other gram-positive cocci of medical importance (exercises 53–56) is through the use of blood agar plates. Hemolysis patterns are particularly useful in identification. In α-**hemolysis,** a zone of greenish coloration with an indistinct margin forms around colonies growing on blood agar (*see figure 55.2*). The color results from partial decomposition of hemoglobin. β-**hemolysis** is a sharply defined zone of clear hemolysis with no greenish tinge surrounding the colony (figure 54.4). Of course, some gram-positive cocci such as *S. salvarius* produce no hemolysis on blood agar.

Mannitol salt agar also can be used for the selective isolation, cultivation, and enumeration of staphylococci from clinical and nonclinical specimens (*see exercise 16*). Mannitol-using bacteria (e.g., *S. aureus*) turn the medium yellow (figure 54.5), whereas those that do not use mannitol (e.g., *S. epidermidis, S. saprophyticus*) produce no color change (figure 54.6).

This exercise is designed to illustrate (1) methods of culturing the staphylococci from certain areas of the human body and (2) the clinical laboratory techniques that are used to distinguish pathogenic from nonpathogenic species.

Procedure

First Period

1. Obtain a nose culture by rotating a moistened (0.85% saline) swab thoroughly around the perimeter of both nares (nostrils). Avoid touching the outside skin area. Seed one half of a Vogel-Johnson (VJ) plate (figure 54.7).

Figure 54.4 Culture of Staphylococci on Blood Agar. Blood agar plate on which large, smooth, β-hemolytic colonies of *S. aureus* are growing. The lysis of the RBCs is due to alpha toxin production.

Figure 54.5 Culture of *Staphylococcus aureus* on Mannitol Salt Agar. Notice that the medium has turned yellow around the growing bacteria since the bacteria are able to ferment the mannitol and produce an acid pH.

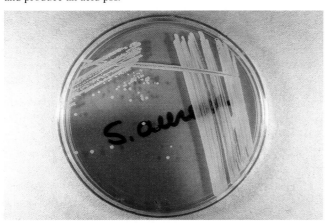

2. Obtain a skin culture by rolling a moistened swab up and down the arm. An alternative method is to swab underneath the fingernails of one hand. Seed the other half of the VJ plate (figure 54.7).
3. Incubate the plate at 35°C for 24 to 48 hours.
4. Sometimes the isolation of staphylococci is not successful; therefore, one can use enrichment media and selective agars such as m-staphylococcus broth and mannitol salt agar. In order to enrich for staphylococci, the swab is placed in a tube of m-staphylococcus broth and incubated at 35°C un-

Figure 54.6 Culture of *Staphylococcus epidermidis* on Mannitol Salt Agar. Notice the small white colonies that do not use the mannitol; that is, no color change is observed since no acid has been produced.

Figure 54.7 VJ Plate Showing Streak Patterns.

til the medium becomes turbid (24 to 48 hours). The culture can then be used to inoculate the VJ plates. Mannitol salt agar can be used either with, or as a replacement for, VJ agar.

Second Period

1. On the VJ plate, *S. aureus* will form small, black colonies surrounded by a yellow zone due to the fermentation of mannitol (figure 54.2). *S. epidermidis* will form smaller black colonies with no yellow zone due to the inability to ferment mannitol. These colonies may even be surrounded by a deep red color. On a mannitol salt agar plate, *S. aureus* will form large golden colonies surrounded by wide yellow zones.
2. Select one presumptive colony of *S. aureus* and one presumptive colony of *S. epidermidis* for further characterization. If you did not obtain one of each from your body, use a plate from one of the other students in the laboratory and inoculate the following:
 a. Blood agar plate. Use a loop and streak for isolation. Incubate for 24 hours at 35°C.

Figure 54.8 Coagulase Test. Coagulase producing strains of *S. aureus* form a clot (solid fibrin gel) when grown in plasma (tube on the left), whereas coagulase negative staphylococci (*S. saprophyticus* [middle tube] and *S. epidermidis* [tube on the right]) do not form a clot.

b. Citrated rabbit plasma. Use a loop for inoculation of 0.5 ml of the plasma with suspected *S. aureus*. Incubate 4 to 6 hours in a 37°C water bath. A positive test (coagulase activity) is indicated by the formation of a clot (figure 54.8).

c. DNase test agar. Use a loop for inoculation. Incubate for 18 to 24 hours at 35°C.

d. Tryptic soy broth. Use a loop for inoculation. (This broth culture will be used later to inoculate a Mueller-Hinton plate.)

e. β-lactamase. To test for β-lactamase, aseptically place two β-lactamase disks into an empty petri plate and wet each with 2 drops of sterile water. Transfer a large inoculum of *S. aureus* and *S. epidermidis* from the above cultures to each disk, respectively. Rub onto the surface. The development of a red color is a positive test for β-lactamase.

f. Gelatinase activity. Test for gelatin hydrolysis as described in exercise 27. *S. aureus* is gelatinase positive (+).

3. Use the STAPHAUREX rapid latex test for detection of *S. aureus*.

a. Shake the latex reagent bottle to obtain an even suspension, and dispense a drop into a circle on the reaction card for each culture to be tested (figure 54.9*a*).

b. Take a mixing stick and pick up some of the culture from the VJ plate by touching it with the flat end of the stick (figure 54.9*b*).

c. Emulsify the sample of culture in a drop of latex by rubbing with the flat end of the stick. Discard the mixing stick in a proper receptacle.

d. Rotate the card gently (figure 54.9*c*) and examine for agglutination for approximately 20 seconds, holding the card at normal reading distance (25 to 35 cm) from the eyes.

Figure 54.9 STAPHAUREX Rapid Latex Test. (a) Add 1 drop of reagent to each circle. **(b)** Select an isolated colony and pick it up with a stick. **(c)** Rotate the card gently after inoculation. **(d)** Positive and negative results. (*d: Wellcome Diagnostics*)

(a)

(b)

(c)

Positive Negative

(d)

e. A positive result (figure 54.9d) is indicated by the development of an agglutination pattern showing clearly visible clumping of the latex particles with clearing of the milky background. Most positive reactions will be almost instantaneous, indicating the presence of *S. aureus*.

f. A negative result is indicated when the latex does not agglutinate and the milky appearance remains substantially unchanged throughout the test (figure 54.9d). A negative test indicates the absence of *S. aureus*.

Third Period

1. a. Blood agar plate. Observe for pigmentation and β-hemolysis (figure 54.4). *S. aureus* should form large golden-yellow colonies surrounded by wide zones of β-hemolysis.

 b. DNase test agar. Flood the plate with 1 N HCl. A distinct zone around the streak indicates that the bacterium is DNase positive (i.e., it breaks down the DNA in the agar medium) (*see figure 29.2*).

 c. Nutrient gelatin deep. Place in the refrigerator for 30 minutes. If the gelatin remains liquid, it has been broken down by the enzyme gelatinase. If the medium hardens, gelatinase was not produced by the bacteria (*see figure 27.1*).

 d. Tryptic soy broth. Inoculate the surface of a Mueller-Hinton plate with a swab for susceptibility testing (*see exercise 43 and figure 43.1*). Incubate for 16 to 18 hours at 37°C. Three of the antibiotics should be methicillin, novobiocin (figure 54.10), and penicillin G. (Test any other antibiotics assigned by your instructor.) Measure and record the diameter of the zone of inhibition (figure 54.10). *See table 43.1 to determine antibiotic sensitivity.*

 e. Select organisms for Gram staining from the same colonies that were used for the blood agar plates.

Fourth Period

1. Observe the susceptibility plate and record in the report for exercise 54 whether the staphylococci were susceptible (+) or resistant (−) to the antibiotics tested.

Penicillin- or Methicillin-Resistant *S. aureus*

1. If you are successful in isolating *S. aureus*, especially if it is penicillin or methicillin resistant, you may wish to consider the reasons for its presence. Questions to ask are the following:

Figure 54.10 Novobiocin Susceptibility Test. (Left on plate) Novobiocin resistance evidenced by lack of zone of inhibition (or a zone less than 17 mm) surrounding a novobiocin disk. Resistance is typical of *Staphylococcus saprophyticus*. (Right on plate) Novobiocin susceptibility evidenced by a zone of inhibition greater than 16 mm surrounding the novobiocin disk. Sensitivity is typical of *Staphylococcus epidermidis* and other coagulase-negative staphylococci, other than *S. saphrophyticus*.

a. Sex of individual

b. Age of individual

c. Have you worked in a hospital environment? For how long?

d. Do you have patient contact?

e. Have you had penicillin or methicillin within the last 6 months?

f. Have you had any illness caused by *S. aureus*? If so, what was it?

g. Has any member of your immediate family been infected with *S. aureus*?

2. The instructor may wish to pass out a questionnaire to the class to determine the possible reasons for finding *S. aureus* in the anterior nares as well as for penicillin or methicillin resistance.

HINTS AND PRECAUTIONS

(1) When working with the HCl, use the fume hood.

(2) Be sure to carefully measure the diameter of inhibition zones at the proper time to get accurate results.

Laboratory Report 54

Name: _____

Date: _____

Lab Section: _____

Staphylococci

1. Biochemical Tests

	S. aureus	*S. epidermidis*
Mannitol fermentation from VJ plate or mannitol salt plate	_____	_____
Hemolysis (blood agar plate)	_____	_____
Coagulase activity	_____	_____
DNase activity	_____	_____
Gelatinase activity	_____	_____
Pigmentation (blood agar of TSA)	_____	_____
β-lactamase	_____	_____

2. Antibiotic Susceptibility Test

Antibiotics	**+ or −**
Novobiocin zone diameter	_____
Penicillin G zone diameter	_____
_____	_____
_____	_____
_____	_____
_____	_____
_____	_____
_____	_____
_____	_____
_____	_____

3. Number of individuals in the class who are "staph" carriers? _____

4. Number of individuals who are harboring coagulase-positive staph? _____

Review Questions

1. What precautions should be taken by a nurse who is a "staph" carrier?

2. What test can differentiate the three major species of *Staphylococcus*?

3. How would you treat a person who has a carbuncle?

4. Describe the cellular morphology and arrangement of staphylococci?

5. What is the principle behind the rapid STAPHAUREX test?

6. How could you distinguish *S. epidermidis* from *S. saprophyticus*?

7. Why are staphylococcal infections frequent among hospital patients?

8. In summary, what is the major purpose of this experiment?

Pneumococci

MATERIALS PER GROUP OF STUDENTS

1 sample of sputum containing pneumococci, or a simulated sputum sample can be prepared by using saliva inoculated with *Streptococcus pneumoniae* (ATCC 6303). The specimen may be placed in sputum cups or petri plates.

3 blood agar plates

1 tuberculin syringe with a 22-gauge, 1/2-inch needle

culture of viridans streptococci (e.g., *Streptococcus bovis*, ATCC 9809)

1 mouse (optional)

3 clean glass slides

toothpicks

inoculating loops and needles

specific pneumococcus typing sera (types I–III) for the quellung reaction (Difco)

optochin or Taxo P (BBL) disks

Erlenmeyer flask with 1% aqueous methylene blue

wax pencil

2 serological test tubes

phenol red indicator

0.05N NaOH

2% bile (sodium deoxycholate) or 10% sodium deoxycholate (SpotTest Ampules by Difco)

0.85% saline (NaCl)

serological pipettes with pipettor

37°C water bath

Gram-stain material

Bunsen burner

pH meter or paper

disinfectant for slides

Learning Objectives

Each student should be able to

1. Culture pneumococci from sputum
2. Perform biochemical and serological diagnostic tests to identify pneumococci and differentiate them from other α-hemolytic isolates
3. Perform an animal susceptibility test

SUGGESTED READING IN TEXTBOOK

1. *Streptococcus,* section 23.4; see also tables 23.4 and 23.5.
2. Streptococcal Pneumonia, section 39.1.

Pronunciation Guide

Streptococcus bovis (bo-VIS)
S. oralis (OH-ral-is)
S. pneumoniae (new-MOH-nee-eye)

PRINCIPLES

S. pneumoniae (pneumococci) is the causative agent of many types of diseases (e.g., bacterial lobar pneumonia, conjunctivitis, otitis media, meningitis, peritonitis). Thirty to seventy percent of normal individuals harbor this bacterium in their pharynx. *S. pneumoniae* is a gram-positive coccus (the distal ends are lancet-shaped), usually arranged in pairs or short chains (figure 55.1). It also occurs singly. The pneumococci are somewhat fastidious in their nutritional needs and do not survive well in competition with other microbiota. The medium of choice for culturing pneumococci is blood agar (figure 55.2). Colonies that form on this medium appear small and shiny with a dense center

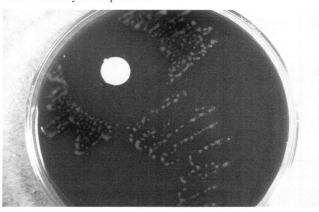

and a lighter, raised margin, giving them a dimpled appearance. The hemolytic pattern observed on blood agar is α-**hemolysis** (a zone of greenish coloration occurs around the colonies). In contrast, *S. oralis* colonies are smaller, gray to whitish gray, and opaque with smooth edges.

The **Neufeld** (named after Fred Neufeld, German bacteriologist, 1861–1945) or **quellung reaction** (German for swelling) is a serological test used to determine the specific capsular type around the bacteria. More than 80 types of pneumococci have been differentiated by the immunologically distinct polysaccharide structure in their capsules determined by this reaction. The test consists of combining the encapsulated bacteria with antibody directed against the capsular structure. If the mixture is homologous (type I bacteria + anti-type-I serum), an apparent swelling of the capsular structure is observed microscopically at 1,000X. If the reaction is nonspecific (type I bacteria + anti-type-III serum), no appearance of swelling will be seen. The quellung reaction is now only performed for research interest or epidemiological surveys. Since the advent of antibiotics, there is no longer a need to determine specific capsular type.

More presumptive tests for pneumococci are **bile solubility** and **optochin tests.** In the bile solubility test, pneumococci will dissolve in bile (2 to 10% sodium deoxycholate), whereas alpha streptococci will not. In the optochin test, the pneumococci exhibit sensitivity to optochin (ethylhydrocupreine) and will lyse, but alpha streptococci are unaffected (figure 55.2).

This exercise is designed to acquaint the student with methods of culturing the pneumococci from sputum, performing biochemical and serological diagnostic tests, and experimenting with animal susceptibility.

Procedure

First Period

1. Using an inoculating loop, inoculate one-third of a blood agar plate with the sputum sample. Restreak the plate with a sterile loop to obtain isolated colonies. Do the same for *S. bovis.*

2. The following can be done as a demonstration by the instructor. Inoculate a mouse intraperitoneally with 0.5 ml of sputum using a sterile tuberculin syringe. To do this, place the mouse on the top of a wire basket or cage. Hold the tail firmly with one hand and grasp the skin over the nape of the neck with the thumb and forefinger. Place the tail between the little finger and palm of the same hand. Pull the tail so that the mouse is held firmly. Take the syringe and place the needle between the forefinger and thumb. Gently pass the needle through the skin into the abdominal cavity, inject the contents, and then withdraw the syringe. Observe the animal daily. If it dies before the next laboratory period, place it in the refrigerator.

3. Using the quellung reaction, type the pneumococcus from the sputum with the type I–III antisera provided as follows:
 a. Place a loopful of sputum on three clean slides.
 b. Transfer 3 small drops of specific antisera to the slides next to the sputum.
 c. Transfer 2 loopfuls of methylene blue to the slides and then mix the three materials together in a small area with a toothpick. Cover with a coverslip and examine, first under high-dry and then under oil. Compare one antiserum preparation with the other. The enlargement of one of the capsular structures around the diplococci is a positive test for the specific type.

4. The bile solubility test is performed as follows:
 a. Using a wax pencil, mark one of the serological tubes "bile" and the other "saline." Place your name on the tubes. Pipette 0.5 ml of bile and saline into the respective tubes.
 b. With the inoculating loop, transfer 2 loopfuls of sputum into each tube.
 c. Add 2 drops of phenol red indicator to each tube. Adjust the pH to 7.0 with the 0.05 N NaOH.
 d. Place both tubes in a 37°C water bath and examine at 30-minute intervals for 2 hours. If the turbidity clears in the bile tube, it indicates that the bacteria have disintegrated and the test is positive for *S. pneumoniae*.
 e. The bile solubility test can also be carried out on isolated colonies. Simply place 1 or 2 drops of 10% deoxycholate on the suspected colony and observe it carefully. If the colony dissolves or flattens and disappears in about 5 minutes, it is bile soluble.

Second Period

1. Observe for colonies of pneumococci on the blood agar, if possible, with a dissecting scope. Inoculate a colony believed to be pneumococci from the sputum sample over one half of a blood agar plate. Keep the streaks close together. Streak the other half with a broth culture of the α-hemolytic streptococcus provided. Place an optochin-impregnated disk or Taxo P disk in the center of each streaked area and incubate at 35°C for 24 hours.

2. If a mouse was infected with *S. pneumoniae* during the first period, kill it and make cultures of blood from the heart and/or peritoneal exudate on a blood agar plate. Smear and Gram stain the blood and/or peritoneal exudate, and examine for white cells and pneumococci. Perform a quellung reaction on the peritoneal exudate.

Third Period

1. Examine both the alpha streptococci and pneumococci for optochin susceptibility. *S. pneumoniae* should have a zone of inhibition ≥ 14 mm in width when 1/4-inch disks are used. Also examine the blood and/or peritoneal exudate.
2. Complete the report for exercise 55.

HINTS AND PRECAUTIONS

(**1**) If the student or instructor does not have experience working with mice, it is advisable to obtain commercially available mouse restrainers for inoculation of mice.

(**2**) A mouse is most rapidly and efficiently killed using cervical dislocation or with ether in a fume hood.

(**3**) A candle jar (*see figure 57.4*) can also be used to create a microaerophilic environment preferred by many pneumococci.

Name: _____

Date: _____

Lab Section: _____

Pneumococci

1. Description of a *S. pneumoniae* colony. _____

2. Description of a *S. bovis* colony.

3. Optochin susceptibility: + −

 a. *S. pneumoniae* _____ _____

 b. *S. bovis* _____ _____

4. Quellung reaction:

 a. Type I _____

 b. Type II _____

 c. Type III _____

5. Clinical observations from the mouse experiment:

6. Pneumococcus suspension with bile _____

7. Pneumococcus suspension without bile _____

8. In summary, what is the major purpose of this experiment?

Review Questions

1. What differentiates sputum from saliva?

2. What does the optochin test illustrate?

3. Describe the quellung reaction.

4. How can you differentiate between *S. bovis* and *S. pneumoniae?*

5. What does the bile solubility test indicate?

6. What type of hemolysis is produced by *S. pneumoniae?*

7. What role does a bacterial capsule play in an infection?

Streptococci

MATERIALS PER GROUP OF STUDENTS

20 ml streptolysin O buffer
6 ml sheep RBCs in 5% buffer
5 ml streptolysin O reagent
0.5 ml test antisera for streptolysin O antibody
9 serological test tubes
5 1-ml pipettes with pipettor
3 5-ml pipettes
1 10-ml pipette
1 tryptic soy broth culture of *Streptococcus equi* (Group C β-hemolytic, ATCC 33398)
1 tryptic soy broth culture of *Streptococcus mitis* (Group D, α-hemolytic, ATCC 903)
1 tryptic soy broth culture of *Streptococcus pyogenes* (Group A β-hemolytic streptococci, ATCC 14299)
1 tryptic soy broth culture of a non-Group A β-hemolytic streptococci (e.g., *Streptococcus agalactiae*, ATCC 13813). *S. agalactiae* is a member of group B.
1 tryptic soy broth culture of *Staphylococcus aureus* (ATCC e25923)
sheep blood agar plates
bacitracin disks
SXT disks (Difco)
4 bile esculin agar slants
Gram-stain material
Bunsen burner
inoculating loop
disinfectant for slides
forceps

Learning Objectives

Each student should be able to

1. Become familiar with morphological and colonial characteristics of the streptococci as well as their medical significance
2. Perform several biochemical and serological tests to identify *S. pyogenes*
3. Compare and contrast the alpha streptococcus *S. mitis* with *S. pyogenes*
4. Show how bacitracin sensitivity, the CAMP reaction, and SXT sensitivity are used to differentiate the Lancefield groups of β-hemolytic streptococci.

SUGGESTED READING IN TEXTBOOK

1. *Streptococcus,* section 23.4; see also tables 23.4 and 23.5.
2. Streptococcal Diseases, section 39.1.

Pronunciation Guide

Staphylococcus aureus (staf-il-oh-kok-kus ORE-ee-us)
Streptococcus (strep-to-KOK-us)
S. agalactiae (a-gal-ACT-e-a)
S. bovis (bo-VIS)
S. equi (e-qy)
S. mitis (MY-tiss)
S. mutans (MYOO-tans)
S. pyogenes (pi-OJ-e-neez)
S. salvarius (sal-vah-REE-uss)

PRINCIPLES

The streptococci are gram-positive cocci arranged in chains (figure 56.1). Streptococci are classified according to their hemolytic activity, immunologic properties (the serological classification of Lancefield), and resistance to chemical and physical factors. The bacterium you will study in this exercise is *S. pyogenes.*

Figure 56.1 Morphology of Streptococci. Gram stain smear of exudate illustrating a long chain (since cell division occurs in one plane) of gram-positive cocci characteristic of streptococci.

Figure 56.2 Streak Plate of Streptococci on Blood Agar. This blood agar plate illustrates the small, semitranslucent gray-white colonies of streptococci surrounded by zones of β-hemolysis.

Streptolysins produced by *S. pyogenes* (Gr. *pyum,* pus) cause the lysis of red blood cells in vitro, producing **β-hemolysis** (a clear zone of hemolysis with no color change) on blood agar (figure 56.2). Two types of beta lysins are produced: streptolysin O and streptolysin S. The former is oxygen-labile, while the latter is oxygen-stable. Streptolysin O is demonstrated only in deep colonies on the blood agar medium. Since most strains of *S. pyogenes* produce both types of lysins, surface hemolysis is generally observed.

Streptolysin O is also immunogenic, and anti-streptolysin O can neutralize the effect of this lysin. The measurement of the serum level of this antibody is called **anti-streptolysin O titration (ASO).** Streptolysin S is nonimmunogenic. The ASO test is routinely used in the diagnosis of recent streptococcal infections that may have caused diseases such as rheumatic fever or glomerulonephritis. Following an infection due to Group A streptococci, the antibody in the individual's serum will begin to rise in 2 to 3 weeks and level off in 5 to 6 weeks. The titer of antibody is measured in **Todd units** (named after American bacteriologist E. W. Todd in 1932). A Todd unit is the reciprocal of the final dilution of serum used in the test that shows no hemolysis. A titer greater than 125 Todd units suggests a recent β-hemolytic, Group A streptococcal infection.

S. pyogenes is classified as Group A, based on a chemical substance known as C carbohydrate (an antigenic, group-specific hapten) found in the cell wall. There are 18 to 20 different immunologic groups of streptococci (A to O, excluding I and J) based on the presence of C factor. The specific type is determined by precipitin analysis. Members of Group A, such as *S. pyogenes,* are the streptococci most often responsi-

ble for human infections. Some diseases in humans caused by Group A streptococci include tonsillitis, septic sore throat, scarlet fever, otitis media, rheumatic fever, meningitis, erysipelas, acute endocarditis, and glomerulonephritis. *S. pyogenes* produces many enzymes and toxins such as streptokinase, leukocidins, streptodornase, hyaluronidase, hemolysins (streptolysin O and S), nucleases, and erythrogenic toxin.

The medium of choice for *S. pyogenes* is blood agar. On blood agar, the colonies are opaque, domed, about 0.5 mm in diameter, and surrounded by a zone of β-hemolysis. The optimal temperature for growth is 35°C.

S. pyogenes may be found in the throat and nasopharyngeal areas of humans. Also found in these areas are the viridans streptococci (e.g., *S. salvarius, S. mutans*) that produce either α-hemolysis (as evidenced by the greenish to gray pigmentation produced around the colony growth on blood agar) or no hemolysis. The viridans streptococci do not produce C carbohydrate and are usually nonpathogenic opportunists. *S. bovis* (L. *bos,* cow) is an example of a member of the viridans streptococci group.

In addition to hemolysis, *S. pyogenes* can be distinguished from viridans streptococci by means of the **bacitracin sensitivity test.** In this test, a filter paper disk impregnated with 0.04 units of bacitracin is applied to the surface of a blood agar plate that has been previously streaked with the bacteria to be identified. The appearance of a zone of inhibition surrounding the disk is a positive test for Group A streptococci (figures 56.3 and 56.4). An absence of a zone of inhibition suggests non-Group A bacteria (e.g., Group C). *S. pyogenes* is also resistant to SXT disks, which contain a mixture of 1.25 μg trimethoprim and 27.75 μg of sulfamethoxazole. Viridans streptococci are inhibited by these agents.

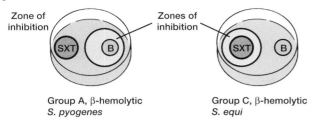

Group A, β-hemolytic
S. pyogenes

Group C, β-hemolytic
S. equi

Figure 56.4 Bacitracin Sensitivity Test. The right half of this blood agar plate was inoculated with a pure culture of a Group A β-hemolytic streptococcus. The left half of the plate was inoculated with *Streptococcus mitis.* The two disks contain 0.04 μg bacitracin. After 24 hours of incubation, the inhibition of growth around the antibiotic (disk on the right) is a positive bacitracin test. The bacteria on the left are bacitracin resistant.

Group B streptococci can be distinguished from other β-hemolytic streptococci by their production of a substance called the **CAMP factor.** CAMP is an acronym for the names of the investigators (**C**hristie, **A**tkins, and **M**unch-**P**etersen) who first described the factor. This factor is a peptide that acts together with the β-hemolysin produced by some strains of *S. aureus,* enhancing the effect of the latter on a sheep blood agar plate (figure 56.5).

Group D streptococci (*S. mitis; L. mitis,* mild) and enterococci can be differentiated from other streptococci by using bile esculin agar slants. Group D streptococci grow readily on the bile esculin agar and hydrolyze the esculin, imparting a dark brown color to the medium. This reaction denotes their bile tolerance and ability to hydrolyze esculin, and constitutes a positive reaction. Gram-positive bacteria (other than group D streptococci and enterococci) are inhibited by the bile salts.

Figure 56.5 The CAMP Test for Distinguishing Group B Streptococci from Other β-Hemolytic Streptococci.

This exercise is designed to familiarize the student with the morphological and colonial characteristics, diagnostic tests (table 56.1), and serological identification of *S. pyogenes.* Also included is the comparison of the usually nonpathogenic alpha streptococci (*S. mitis*) with the pathogenic *S. pyogenes.*

Procedure

First Period

Determination of Anti-Streptolysin O Antibody (ASO): A Demonstration

1. The following reagents will be used:
 a. Streptolysin O buffer
 b. Red blood cell suspension (5% sheep red blood cells in buffer)
 c. Streptolysin O reagent (lyophilized enzyme preparation, which must be reconstituted by adding distilled water to the ampule in the amount indicated on the label. It is prepared only when ready to be added to the diluted serum).
 d. Serum dilution—the following dilutions of serum, heated at 56°C for 30 minutes, will be prepared in buffer:
 1/10, 0.05 ml serum + 4.5 ml of buffer1/100, 1.0 ml of the 1/10 dilution + 9 ml of buffer
2. Your instructor will demonstrate the ASO test according to the protocol illustrated in table 56.2.

Study of Streptococci

1. Each group of students is to inoculate one blood agar plate with Group A, β-hemolytic *S. pyogenes* and another plate with Group C β-hemolytic *S. equi.* Use a loop to inoculate. A viridans streptococcus such as *S. salvarius* can be used in place of *S. equi.*
2. Place a bacitracin disk and an SXT disk on the plates. Be sure to place the disks on opposite sides of the plates, in well-inoculated areas. Incubate the plate for 24 hours at 35°C. Do not invert.
3. With an inoculating loop, streak a strain of *S. aureus* down the center of a blood agar plate. On one

Table 56.1
Diagnostic Tests for Streptococcal Differentiation

Group Species	Hemolysis	Bacitracin Sensitivity	CAMP	Bile Esculin	SXT Sensitivity
A					
S. pyogenes	β	Sensitive	−	−	Resistant
B					
S. agalactiae	β	Resistant	+	−	Resistant
C					
S. equi	β	Resistant	−	−	Sensitive
D					
E. faecalis	α, β, or none	Resistant	−	+	Resistant (some sensitive)
S. bovis					
Viridans					
S. salvarius	α or none	Resistant (some sensitive)	−	−	Sensitive
S. mitis					
S. mutans					

Table 56.2
Protocol for the ASO Test

Tube No.	1	2	3	4	5	6	7	Buffer-Red Cell Control 8	Streptolysin Control 9
ml serum dilution 1/10	0.8	0.2	−	−	−	−	−	−	−
ml serum dilution 1/100	−	−	1.0	0.8	0.6	0.4	0.3	−	−
ml of buffer	0.2	0.8	0	0.2	0.4	0.6	0.7	1.5	1.0
Shake gently									
Rehydrate streptolysin	0.5	0.5	0.5	0.5	0.5	0.5	0.5	−	0.5
ml streptolysin O reagent									
Shake gently									
Incubate at 37°C for 15 minutes									
ml 5% RBC	0.5	0.5	0.5	0.5	0.5	0.5	0.5	0.5	0.5
Shake gently									
Incubate at 37°C for 45 minutes									
Shake once after 15 minutes									
Centrifuge tubes for 1 minute at 1,500 rpm									
Examine									
Degree of hemolysis 0 to 4+ ; 0 = no hemolysis; 4 + = complete hemolysis									
Unit value of each tube in Todd units	12	50	100	125	166	250	333	−	−

side of the plate, inoculate a strain of Group B streptococci (*S. agalactiae*) by making a streak at a 90° angle, starting 5 mm away from the *S. aureus* and extending outward to the edge of the agar. On the other side of the plate, inoculate the strain of Group A streptococci (*S. pyogenes*). This streak should not be directly opposite the Group B streak.

4. Incubate the plates at 35°C for 24 hours.
5. Inoculate one bile esculin agar slant with *S. equi,* one with *S. mitis,* one with *S. agalactiae,* and one with *S. pyogenes.* Incubate the slants at 35°C for 24 hours.

Second Period

1. Examine the plates for bacitracin and SXT susceptibility. A zone of bacitracin inhibition greater than or equal to 14 mm in diameter indicates the presence of Group A streptococci. There will not be a zone around the non–Group A streptococci. Positive SXT results for the presence of Group C streptococci are indicated by a zone of inhibition greater than or equal to 16 mm in diameter. There will not be a zone around the non–Group C.
2. Observe the area of hemolysis around the *S. aureus* streak. At the point adjacent to the streak of Group

B streptococci, you should be able to see an arrowhead-shaped area of increased hemolysis indicating the production of the CAMP factor. There should be no change in the hemolytic zone adjacent to the streak of the Group A streptococci since most strains do not produce the CAMP factor.

3. Gram stain each group of streptococci from the blood agar plates.
4. Examine the bile esculin agar slants for any color change. A dark brown color change is a positive reaction for group D streptococci. No color change is a negative reaction.
5. Complete the report for exercise 56.

HINTS AND PRECAUTIONS

(1) It should be noted that the zone of hemolysis around the bacitracin disk is so distinct because a pure culture is being used. If this is not so, then your stock culture is probably contaminated.

(2) Sometimes it is easiest to observe hemolysis by examining the zone with the 4× objective of a compound microscope. Any intact erythrocytes are readily seen.

Laboratory Report 56

Name: _____

Date: _____

Lab Section: _____

Streptococci

1. Serum titer in Todd units: _____
2. Streptococci.

Bacterium	Gram Stain	Group	CAMP Factor	Colony on Blood Agar	Hemolysis	Bacitracin Sensitivity	SXT Sensitivity	Bile Esculin Hydrolysis
S. mitis	_____	_____	_____	_____	_____	_____	_____	_____
S. pyogenes	_____	_____	_____	_____	_____	_____	_____	_____
S. equi	_____	_____	_____	_____	_____	_____	_____	_____
S. agalactiae	_____	_____	_____	_____	_____	_____	_____	_____

3. In summary, what is the major purpose of this experiment?

Review Questions

1. How could you determine if a sore throat was caused by *S. pyogenes* (Group A, β hemolytic)?

2. What types of diseases are caused by *S. pyogenes?*

3. What is a Todd unit?

4. Do all Group A streptococci produce erythrogenic toxin? Explain your answer.

5. What is the purpose of the ASO test? When would you use it in a clinical laboratory?

6. What is SXT sensitivity? CAMP factor? What organisms are identified by those tests?

7. How would you differentiate between α- and β-hemolysis?

Exercise 57

Neisseriae

MATERIALS PER GROUP OF STUDENTS

1 tryptic soy agar plate
1 modified Thayer-Martin plate
3 tubes enriched nitrate broth containing Durham tubes
candle jar
1 stock culture of *Neisseria gonorrhoeae* (ATCC 9793)
1 stock culture of *Neisseria flavescens* (ATCC 13120)
1 stock culture of *Moraxella nonliquefaciens* (ATCC 19975)
3 CTA (cysteine tryptic agar) glucose medium tubes
3 CTA fructose medium tubes
3 CTA maltose medium tubes
3 CTA sucrose medium tubes
indophenol oxidase reagent or Difco SpotTest Oxidase Reagent
catalase reagent (3% hydrogen peroxide)
nitrate test reagents
DNase test agar plate
1-ml pipettes with pipettor
Gram-stain reagents
glass slides
wax pencil
37°C incubator
prepared slides of *Neisseria gonorrhoeae* exudate
1 N HCl
Bunsen burner

Learning Objectives

Each student should be able to

1. Become familiar with the clinical isolation and identification of *Neisseria* species
2. Perform isolation and biochemical tests to differentiate *N. gonorrhoeae, N. flavescens,* and *M. nonliquefaciens*

SUGGESTED READING IN TEXTBOOK

1. Order *Neisseriales,* section 22.2.
2. Gonorrhea, section 39.3.

Pronunciation Guide

Moraxella nonliquefaciens (mo-rak-SEL-ah non-li-kwe-FA-sens)
Neisseria flavescens (nis-SE-re-ah flav-ES-sens)
N. gonorrhoeae (go-nor-REE-ah)
N. lactamica (lak-tam-ica)
N. mucosa (MU-cosa)
N. sicca (sik-ah)
N. subflava (sub-FLA-vah)

PRINCIPLES

The family *Neisseriaceae* includes bacteria that are gram-negative cocci occurring in pairs or groups. The two species that are pathogenic in humans are *N. gonorrhoeae* (Gr. *gonorrhoea,* gonorrhea), which causes gonorrhea, and *N. meningitidis* (Gr. *meninx,* the membrane enclosing the brain), which causes cerebrospinal meningitis. Since these bacteria are very fastidious and sensitive to changes in the atmosphere, the following precautions must be taken in isolating and culturing them: (1) highly enriched media must be used, and (2) the bacteria must be incubated in an increased CO_2 and water atmosphere.

Table 57.1

Some Distinguishing Characteristics of *Neisseria* and *Moraxella*

Characteristic	*N. gonorrhoeae*	*N. flavescens*	*M. nonliquefaciens*
		Bacteria	
Growth on soy agar at 35 to 37°C	−	+	+
Growth on modified Thayer-Martin agar, Martin-Lewis agar, or New York City medium at 35–37°C	+	−	−
Acid produced from			
Glucose	+	−	−
Fructose	−	−	−
Maltose	−	−	−
Sucrose	−	−	−
Lactose	−	−	−
Reduction of			
NO_3	−	−	+
NO_2	−	+	−
Colony pigmentation	None	Yellow	None
Colony morphology	Gray to white, smooth	Opaque, smooth	Opaque, smooth
DNase	−	−	+
Catalase	+	+	+
Oxidase	+	+	+

Other nonpathogenic *Neisseria* may be isolated from the respiratory tract of humans. These include *N. sicca*, *N. subflava*, *N. flavescens*, *N. mucosa*, and *N. lactamica*. *Moraxella nonliquefaciens* is also commonly found in the respiratory tract. However, it is a gram-negative rod that produces cocci only in a stationary culture. All of these bacteria can be distinguished from the pathogenic species by their fermentation patterns, nitrate and nitrite reduction characteristics, and growth patterns on enriched and nutrient agars (table 57.1).

The purpose of this exercise is to acquaint the student with methods of clinical laboratory identification of the *Neisseria* species. *M. nonliquefaciens* will also be studied because of its practical laboratory implications—it is the only species that produces DNase.

Procedure

First Period

1. Gram stain the known stock cultures of *N. gonorrhoeae* (figure 57.1), *N. flavescens,* and *M. nonliquefaciens.* Note the resistance of *M. nonliquefaciens* to decolorization.
2. Carefully examine the prepared slide of a *N. gonorrhoeae* exudate. Notice the many white blood cells that are present (figure 57.2).
3. Warm the media to be used to room temperature. With a wax pencil, divide the bottom of a tryptic soy agar plate, modified Thayer-Martin plate, and DNase plate into three sections. Modified Thayer-

Figure 57.1 **Morphology of *Neisseria.*** Gram stain of a culture of *N. gonorrhoeae* showing the gram-negative cocci occurring most in pairs; some have formed clumps during the staining procedure (×1,000).

Figure 57.2 **Gram Stain of a Urethral Discharge Showing Acute Gonococcal Urethritis.** Note the presence of the intracellular gram-negative diplococci within the pale-staining segmented neutrophil.

Figure 57.3 Typical Colonies of *N. gonorrheae* on Modified Thayer-Martin Medium. Notice that the gonococcal colonies are typically small, raised, glistening, and gray-white as shown here.

Figure 57.4 A Candle Jar.

Martin agar is one of the recommended media when *Neisseria* spp. are suspected (table 57.1 and figure 57.3). With the wax pencil, place your name, date, and the name of the respective bacteria to be inoculated on the plates. Streak each section for isolation with the respective bacteria.

4. Incubate the plates at 37°C in a candle jar that contains a dampened paper towel for providing a humid atmosphere (figure 57.4) for 24 to 48 hours. (The candle jar provides an anaerobic environment as the burning candle uses up the O_2 in the sealed jar.)

5. Inoculate the enriched nitrate tubes with the respective bacteria. Label with your name, date, and the name of the bacteria. Incubate for 24 to 48 hours at 37°C (*see figure 34.1*).

6. Heavily inoculate the following CTA carbohydrate-containing media tubes just below the surface (**do not go deep into the medium**): glucose, maltose, fructose, and sucrose. Tighten the caps and incubate at 37°C. Examine daily for 5 days or until acid production is seen (*see figure 20.2*).

7. Add 1 to 2 drops of oxidase reagent to each stock culture. A positive oxidase test is indicated by a darkening of the color of the colony within 20 to 30 seconds. This color will eventually turn purple.

8. To test for catalase, pipette several milliliters of the hydrogen peroxide reagent over select colonies. The appearance of gas bubbles indicates a positive test; the absence of gas bubbles is a negative test (*see figure 28.1*).

Second Period

1. Note the characteristics of any isolated colonies of *N. gonorrhoeae*, *N. flavescens*, and *M. nonliquefaciens* on tryptic soy, modified Thayer-Martin agar, and DNase plates. Describe these colonies.

2. Test and observe the nitrate broth cultures for the reduction of nitrate and/or nitrite (*see figure 34.1*).

3. Read the results of the carbohydrate degradation tubes.

4. Complete the report for exercise 57.

Name: _____

Date: _____

Lab Section: _____

Neisseriae

1. *Neisseria* and *Moraxella* culture data (colony size, form, surface, margin, color).

Agar	*N. gonorrhoeae*	*N. flavescens*	*M. nonliquefaciens*
Tryptic soy agar	_____	_____	_____
Modified Thayer–Martin medium	_____	_____	_____

2. *Neisseria* and *Moraxella* biochemical data.

Test	*N. gonorrhoeae*	*N. flavescens*	*M. nonliquefaciens*
Oxidase ($+$ or $-$)	_____	_____	_____
Catalase ($+$ or $-$)	_____	_____	_____
DNase ($+$ or $-$)	_____	_____	_____
Acid production			
Glucose	_____	_____	_____
Fructose	_____	_____	_____
Maltose	_____	_____	_____
Sucrose	_____	_____	_____
Nitrate and nitrite reduction			
Nitrate	_____	_____	_____
Nitrite	_____	_____	_____
Gram stain			
Reaction	_____	_____	_____
Morphology	_____	_____	_____
Arrangement	_____	_____	_____

3. In summary, what is the major purpose of this experiment?

Review Questions

1. From what type of clinical specimen would one expect to isolate *N. gonorrhoeae*? *M. nonliquefaciens*?

2. Why is the oxidase test useful in the separation of gram-negative, facultatively anaerobic bacteria?

3. Give one example in which gonococci might give a false-negative test in a CTA carbohydrate test.

4. How does a candle jar function?

5. Since the *Neisseria* are fastidious and sensitive to atmospheric changes, what precautions must be taken in isolating and culturing them?

6. How are pathogenic *Neisseria* identified?

7. What are intracellular gram-negative diplococci?

Aerobic and Anaerobic Endospore-Forming Bacteria

MATERIALS PER GROUP OF STUDENTS

1 tryptic soy agar plate

1 24-hour tryptic soy broth culture of *Bacillus subtilis* (ATCC 6051)

1 36-hour tryptic soy agar slant of *Bacillus megaterium*

1 24-hour sodium thioglycollate broth culture of *Clostridium perfringens* (ATCC 3620)

2 blood agar plates

spore-staining material (*see exercise 10*)

anaerobic jar, GasPak System (BBL), or Oxyrase plate

wax pencil

inoculating loop

35°C incubator

clean glass slides

Bunsen burner

Gram-stain reagents

test-tube rack or can

Learning Objectives

Each student should be able to

1. Perform aerobic and anaerobic cultures on endospore-forming bacteria
2. Do an endospore stain

SUGGESTED READING IN TEXTBOOK

1. The Bacilli and Lactobacilli, section 23.4; see also tables 23.2 and 23.3, and figures 23.5, 23.6, and 23.8.
2. Anthrax, section 39.3.
3. Gas Gangrene, section 39.3.
4. Tetanus, section 39.3.
5. Botulism, section 39.4.

Pronunciation Guide

Bacillus subtilis (bah-SIL-lus sub-til-us)
B. megaterium (meg-AH-ter-ee-um)
Clostridium perfringens (klos-STRID-ee-um per-FRIN-jens)

PRINCIPLES

Bergey's Manual lists six gram-positive, endospore-forming genera. The most thoroughly studied and most important are *Bacillus* and *Clostridium*.

The genus *Bacillus* (e.g., *B. subtilis, B. megaterium*) is aerobic, or sometimes facultatively anaerobic, and catalase positive. *Bacillus* species are ubiquitous in soil and are common laboratory contaminants. In older cultures, these bacteria will lose their ability to retain the primary stain (crystal violet) in the Gram reaction. Under adverse conditions, *Bacillus* spp. form endospores (figure 58.1*a,b*). Subgroups of the *Bacillus* genus have been constructed based on endospore shape (oval, spherical), position of the spore in the cell (central, terminal), and the species' ability to ferment sugars. Most members of this group are motile, some produce capsules, some are thermophilic (heat-loving) and, with the exceptions of *B. anthracis* (the etiologic agent of **anthrax**) and *B. cereus* (the etiologic agent of **gastroenteritis**), are nonpathogenic.

Members of the genus *Clostridium* are obligately anaerobic and lack both catalase and a complete electron transport chain. The endospores may be central, subterminal, or terminal in position, and spherical or oval in shape. Like *Bacillus,* bacteria of this genus tend to give a gram-negative stain in older cultures. *Clostridium* bacteria are widely distributed in the soil and in the intestinal tract of humans and other animals.

Figure 58.1 **Bacterial Endospores.** (**a**) *Bacillus subtilis.* Notice that the spores are elliptical and central (×1,000). (**b**) *Bacillus megaterium.* Notice the large elliptical central spores (×1,000). (**c**) *Clostridium perfringens.* Notice that the spores are elliptical, and central or subterminal, cells slightly swollen (×1,000). (**d**) *Clostridium tetani.* Notice that the spores are round and terminal (×1,000). (**e**) *Bacillus subtilis* on blood sugar. The large, irregular, rough and waxy colonies are surrounded by zones of β-hemolysis.

(a)

(b)

(c)

(d)

(e)

Three species are of medical importance: *C. tetani,* which causes tetanus (figure 58.1*d*); *C. botulinum,* which causes botulism; and *C. perfringens,* which causes gas gangrene and food poisoning (figure 58.1*c*). **Tetanus** is caused by *C. tetani* endospores entering tissue with a low oxidation-reduction potential. The spores germinate, and the bacteria excrete exotoxins. **Botulism** is the result of the ingestion of foods containing the botulinum exotoxin, which is produced by *C. botulinum.* **Gas gangrene** results when *C. perfringens* (also *C. novyi, C. septicum,* or *C. histolyticum*) is introduced into tissue that has been injured. When anaerobiosis occurs in the surrounding area, this bacterium ferments the carbohydrates present in the tissue, producing gas.

An anaerobic atmosphere can be obtained by replacing air in a jar with the desired gas. The gases usually include mixtures of CO_2, and/or H_2 in N_2. The H_2 gas aids in decreasing the oxidation-reduction potential by combining with O_2 to form water (in the presence of a catalyst). Chemical reactions can also be used to generate anaerobic environments. The GasPak System (BBL) provides a simple means for creating an anaerobic environment (*see figures 18.4 and 18.5*). Cultures are placed in a GasPak along with a GasPak Disposable Hydrogen + Carbon Dioxide Generator Envelope with a palladium-coated aluminum pellet catalyst, and an anaerobic indicator. Water is added to the envelope, and the container is sealed. The hydrogen gas combines with O_2 to form water and provides the anaerobic conditions.

In this exercise, students will learn the aerobic procedures for culturing *Bacillus* (figure 58.1*e*) and the anaerobic procedures for culturing *Clostridium.* Spore and Gram staining will also be done.

Procedure

First Period

Pure Culture Study

1. With the *B. subtilis* broth culture, perform an isolation streak on the tryptic soy agar plate. Label the plate with your name, date, medium, and bacterium.
2. Incubate the plate inverted for 24 to 48 hours at 35°C.
3. Streak two blood agar plates for isolation with *C. perfringens.* Label as in step 1.
4. Place one of the blood agar plates in either the anaerobic jar supplied or the GasPak. (Alternatively, an Oxyrase plate may be inoculated to grow *C. perfringens* anaerobically.) Your instructor will demonstrate the use of the jar, GasPak Systems, or Oxyrase plate.
5. Place either the jar or the GasPak and the other blood agar plate in the 35°C incubator for 48 hours.

6. Save all cultures for the second period, including the stock cultures.

Endospore Stain

1. Fix a smear from each culture (tryptic soy agar slant and sodium thioglycollate broth) on a clean glass slide.
2. Stain for endospores as described in exercise 10.
3. Draw the endospores in part 1 of the report for exercise 58 (figure 58.1*a,b*).

Second Period

1. Examine the aerobic culture of *B. subtilis* and record the colonial morphology in part 2 of the report for exercise 58.
2. Do Gram stains on well-isolated bacterial colonies from the TSA plate and from the tryptic soy broth tube.

3. Examine the aerobic and anaerobic cultures of *C. perfringens*. Notice that no growth was observed in the aerobic culture due to the fact that this bacterium is a strict anaerobe.
4. Record the colonial morphology of the blood agar plate grown under anaerobic conditions in part 2 of the report for exercise 58. Also notice the double zone of β-hemolysis, which is a characteristic used in the identification of *C. perfringens*.
5. Do Gram stains on well-isolated colonies from the blood agar plate and sodium thioglycollate broth.

HINTS AND PRECAUTIONS

Use the fume hood when heat-staining spores with malachite green.

Name: _____

Date: _____

Lab Section: _____

Aerobic and Anaerobic Endospore-Forming Bacteria

1. Drawing of endospores.
 a. *C. perfringens*

 b. *B. megaterium*

2. Pure culture study.

Bacterium	Colonial Morphology	Position of the Spore	Gram Stain	
			Young Culture	Old Culture
C. perfringens	_____	_____	_____	_____
B. subtilis	_____	_____	_____	_____

3. In summary, what is the major purpose of this experiment?

Review Questions

1. What are several functions of bacterial endospores?

2. Can the position of an endospore help in identifying bacteria? Give examples.

3. What is the major source of *Clostridium* in our environment?

4. Discuss oxidation-reduction potentials in terms of anaerobiosis.

5. How does a GasPak System function?

6. Why is sodium thioglycollate used to culture anaerobic bacteria?

7. Would you expect to find endospores in the blood agar plate culture of *Clostridium?* Why or why not?

Survey of Selected Eucaryotic Microorganisms

Fungi

For 300 years after the birth of Christ rainfall was abnormally heavy. . . . The crops flourished and so did the host of tiny living organisms—the rusts, rots, mildews, molds, smuts, and blights that were to ultimately destroy the grain. . . . Field after field was laid to waste. History is made by a multiplicity of little things, as well as major revolutions, wars, earthquakes, and floods. And so it was that the fungi, by bringing hunger and unrest, contributed to the decline of the Great Roman Empire.

Lucy Kavaler (American author of *Mushrooms, Molds, and Miracles*, 1938–)

Eucaryotic microorganisms include the algae, fungi (yeasts and molds), and the protozoa. All of these eucaryotes consist of single cells (or simple aggregates of cells) that contain a membrane-bound nucleus and intracellular membrane-bound organelles. Free-living, eucaryotic microorganisms are widely distributed in the environment; eucaryotic microorganisms also participate in various types of symbiotic relationships. Many cause human diseases and are of medical importance; others are domestic animal and agricultural plant pathogens; still others benefit society.

This section of the manual contains two laboratory exercises that have been designed to better acquaint the student with the fungi. Students will work with both prepared specimens and living microorganisms in order to better understand the life cycles, salient characteristics, and morphology of selected examples.

After completing one or more of the exercises in part ten, you will, at the minimum, be able to demonstrate an increased understanding of microbial diversity. This will meet the American Society for Microbiology Core Curriculum theme number 1: impact of microorganisms on the biosphere and humans; microbial diversity (*see pp. vi–viii*).

Fungi I: Yeasts

MATERIALS PER STUDENT

7- to 10-day Sabouraud dextrose plate culture of
 Saccharomyces cerevisiae (ATCC 2366) and/or
 Rhodotorula species (ATCC 22078)
dried commercial baker's yeast
iodine solution (3 ml water to 1 ml Gram's iodine)
methylene blue solution
Bunsen burner
inoculating loop
clean glass slides
coverslips
wax pencil
2 Sabouraud dextrose agar plates
35°C incubator
1 glucose fermentation tube with Durham tube
 (Difco: neopeptone + 1% yeast extract + 1%
 glucose)
1 sucrose fermentation tube with Durham tube
 (Difco: neopeptone + 1% yeast extract + 1%
 sucrose)
sterile cotton swabs

Learning Objectives

Each student should be able to

1. Understand the morphological characteristics of yeast cells
2. Culture a typical yeast and study carbohydrate fermentation
3. Stain a typical yeast cell

SUGGESTED READING IN TEXTBOOK

1. Structure (of Yeast), section 25.3; see also figure 25.3.

2. Division *Ascomycota,* section 25.6; see also figures 25.2 and 25.10.
3. Opportunistic Mycoses, section 40.1; see also figure 40.16.

Pronunciation Guide

Rhodotorula spp. (ro-do-TOR-u-lah)
Saccharomyces cerevisiae (sak-ah-ro-MI-seez ser-ah-VEES-ee-eye)

Why Are the Above Fungi Used in This Exercise?

The major objectives of this exercise are to learn yeast morphology and culture some typical yeasts. To achieve these objectives, the authors have chosen the common yeast *Saccharomyces cerevisiae* (or *Rhodotorula* spp.) for this exercise. Because *Candida albicans* is a yeast that is commonly found in the mouth, the student should be able to isolate it.

PRINCIPLES

Yeasts are unicellular fungi that are spherical, ellipsoidal, or oval in shape, and usually (with the exception of the dimorphic yeasts such as *Candida*) do not form hyphae (fungal filaments). They are about 5 to 10 times larger than bacteria. Yeasts commonly reproduce asexually by **budding,** a process in which a new cell (called a daughter cell) is formed by the parent cell from a protuberance called a **bud.** When yeasts reproduce sexually, they produce several types of **sexual spores** (e.g., ascospores). The type of spore produced is very useful in classifying yeasts. Metabolic activities are also used to identify and classify yeasts. For example, the yeast *Saccharomyces cerevisiae* will fer-

ment glucose but not sucrose. In the laboratory, **Sabouraud dextrose agar** is commonly used to isolate yeasts. It is a selective medium containing glucose and peptone, and has a low pH, which inhibits the growth of most other microorganisms.

This exercise illustrates the fermentation ability and colony and cell morphology of the common yeast, *S. cerevisiae*. This yeast is used in making bread and in various alcoholic fermentations. In addition, each student will attempt to isolate and culture a yeast from his/her mouth. A yeast that is most commonly found in the mouth is *C. albicans*. In healthy individuals, *C. albicans* does not produce disease. However, if the normal microbiota of the mouth (or other areas of the body) is upset, or the individual is compromised, the disease **candidiasis** may result.

Procedure

First Period

1. With a wax pencil, draw two circles on a clean glass slide. Place several drops of the water-iodine solution into one circle and several drops of the methylene blue solution into the other circle.
2. Suspend a loopful of yeast culture into each circle. Place a coverslip over each.
3. Examine both yeast cell preparations under low and high power. Note the shape and relative size and the presence or absence of budding (figure 59.1). Look for the small nucleus and larger vacuole.

Figure 59.1 Yeasts. Oval yeast cells (*Saccharomyces cerevisiae*) in a wet mount (×1,000).

4. Draw the representative yeast cells in part 1 of the report for exercise 59.
5. Suspend a pinch of commercial baker's yeast in 2 to 3 ml of warm water. With the inoculating loop, streak onto one half of a Sabouraud dextrose agar plate. Using the inoculating loop, remove some *S. cerevisiae* from the stock plate. Streak the other half of the plate. Label the plate with your name, date, and yeast. Incubate the plate at 35°C until growth is seen.
6. Inoculate about 1 ml of the commercial baker's yeast suspension into a glucose fermentation tube and a sucrose fermentation tube. Incubate these tubes at 35°C and observe daily until growth is seen. Look for fermentation (gas bubbles in the Durham tubes) and growth as indicated by turbidity.
7. Try to isolate *S. albicans* by swabbing the surface of your tongue. Streak a Sabouraud dextrose agar plate. Incubate the plate at 35°C until growth has occurred.

Second Period

1. Smell the Sabouraud dextrose agar plate. Observe the appearance of the yeast colonies. Record your results in part 2 of the report for exercise 59.
2. Examine the glucose and sucrose fermentation tubes. Record your results in part 3 of the report for exercise 59.
3. Examine the Sabouraud dextrose agar plate prepared by swabbing your tongue. Smell the plate. Stain several of the colonies with the methylene blue as in step 1 of the first period procedure. Record your results in part 4 of the report for exercise 59.

HINTS AND PRECAUTIONS

If phase-contrast or ultraviolet microscopes are available, the observation of cellular organelles and inclusions can be particularly revealing when students work with unstained yeast wet mounts.

Fungi I: Yeasts

1. Draw typical yeast cells in the following space.
 a. Water-iodine suspension (Magnification: × _____).

 b. Methylene blue suspension (Magnification: × _____).

2. Draw typical yeast colonies in the following space (Magnification: × _____).

Commercial baker's yeast *S. cerevisiae*

a. color _____ a. color _____

b. smell _____ b. smell _____

3. Fermentation.
 a. What is the color of the glucose broth after _____ days incubation? _____
 Was gas produced? Explain. _____
 b. What is the color of the sucrose broth after _____ days incubation? _____
 Was gas produced? Explain. _____

4. Draw the appearance of the yeast colony from your mouth in the following space. Do the same for the cell morphology.

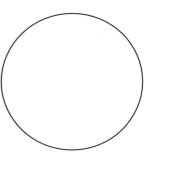

Magnification × _____

Review Questions

1. Define:

 a. budding

 b. hyphae

 c. yeast

2. Compare and contrast the characteristics of baker's yeast and *Saccharomyces cerevisiae.*

3. Why are yeast colonies larger than bacterial colonies?

4. Why do yeasts generally have to be cultured for longer periods than most bacteria?

5. What are some similarities and differences of the yeasts that were cultured from your mouth?

6. Why are stains not required for yeast identification?

7. Why is Sabouraud dextrose agar used to cultivate yeasts?

Fungi II: Phycomycetes, Ascomycetes, and Basidiomycetes

MATERIALS PER GROUP OF STUDENTS

7- to 10-day Sabouraud dextrose plate cultures of *Penicillium notatum* (ATCC 9178) and *Aspergillus niger* (ATCC 10535)
preserved mushrooms (*Agaricus* or *Coprinus*)
plus and minus strains of *Rhizopus stolonifer* (ATCC 12938$^+$, 12939$^-$)
1 Sabouraud dextrose agar deep
2 potato dextrose agar deeps
48° to 50°C water bath
3 sterile petri plates
methylene blue in 90% methanol
inoculating loop
wax pencil
clean microscope slides
22 × 40-mm coverslips
tweezers
Bunsen burner
syringes containing paraffin and oil, or tubes of silicone tub caulk
wooden swab sticks (or a syringe)
1-ml pipettes with pipettor
glass or wooden support rods
sterile Pasteur pipettes and bulbs
prepared slides of Phycomycetes (e.g., *Rhizopus, Saprolegnia*), Ascomycetes (e.g., *Penicillium, Aspergillus, Morchella*), Basidiomycetes (e.g., *Polyporus, Lycoperdon, Coprinus, Puccinia*)

Learning Objectives

Each student should be able to

1. Describe a typical mold
2. Grow colonies of several molds on Sabouraud agar plates
3. Prepare culture slides of several molds
4. Identify the morphology and reproductive structures of representative molds
5. Observe a complete mold life cycle

SUGGESTED READING IN TEXTBOOK

1. The Fungi, chapter 25. It is suggested that each student bring his or her textbook to the laboratory since all of the morphology data, life cycles, and terminology covered in this exercise can be found in chapter 25 and will not be repeated here.

Pronunciation Guide

Aspergillus niger (as-per-JIL-us NI-jer)
Penicillium notatum (pen-a-SIL-ee-um no-TAY-tum)
Rhizopus stolonifer (ri-ZO-pus sto-LON-a-fer)

Why Are the Above Molds Used in This Exercise?

In this exercise, the student will become familiar with several aspects of fungus biology. The authors have chosen three of the most common molds (*Penicillium notatum, Rhizopus stolonifer,* and *Aspergillus niger*) for the student to culture and study. Other fungi are introduced using prepared slides and preserved materials.

PRINCIPLES

Molds are multicellular, filamentous fungi. The techniques of culturing and observing fungi differ from the methods used to study bacteria. Fungi grow at comparatively slow rates, often requiring several days to weeks to form macroscopically visible colonies. Usually, the growth will spread over the entire culture plate. Molds produce spores on brightly colored aerial hyphae. Most molds grow best at room temperature (25°C) rather than at 35°C.

The basic medium for the culture of many molds is Sabouraud dextrose agar. The high sugar concentration and low pH (5.6) of this medium make it unsuitable for the growth of most bacteria, thus guarding against contamination. Most fungi grow well at pH 5.6.

Mold colonies may be examined directly in culture with a dissecting microscope. However, it is better to tease away a portion of the growth and place it on a slide with a drop of water or stain (a wet-mount). An alternative method is a **slide culture.** The latter is more advantageous because one is able to study the growing mold and identify specific structures without disturbing the culture. In this way, the reproductive structures can be easily observed. Fungi are identified primarily by examining their reproductive structures, morphological characteristics, and colony growth.

The macroscopic aggregation (colony) of mold cells is called a **thallus.** A thallus is composed of a mass of strands called a **mycelium;** each strand is a **hypha. Vegetative hyphae** grow on the surface of culture media. They form aerial hyphae, called **reproductive hyphae,** that bear asexual reproductive **spores** or **conidia.** The hyphae that grow below the surface of culture media are called **rhizoidal hyphae.** The hyphal strand of some molds may be separated by a crosswall called a **septum.** Hyphae that contain septa are called **septate hyphae.** Molds with hyphae that lack septa are called **coenocytic hyphae.**

Molds are classified primarily by their sexual stages of reproduction. They are also characterized and classified according to the appearance of the colony (e.g., color, size), organization of the hyphae (e.g., septate or coenocytic), and structure and organization of the spores (e.g., sporangiospores, conidiospores). Molds are important clinically, industrially, and as decomposers (saprophytes) in the environment. Mold spores are a very common source of contamination of laboratory cultures.

In this exercise, the methods used to study molds will include the preparation of colonies on petri plates, the preparation of special culture slides, the study of the life cycle of *Rhizopus* (the common bread mold), the

dissection of a mushroom, and the examination of commercially prepared slides of preserved molds.

Procedure

Preparation and Observation of Colonies

1. Melt one tube of Sabouraud dextrose agar and one of potato dextrose agar.
2. Cool to 48° to 50°C in a water bath.
3. Pour into two petri plates respectively and allow to harden.
4. Using the wax pencil, label the Sabouraud dextrose agar plate *Aspergillus* and the potato dextrose agar plate *Penicillium.* Add your name and date to each plate.
5. Using aseptic technique, inoculate the plates as labeled with a single loopful of the mold suspension. Place the loopful of mold inoculum in the center of the plate. Do not spread the inoculum. Handle plates carefully so that they are not jostled.
6. **Do not invert the petri plates.** Incubate them at room temperature for 2 to 7 days.
7. After the colonies have developed properly (figure 60.1), sketch and describe the macroscopic appearance (e.g., color, texture) of each in part 2 of the report for exercise 60. If dissecting microscopes are available, examine the hyphae and conidia under the microscopes and draw the conidia.

Preparation of Slides for Microscopic Examination of Molds

1. Obtain two clean glass slides and coverslips.
2. Flame the surface of the slides and coverslips in order to sterilize them. Use tweezers to hold the slides and coverslips.

Figure 60.1 **Molds.** *Aspergillus* showing black conidiospores.

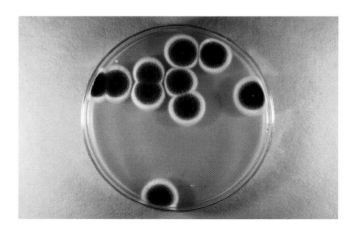

3. Use the wooden sticks (or a heated syringe) to transfer sufficient melted paraffin or silicone caulk to each slide to support the coverslip about 1 mm over the surface of the slide. Let the paraffin or silicone harden. Each slide should appear as in figure 60.2*a*.

4. Heat the coverslips sufficiently so that they will form a seal when you set them over the hardened paraffin or silicone. Each slide should look like figure 60.2*b*.

5. Melt and cool to 48° to 50°C a tube of Sabouraud dextrose agar (for *Aspergillus*) and a tube of potato dextrose agar (for *Penicillium*). Label the tubes accordingly.

6. Using a sterile pipette, transfer and mix 0.5 ml of the proper mold suspension with the proper agar tube.

7. For each mold, use a sterile Pasteur pipette and quickly let sufficient inoculated agar run under the coverslip of the prepared slide to half fill the chamber. Each slide should now appear as shown in figure 60.2*c*. This procedure must be completed before the agar hardens.

8. Moisten two circles of paper towel that just fit the bottom of the petri plates, and place one in each plate. Each slide should then be placed in its own petri plate. Label the plates accordingly. The slide should be supported above the moist paper by two wooden sticks or a piece of a V-shaped glass rod (figure 60.2*d*). Place the lid on the petri plate and incubate at room temperature for 2 to 4 days.

Figure 60.2 Mold Culture Slide.

9. After incubation, observe the slides with a microscope, using the low-power objective. Add a few drops of methylene blue in methanol to stain the various structures. (The alcohol is necessary to soften the cell wall and allow the stain to enter.)

Commercially Prepared Slides

1. Obtain commercially prepared slides illustrating the various classes of fungi, their morphology, and reproduction. Carefully study each of these slides. Draw and label the indicated structures in part 3 of the report for exercise 60. These slides should supplement your observations of live fungi and give you a good picture of fungus morphology and the various forms of fungal reproduction.

Phycomycetes:
 Rhizopus (sporangia and zygotes)
 Saprolegnia (fruiting bodies, sex organs, sporangia)
Ascomycetes:
 Penicillium (sections, conidia)
 Aspergillus (conidia, cleistothecia)
 Morchella (sections with asci)
Basidiomycetes:
 Polyporus (sections with basidia)
 Lycoperdon (longitudinal section)
 Coprinus (pileus, medium longitudinal section, cross section)
 Puccinia (aecia, telia, uredia)

Dissection of Preserved Mushrooms

1. Examine a specimen of *Agaricus* or *Coprinus*. Note and draw its major anatomical features (cap, stipe, annulus, and mycelia at the base of the stipe).

2. Carefully dissect out a gill section, mount it in water, cover it with a coverslip, and gently crush it. Examine your specimen under the microscope. At the gill surface, you should be able to see the basidia, sterigmata, and basidiospores.

3. Make your drawings in part 4 of the report for exercise 60.

Rhizopus Morphology and Reproduction

1. The life cycle of *Rhizopus stolonifer* can be easily studied.

2. Inoculate a potato dextrose agar plate with the plus and minus culture strains provided.

3. A spot of each spore type should be placed at opposite sides of the plate, about 4 cm or more apart.

4. Within 4 to 7 days, the two strains will have grown together and a line of zygospores will be formed in the center of the plate. Carefully remove the cover

of the petri plate and observe the morphology of the zygomycete with a dissecting scope. You should be able to see the hyphae and sporangia. Zygospores can be studied by carefully teasing apart the mycelium in the center of the plate. Zygospores in various stages of development will be seen beneath the surface. You should also be able to observe gametangia as well. Portions of the thallus may also be removed and pressed under a coverslip for observation at higher magnifications.

HINTS AND PRECAUTIONS

(1) When preparing slides for fungal culture and observation, it is essential to heat the coverslip properly (step 4). If the coverslip is heated too little, it will not stick to the paraffin. However, be careful not to overheat the coverslip, or the paraffin will liquify and run over the slide surface.

(2) After the fungi are incubated for several days, desiccation of the agar and cultures can be avoided if the petri plates are taped for a tighter seal.

Laboratory Report 60

Name: _____

Date: _____

Lab Section: _____

Fungi II: Phycomycetes, Ascomycetes, and Basidiomycetes

1. Mold plate colonies.

 Aspergillus

 a. extent of growth _____

 b. pigmentation _____

 c. aerial hyphae present _____

 Penicillium

 a. extent of growth _____

 b. pigmentation _____

 c. aerial hyphae present _____

2. Mold slide cultures.

Aspergillus

Penicillium

3. Commercially prepared slides.

Rhizopus

Saprolegnia

Penicillium

Aspergillus

Morchella

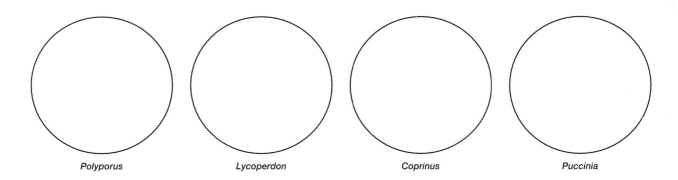

| Polyporus | Lycoperdon | Coprinus | Puccinia |

4. Dissection of preserved mushrooms.
 a. Drawing of the major anatomical features of a mushroom.

 b. Drawing of a mushroom gill section.

5. In summary, what is the major purpose of this experiment?

Review Questions

1. What is the difference between vegetative and aerial mycelia?

2. Can bacteriological media be used for the cultivation of molds? Explain your answer.

3. In the common bread-mold life cycle, what do the plus and minus signs mean?

4. Why aren't molds streaked for isolation like previous exercises have done for bacteria?

5. What is the difference between reproductive and rhizoidal hyphae?

6. Are *Rhizopus* hyphae coenocytic or septate?

7. How would you describe the fruiting bodies of *Aspergillus? Penicillium?*

Microbial Genetics and Genomics

The Double Helix is indeed a remarkable molecule. Modern man is perhaps 50,000 years old, civilization existed for scarcely 10,000 years, and the United States for only just over 200 years; but DNA and RNA have been around for at least several billion years. All that time the double helix has been there, and active, and yet we were the first creatures on earth to become aware of its existence.

Francis Crick (1916–)
What Mad Pursuit

All of the characteristics of microorganisms (e.g., their chemical composition, growth patterns, metabolic pathways, and pathogenicity) are inherited. These characteristics are transmitted from parent cell to daughter cells through genes. The science of **genetics** is concerned with the study of what genes are, how they carry genetic information, how they replicate, and how their information is expressed within a microorganism to determine its particular characteristics.

Bacteria are especially useful in the study of genetics because large numbers of them can usually be cultured in a short period, and their relatively simple genetic composition facilitates studying the structure and function of specific genes.

Genetic engineering involves the joining together of DNA (genes) from different sources and the introduction of the joined material into a suitable cell in which it will be both replicated and expressed. Several steps are usually part of the process: (1) provision of a suitable carrier such as a **plasmid;** (2) the formation of composite DNA molecules, or **chimeras;** (3) introduction of the composite DNA into a functional **recipient cell;** and (4) an appropriate **selection method** to isolate the desired recombinant bacterium. The first three exercises in this part of the manual use some of these genetic techniques.

DNA is the genetic material that gives a microorganism its inherited characteristics. The basic structure of DNA is identical in every organism. One experiment in this part of the manual is thus designed to understand the principles and steps necessary to isolate DNA from yeast.

Joshua Lederberg (1925–)

Joshua Lederberg, an American biochemist, was cowinner with George Beadle and Edward Tatum of the Nobel Prize for Physiology or Medicine in 1958 for discoveries concerning genetic recombination and the organization of genetic material in bacteria. In 1946, Joshua Lederberg along with E. L. Tatum published a paper in *Nature* (158:558) demonstrating the first sexual process in a bacterium (*E. coli*).

> Analysis of mixed cultures of nutritional mutants has revealed the presence of new types which strongly suggests the occurrence of a sexual process in the bacterium, *Escherichia coli*. . . . These experiments imply the occurrence of a sexual process in the bacterium. . . . As a result, genetic recombination in *E. coli* appears to imply a sexual mechanism.

Lederberg went on to show that bacteria are able to transmit genetic elements in several ways. He also showed his versatility by not limiting himself to bacterial genetics. Lederberg contributed to understanding penicillin's mode of action and extended his activities to space microbiology and the use of computers in biology.

Genomics is the study of the molecular organization of genomes, their information content, and the gene products they encode. It is a broad discipline, which may be divided into at least three general areas. **Structural genomics** is the study of the physical nature of genomes. Its primary goal is to determine and analyze the DNA sequence of the genome. **Functional genomics** is concerned with the way in which the genome functions. That is, it examines the transcripts produced by the genome and the array of proteins they encode. The third area of study is **comparative genomics,** which compares genomes from different organisms to look for significant differences and similarities. This helps identify important, conserved portions of the genome, and discern patterns in function and regulation. The data also provide much information about microbial evolution, particularly with respect to phenomena such as horizontal gene transfer. The last two experiments in this lab manual were designed to help illustrate the usefulness of genomics in the microbial world.

After completing one or more of the exercises in part eleven, you will, at the minimum, be able to demonstrate an increased understanding of microbial genetics, including mutation and genomics. This will meet the American Society for Microbiology Core Curriculum theme number 3: microbial genetics, including mutations (*see pp. vi–viii*).

Bacterial Mutation

Exercise 61

SAFETY CONSIDERATIONS

Staphylococcus aureus is a pathogen. Follow aseptic procedures. Alcohol is extremely flammable. Keep the beaker of ethyl alcohol away from the Bunsen burner. Do not put a flaming rod back into the alcohol. No mouth pipetting. Keep all culture tubes upright in a test-tube rack or can.

MATERIALS PER GROUP OF STUDENTS

sterile petri plate
glass rod or wooden dowel stick (1.6 mm in
 diameter)
2 tryptic soy agar tubes containing 6 to 7 ml of
 agar
1 tryptic soy agar tube containing 0.05 mg of
 streptomycin in 6 to 7 ml of agar
1-ml pipette with pipettor
24-hour tryptic soy broth culture of
 Staphylococcus aureus (ATCC 25903);
 Serratia marcescens (ATCC 13880) can also be
 used
95% ethyl alcohol
spreading rod
3 tryptic soy broth tubes
2 tryptic soy broth tubes containing 0.01 mg of
 streptomycin in 6 to 7 ml of broth
2 tryptic soy broth tubes containing 0.05 mg of
 streptomycin in 6 to 7 ml of broth

Learning Objectives

Each student should be able to

1. Define a bacterial mutation
2. Isolate a streptomycin-resistant mutant

SUGGESTED READING IN TEXTBOOK

1. Mutations and Their Chemical Basis, section 11.6.
2. Streptomycin, section 35.6.
3. Drug Resistance, section 35.7.

Pronunciation Guide

Staphylococcus aureus (staf-il-oh-KOK-kus ORE-ee-us)

Why Is the Above Bacterium Used in This Exercise?

The student will isolate a streptomycin-resistant mutant bacterium in this experiment. For this purpose, the authors have chosen the common bacterium *Staphylococcus aureus*. Some mutant strains of *S. aureus* are resistant to high concentrations of streptomycin. These strains are present in a ratio of only about one streptomycin-resistant cell to several million sensitive cells. However, these rare *S. aureus* mutants can be readily detected by plating a large number of cells on a medium containing a high concentration of streptomycin. Because the bacteria have not been exposed to any mutagenic agent, the isolated mutants are spontaneous mutants.

PRINCIPLES

A **mutation** is the result of a stable, heritable change in the nucleotide sequence of DNA. By manipulating the chemical or physical environment of a bacterium, one can increase the frequency of mutations. If a mutant bacterium, by virtue of the change that it has undergone, is suited to the environment in which it is formed, its growth can be greater than the parent bacterium. In such a case, the mutant will quickly become the dominant bacterium in a culture.

Mutations occur in one of two ways. (1) Spontaneous mutations arise occasionally in all bacteria and develop in the absence of any added agent. (2) Induced mutations are the result of the bacterium's exposure to a **mutagen,** which is a physical or chemical agent.

Spontaneous mutations resistant to antibiotics such as streptomycin are easily detected because they grow

389

Figure 61.1 Preparation of a Gradient-Agar Plate. (a) Place a 1.6-mm glass rod or dowel under a petri plate. Add tryptic soy agar and allow to harden. (b) Remove the rod and add streptomycin agar.

in the presence of antibiotic concentrations that inhibit the growth of normal bacteria. This introductory exercise in microbial genetics employs the gradient-agar plate method to isolate and select streptomycin-resistant mutant *Staphylococcus aureus* strains.

Procedure

First Period

1. Prepare a gradient-agar plate as follows:
 a. Place a sterile petri plate on an angle by placing either a glass rod or dowel stick under one side (figure 61.1*a*).
 b. Aseptically pour a tube of melted tryptic soy agar into the petri plate, immediately cover the plate and allow the agar to harden. Label the plate with your name and date.
 c. After the tryptic soy agar has hardened, remove the glass rod or dowel and place the plate flat on the table.
 d. Aseptically pour a tube of tryptic soy agar containing 0.05 mg of streptomycin on the surface of the gradient-agar plate (figure 61.1*b*). Cover the plate and allow the new agar to harden.

2. Pipette 0.3 ml of the *Staphylococcus aureus* culture onto the agar surface. Using the sterile glass spreader (sterilize by dipping into 95% alcohol, flaming, and cooling on the sterile surface of the gradient plate; *see figure 15.2*), spread the culture over the surface of the plate.

3. Allow the surface of the agar plate to dry with the lid on prior to incubating. Incubate inverted at 37°C for 24 to 48 hours.

Second Period

1. Observe the gradient-agar plate for the development of resistant *S. aureus* colonies in the areas of higher streptomycin concentration.

2. Record your results in the report for exercise 61.

3. Using the initial tryptic soy broth culture, inoculate one control tube of tryptic soy broth, one tube containing 0.01, and one tube containing 0.05 mg of streptomycin. (These represent the original *S. aureus* bacteria.)

4. In like manner, with a sterile inoculating loop, make a bacterial suspension in tryptic soy broth from colonies growing in the region of highest antibiotic concentration on the gradient-agar plate. (These represent the resistant or mutant *S. aureus.*)

5. From this broth suspension, inoculate a loopful of bacteria into a tube of tryptic soy broth containing 0.01 mg of streptomycin and one containing 0.05 mg of streptomycin. Do the same for a tube of regular tryptic soy broth (this serves as your control).

6. Incubate all tubes at 37°C for 24 to 48 hours.

Third Period

1. Observe the four streptomycin tubes for growth and compare them with the two control tubes.

2. Record your results in the report for exercise 61.

HINTS AND PRECAUTIONS

It is extremely important to be certain that the regular tryptic soy agar is completely cool and hard before pouring the streptomycin agar over it.

Bacterial Mutation

1. Draw the number of *Staphylococcus aureus* colonies growing on the different areas of the gradient-agar plate.

2. Record the relative growth of *S. aureus* in the three tryptic soy broth tubes. Use a scale from 0 for no growth to + + + + for heavy growth.

Inoculum	Streptomycin		
	Control	0.01 mg	0.05 mg
Original *S. aureus*	_____	_____	_____
Resistant *S. aureus*	_____	_____	_____

3. In summary, what is the major purpose of this experiment?

Review Questions

1. How would you define a bacterial mutation?

2. How would you describe a streptomycin-dependent mutant bacterium?

3. How can bacterial mutations occur?

4. What is a mutagen? Give several examples.

5. What methods could be used to increase the frequency of the appearance of mutant *Staphylococcus aureus?*

6. Could a gradient-agar plate be used for any antibiotic? Explain your answer.

Exercise *62*

Bacterial Transformation

SAFETY CONSIDERATIONS

Chloroform is very volatile and flammable. Keep away from fire or flames. Inhalation of large doses of chloroform may cause hypotension, respiratory and myocardial depression, and death; it is also carcinogenic. No mouth pipetting. Make sure the centrifuge tubes are balanced. Keep all culture tubes upright in a test-tube rack or can.

MATERIALS PER GROUP OF STUDENTS

24-hour nutrient broth culture of *Bacillus subtilis* strain SB 100 (ATCC 29056)

24-hour nutrient broth culture of *Bacillus subtilis* wild type (ATCC 6051)

lysozyme solution (2 mg/ml)

1 glucose—minimal salts agar plate containing tyrosine and histidine

1 glucose—minimal salts agar plate containing tyrosine and tryptophan

2 glucose—minimal salts agar plates containing tyrosine, tryptophan, and histidine

1 glucose—minimal salts agar plate containing histidine

sterile test tubes

sterile phosphate buffer (0.1 M, pH 6.2)

chloroform

sterile centrifuge tubes

sterile toothpicks

centrifuge

ice-water bath

sterile distilled water

2 sterile 250-ml flasks

50-ml Difco Antibiotic Assay Medium 3 in a 250-ml flask

15 ml of MM1 medium

15 ml of MM2 medium

5-ml pipettes with pipettor

1-ml pipettes with pipettor

thermometer
glass spreader
70% ethanol
Bunsen burner

Learning Objectives

Each student should be able to

1. Describe the process of transformation
2. Perform a DNA extraction procedure
3. Demonstrate the presence of competent bacterial cells in a bacterial culture

SUGGESTED READING IN TEXTBOOK

1. DNA transformation, section 13.5; (see also figure 13.6).

Pronunciation Guide

Bacillus subtilis (bah-SIL-lus SUB-til-us)

Why Is the Above Bacterium Used in This Exercise?

In this exercise, the student will learn how to form recombinant bacterial cells by transformation. To accomplish this objective, the authors have chosen a wild-type strain of *Bacillus subtilis* as the donor strain and the SB 100 *B. subtilis* strain as the recipient. As a result of multiple mutations, SB 100 *B. subtilis* is an auxotroph that requires the amino acids tyrosine, tryptophan, and histidine for growth. It has lost the ability to synthesize these amino acids. The wild-type strain of *B. subtilis* can synthesize these amino acids and grow on a glucose–minimal salts medium lacking them. By mixing DNA extracted from the wild type with viable cells of the SB 100 auxotrophic strain, transformants will be produced that now have the ability to grow in the absence of tryptophan or tyrosine.

393

PRINCIPLES

Transformation is the uptake by a recipient bacterium of a naked DNA molecule or a fragment from a culture medium and the incorporation of this molecule (or fragment) into the recipient chromosome in a heritable form. In natural transformation, the DNA comes from a donor bacterium. The process is random, and any portion of the genome may be transferred between bacteria.

When bacteria lyse, they release considerable amounts of DNA into the surrounding environment. These fragments may be relatively large and contain several genes. If a fragment contacts a **competent bacterium,** one able to take up DNA and be transformed, it can be bound by the cell and taken inside. Transformation may be an important route of genetic exchange in nature.

In this introductory transformation exercise, a wild-type strain of *Bacillus subtilis* will be used as the donor strain and *B. subtilis* strain SB 100 as the recipient strain. This recipient strain is an **auxotroph** (a mutated bacterium that lacks the ability to synthesize an essential nutrient and must obtain it from its surroundings) that requires the amino acids tyrosine, tryptophan, and histidine to grow because it has lost the ability to synthesize these amino acids. The wild-type strain, however, can synthesize these amino acids and grow on a glucose–minimal salts medium lacking these amino acids. By mixing DNA extracted from the wild strain with viable bacteria of the auxotrophic strain, you will (1) look for transformants that have gained the ability to grow in the absence of tryptophan, and (2) look for those that have acquired the ability to synthesize tryptophan and/or histidine and no longer require that the particular amino acid be supplied to the medium.

Procedure

First Period
Preparation of Donor DNA

1. Aseptically transfer 5 ml of the wild-type *Bacillus subtilis* to a centrifuge tube and centrifuge at 5,000 g for 10 minutes.
2. After centrifugation, using a new pipette, discard the supernatant and resuspend the bacteria in 5 ml of sterile phosphate buffer.
3. Add 0.05 ml of the lysozyme solution and incubate (with gentle agitation) for 30 minutes at room temperature.
4. Cool the centrifuge tube to 5°C in an ice-water bath.
5. Add 5 drops of chloroform to lyse the bacteria whose walls have been weakened by lysozyme digestion. Agitate the tube to mix the chloroform and medium. This is the source of transforming DNA for the second period of this procedure. This preparation can be stored at 5°C for several days.

Preparation of Competent Bacteria

1. Aseptically inoculate a culture of *B. subtilis* SB 100 into a 250-ml flask containing 50 ml of Difco Antibiotic Assay Medium 3.
2. Incubate for 24 hours at 37°C. The culture should be gently agitated.

Second Period
Preparation of Competent Bacteria (continued)

1. Harvest the bacteria by centrifugation at 5,000 g for 10 minutes and resuspend them in 15 ml of MM1 medium.
2. Add 12.5 ml of the resuspended culture to a sterile 250-ml culture flask and incubate with shaking for 5 hours.
3. Harvest these bacteria by centrifugation and dilute them fivefold by resuspending in 15 ml of MM2 (transformation medium). These are the competent bacteria that will be used in the next part of this experiment.

Transformation of Competent Bacteria

1. Aseptically transfer 0.9 ml of the competent bacteria (*B. subtilis* SB 100) into a sterile test tube. Carefully, without adding any chloroform, add 0.1 ml of the DNA solution (the donor DNA) prepared from the wild-type *B. subtilis*. The white DNA layer is at the chloroform-water interface.
2. Incubate the test tube at 37°C for 30 minutes with gentle shaking.
3. Pipette 0.1 ml of the transformation mixture onto a glucose–minimal salts agar plate containing tryptophan, tyrosine, and histidine. Spread the bacteria with a flame-sterilized glass spreader. Label the plate with your name, date, and amino acid mixture.
4. Dilute the transformation mixture 1/10 using sterile distilled water. Pipette 0.1 ml of this dilution onto a glucose–minimal salts agar plate containing a mixture of tryptophan, tyrosine, and histidine. Spread out the bacteria as before. Label with your name, date, and amino acid mixture.
5. Incubate the plates at 37°C for 48 to 72 hours.
6. Using sterile toothpicks, transfer 59 well-isolated colonies from a transforming petri plate to four glucose–minimal salts plates containing (1) tyrosine and histidine, (2) tyrosine and tryptophan, (3) tyrosine, tryptophan, and histidine, and (4) histidine alone. Use the grid pattern given in

Figure 62.1 Grid Pattern for a 90-mm Petri Plate.

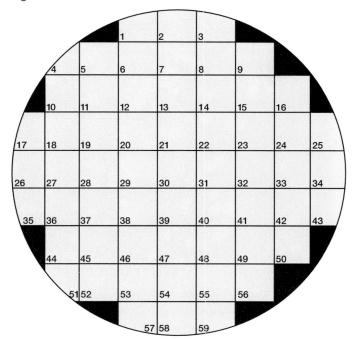

figure 62.1 for a 90-mm petri plate as a template for each of the above three plates. Transfer each well-isolated colony by stabbing the colony with the toothpick and inoculating one grid in each of the above three plates (e.g., colony 1 will be inoculated into grid area 1 of all three plates, then colony 2 into grid area 2, and so forth until all 59 grids have been streaked). Be careful when doing this transfer not to go into adjacent grids. Keep the plates covered as much as possible during these transfers.

7. The tyrosine/tryptophan/histidine plate serves as the control plate. If growth does not occur on this plate, data from the other plates should be discarded.

8. Incubate the four plates for 48 hours at 37°C.

9. Record the grid areas and indicate in the report for exercise 62 whether a particular bacterial transformant grows (+) or does not grow (−) on each kind of medium.

Name: _____

Date: _____

Lab Section: _____

Bacterial Transformation

Grid Number	Growth in Minimal Salts–Glucose Agar Plate Containing			
	Tyrosine/Histidine	Tyrosine/Tryptophan	Tyrosine/Tryptophan/ Histidine	Histidine
——	——————	——————	——————	——————
——	——————	——————	——————	——————
——	——————	——————	——————	——————
——	——————	——————	——————	——————
——	——————	——————	——————	——————
——	——————	——————	——————	——————
——	——————	——————	——————	——————
——	——————	——————	——————	——————
——	——————	——————	——————	——————
——	——————	——————	——————	——————
——	——————	——————	——————	——————
——	——————	——————	——————	——————
——	——————	——————	——————	——————
——	——————	——————	——————	——————
——	——————	——————	——————	——————
——	——————	——————	——————	——————
——	——————	——————	——————	——————
——	——————	——————	——————	——————
——	——————	——————	——————	——————
——	——————	——————	——————	——————
——	——————	——————	——————	——————
——	——————	——————	——————	——————
——	——————	——————	——————	——————
——	——————	——————	——————	——————
——	——————	——————	——————	——————
——	——————	——————	——————	——————
——	——————	——————	——————	——————
——	——————	——————	——————	——————
——	——————	——————	——————	——————
——	——————	——————	——————	——————
——	——————	——————	——————	——————
——	——————	——————	——————	——————
——	——————	——————	——————	——————
——	——————	——————	——————	——————
——	——————	——————	——————	——————

Number of transformants growing on the control plate (tyrosine/tryptophan/histidine): _____

Number of transformants growing on the tyrosine/tryptophan plate: _____

Number of transformants growing on the tyrosine/histidine plate: _____

Number of transformants growing on the histidine plate: _____

The genes transferred by transformation: _____

Review Questions

1. How would you define bacterial transformation?

2. What is a competent bacterium?

3. What is an auxotroph?

4. Could phenylalanine replace the amino acid tyrosine in this experiment? Explain your answer.

5. What are several factors upon which the successful development of a transformation system depends?

6. What are some important uses of transformation?

Bacterial Conjugation: The Transfer of Antibiotic-Resistant Plasmids

SAFETY CONSIDERATIONS

In this experiment, students will be working with antibiotic-resistant strains of *Escherichia coli*. It is essential that the bacterium and its plasmid **not** be spread around the laboratory. It is also imperative to ensure that one is working with a **pure culture.** All manipulations must, therefore, be carried out in such a way as to avoid contamination of the culture. The following standard **aseptic procedures** must be followed:

1. Sterilize all loops before use; cool slightly before inoculum pick-up to avoid killing the bacteria on contact.
2. Resterilize the loop after use by gradual heating. This is recommended to minimize spattering and the formation of aerosols.
3. Flame the mouths of tubes and flasks momentarily after removing the cap or plug, and again after inoculation.
4. Never leave the cap or plug on the bench during inoculation. Hold it in one hand and replace it in the tube or flask after flaming the mouth when the inoculum has been transferred.
5. Cotton wool plugs removed from tubes and flasks should be held by their projection surfaces only. Handling of the portion of the plug that is inserted into the flask or tube will result in contamination.
6. Open tubes should be held in an inclined position to minimize contamination.
7. Lids of petri plates should be held over the plate while inoculating and should be replaced as soon as possible. Never place the lids on the bench surface.
8. When finished, all cultures and tubes will be autoclaved at 121°C for 15 minutes. After autoclaving, the tube contents will be poured down the sink. The petri plates are disposable and the plates plus agar contents will be autoclaved and discarded.
9. In the event that a culture is accidentally spilled during handling, it is imperative to flood the immediate area with Lysol solution. Allow 3 hours for disinfection and then wash thoroughly with tap water.

MATERIALS PER GROUP OF STUDENTS

overnight tryptic soy broth culture of *Escherichia coli* recipient strain J-53R (Wards number 85W1686)

overnight tryptic soy broth culture of *Escherichia coli* donor strain HT-99 (Wards number 85W1684). These strains may be obtained from Wards, Natural Science Establishment, Rochester, NY by special order or calling Wards at 1-800-962-2660.

5 ml sterile nutrient broth in a screw-cap tube

glass spreader in 80% ethanol

2 sterile 1-ml pipettes with pipettors

3 sterile Pasteur pipettes with pipetting bulbs

1 tryptic agar plate containing 25 μg chloramphenicol per ml

1 tryptic agar plate containing 100 μg rifampin per ml

1 tryptic agar plate containing chloramphenicol (25 μg/ml) and rifampin (100 μg/ml)

Bunsen burner

petri plate cans

37°C incubator

wax pencil

Learning Objectives

Each student should be able to

1. Perform the transfer of a conjugative plasmid coding for chloramphenicol resistance from a donor to a suitable *E. coli* recipient
2. Describe the method in which chimeric plasmids, formed by in vitro techniques, may be introduced into a suitable recipient and may be both replicated and expressed with the recipient bacterium

SUGGESTED READING IN TEXTBOOK

1. Chloramphenicol, section 35.6; see also tables 35.1, 35.2, and 35.4.
2. The Origin and Transmission of Drug Resistance, section 35.7.
3. Bacterial Plasmids, section 13.2.
4. Bacterial Conjugation, section 13.4.
5. Transduction, section 13.6.
6. Cloning Vectors, section 14.5.
7. Expression of Foreign Genes in Bacteria, section 14.7.

Pronunciation Guide

Escherichia coli (esh-er-I-ke-a KOH-lee)

Why Is the Above Bacterium Used in This Exercise?

The student will perform a bacterial conjugation experiment in this exercise. The authors have chosen two strains of *Escherichia coli* for this purpose. The HT-99 donor strain of *E. coli* is resistant to chloramphenicol. The J-53R recipient strain of *E. coli* cannot synthesize methionine and proline and is rifampin resistant. Recombinants of these two strains will have different combinations of either the donor or recipient strains' characteristics and can be readily detected by using selective plating media.

PRINCIPLES

Plasmids that can affect the conjugation between individual bacteria are termed **conjugative plasmids.** These plasmids carry genes that code for the ability to transfer themselves from the **donor** host bacterium to a suitable **recipient** bacterium by conjugation. **Conjugation** describes the "mating" between two bacteria, when (part of) the chromosome is transferred from one to the other. In order for conjugation to occur, there must be direct physical contact between bacteria. The actual "mating" between donor and recipient is dependent on the production of special filaments called sex pili by the donor bacteria. Since the genes for sex pili formation are an integral part of the conjugative plasmid DNA, the plasmids themselves determine their own transmissibility between bacteria.

Plasmids are small, circular, double-stranded DNA molecules that can act as extrachromosomal elements (i.e., they can carry genetic information and exist apart from the chromosome). These small DNA circles can reproduce independently of the bacterial genome, and many can be incorporated into it and reproduce when the genome does. The plasmid used in this experiment carries a gene for chloramphenicol resistance and can be transferred during bacterial conjugation.

This experiment employs a broth procedure for in vitro mating. Since conjugation requires cell-to-cell contact, it can be described as a collision-dependent process. In the experimental procedure, sufficient time must be provided for the mating cells to make contact.

In this experiment, students will work in groups using two *E. coli* strains: *E. coli* J-53R (recipient) and *E. coli* HT-99 (donor). J-53R is an **auxotrophic** strain (a mutant that has a growth factor requirement) for the amino acids methionine and proline. It also carries a chromosomal gene for rifampin resistance. HT-99 is a wild-type isolate that is resistant to chloramphenicol, and the resistance determinant is located on a conjugative plasmid. By this simple experiment, the basic goal of genetic engineering research (i.e., the creation, replication, and expression of hybrid strains) is illustrated.

Procedure

First Period

1. Prepare the mating mixture by aseptically transferring 0.1 ml of an overnight tryptic soy broth culture of *E. coli* HT-99 (donor) and 0.9 ml of an overnight tryptic soy broth culture of *E. coli* J-53R (recipient) into 5 ml of sterile tryptic soy broth (figure 63.1).
2. Mix and incubate at 37°C for 3 hours or overnight.
3. Confirm the resistance status of the donor and recipient by culturing on antibiotic-containing petri plates as follows:
 a. Using the wax pencil, divide the chloramphenicol and the rifampin plates in half by drawing a line on the bottom surface of the plate.
 b. Using the same sterile Pasteur pipette, transfer a drop of donor culture to one half of the chloramphenicol plate and one drop to one half of the rifampin plate. Carefully spread over one half of the plate using the glass spreader sterilized by flaming (*see exercise 15 and figure 15.2*). Sterilize the glass spreader between spreadings. Using another sterile Pasteur pipette, transfer a drop of recipient culture to the other half of the chloramphenicol and rifampin plates. Spread as above. Label the plates with your name and date. Allow them to dry thoroughly and incubate inverted overnight at 37°C in a petri plate can.

Figure 63.1 Plasmid Transfer Procedure.

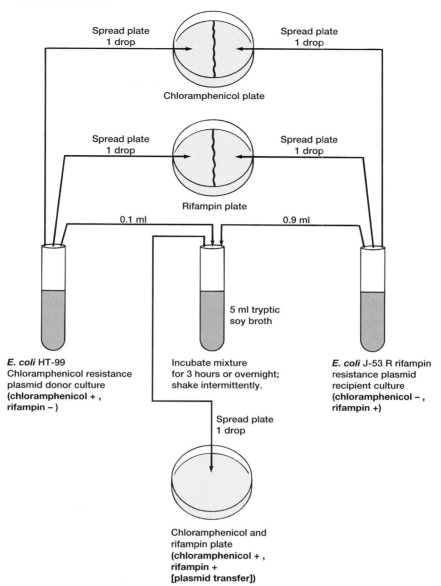

Spread plate
1 drop

Spread plate
1 drop

Chloramphenicol plate

Spread plate
1 drop

Spread plate
1 drop

Rifampin plate

0.1 ml 0.9 ml

5 ml tryptic
soy broth

E. coli HT-99
Chloramphenicol resistance
plasmid donor culture
(**chloramphenicol + ,
rifampin –)**

Incubate mixture
for 3 hours or overnight;
shake intermittently.

E. coli J-53 R rifampin
resistance plasmid
recipient culture
(**chloramphenicol – ,
rifampin +)**

Spread plate
1 drop

Chloramphenicol and
rifampin plate
(**chloramphenicol + ,
rifampin +
[plasmid transfer])**

4. Remove the mating mixture from the incubator after the desired period of mating and transfer 1 drop to the center of the petri plate containing both chloramphenicol and rifampin. Spread as described above. Label the plate with your name and date. Incubate inverted at 37°C overnight in a petri plate can.

Second Period

1. Examine all plates for growth after overnight incubation, and record the results in the report for exercise 63.

2. Since the donor (*E. coli* HT-99) is resistant to chloramphenicol and sensitive to rifampin, it will be able to grow in the presence of 25 μg/ml of chloramphenicol but unable to grow in the presence of

100 μg/ml of rifampin. An examination of the plate from step 3b in the first period procedure should confirm this result.

3. The recipient (*E. coli* J-53R) is sensitive to chloramphenicol and resistant to rifampin; thus, it will be able to grow on the rifampin plate but unable to grow on the chloramphenicol plate (plate from step 3b).

4. Neither the donor nor the recipient can grow in the presence of chloramphenicol plus rifampin because both strains are inhibited by one or the other of the antibiotics. Colonies appearing on the double antibiotic plate prepared in step 4 during the first period must, therefore, result from growth of

exconjugant bacteria arising by conjugal transfer of the chloramphenicol resistance plasmid from a donor to a recipient bacterium. Such bacteria are the result of in vivo natural genetic engineering and are genetically new bacteria.

5. Count the number of doubly resistant bacterial colonies that have been formed. If desired, each team may use a different time period of mating (e.g., 1 to 24 hours) and observe the effect of mating time on the number of exconjugants appearing on the plates.

Bacterial Conjugation: The Transfer of Antibiotic-Resistant Plasmids

1. Sketch the growth on the following plates:

Chloramphenicol Rifampin Chloramphenicol + rifampin

2. What does this growth or lack of it indicate on each side of the following plates?
 a. chloramphenicol (HT-99)

 chloramphenicol (J-53R)

 b. rifampin (HT-99)

 rifampin (J-53R)

3. How many doubly resistant colonies were formed? _____

 Assuming that a drop is approximately 0.05 ml, about how many plasmid transfers took place in the broth culture? Should this number increase with incubation time? Why or why not?

4. Describe how *E. coli* is able to grow on the plate containing both chloramphenicol and rifampin.

5. In summary, what is the major purpose of this experiment?

Review Questions

1. Why is the aseptic technique so important in this experiment?

2. Why is *E. coli* HT-99 able to grow in the presence of chloramphenicol but not in the presence of rifampin? Why is *E. coli* J-53R able to grow on the rifampin plate but unable to grow on the chloramphenicol plate?

3. Would you expect that the new doubly antibiotic-resistant strain could grow well if it were released into the environment? Explain.

4. Is it likely that rifampin resistance will be frequently transferred from the J-53R strain to HT-99? Explain your reasoning.

5. How do plasmids function?

6. Do bacteria reproduce sexually? Explain.

7. How would you define genetic engineering?

Exercise 64

Isolation of Genomic DNA from *Saccharomyces cerevisiae*

MATERIALS PER GROUP OF STUDENTS

glass stirring rods
baker's yeast, dehydrated or *Saccharomyces cerevisiae* YPH499 (ATCC #76625)
95% ethanol
plastic 10 ml transfer pipettes
250-ml beaker
50–60° tap water
temperature controlled hot plate
thermometer
liquid kitchen dishwashing detergent (concentrated)
Adolph's meat tenderizer
microbalance

Learning Objectives

Each student should be able to

1. Understand the basics of DNA structure and function
2. Understand the principles and steps needed to isolate genomic DNA from the yeast, *Saccharomyces cerevisiae*

SUGGESTED READING IN TEXTBOOK

1. All of chapter 15 on Microbial Genomics.
2. DNA as genetic material, section 11.1.
3. The genetic code, section 11.4.

Pronunciation Guide

Saccharomyces cerevisiae (sak-ah-ro-MI-seez SER-ah-VEES-ee-eye)

Why Is the Above Yeast Used in This Exercise?

Isolation and purification of DNA are often the first steps in starting genetic analysis. DNA is basically the same from yeast to bacteria to humans. However, DNA packaging varies greatly. The structure of each organism's cells and the application for which the extracted DNA is required must be taken into consideration when selecting an extraction procedure. Do you want plasmid DNA or genomic DNA? Are you isolating DNA from a bacterium or a multicellular organism? Do you want to use DNA for Southern blotting, the polymerase chain reaction, restriction enzyme analysis, cloning, or other types of analysis? How does this affect isolation and purification schemes? The simple procedure described in this experiment uses yeast genomic DNA; it omits purification steps, is quick, easy, and an inexpensive method that can be used in any microbiology laboratory with minimum equipment and time. The resulting DNA is visible and provides a source of student discussion for the DNA molecule, genetics, evolution, and biotechnology.

PRINCIPLES

DNA is the genetic material that gives an organism its inherited characteristics. The basic structure of DNA is the same in every organism. DNA is composed of two strands of nucleotides wound together in a double helix. Each nucleotide consists of a deoxyribose sugar, a phosphate group, and one of four possible nitrogenous bases: adenine pairing with thymine and cytosine pairing with guanine. The two strands are attached to each other by complementary base pairing. This base pairing involves hydrogen bonds between opposing adenine and thymine residues and opposing cytosine and guanine residues.

Figure 64.1 DNA Spooled onto a Glass Stirring Rod.

Spooling DNA onto a glass stirring rod (figure 64.1) is a common way to isolate DNA once it has been precipitated out of an aqueous solution using ethanol. One of the reasons the yield of yeast DNA is so large is that the size of the yeast genome averages 1.3×10^7 base pairs compared to bacteria that average 4.7×10^6 base pairs.

Procedure

1. Measure 100 ml of 50–60°C tap water into a 250-ml beaker.
2. Add one-half package of dry yeast to the tap water and stir until completely suspended.
3. Incubate for 5–15 minutes at 50–60°C. The warm water activates the yeast.
4. Add 5 ml of concentrated detergent; stir occasionally for 5 minutes while keeping the suspension warm. (The detergent lyses the yeast cell plasma membrane, releasing the contents of the cell, including the DNA into water solution.)
5. Add 3 grams of meat tenderizer and stir. (The meat tenderizer contains the proteolytic enzyme papain that digests the cellular proteins.)
6. Incubate at room temperature for 20 minutes.
7. Using a plastic 10-ml transfer pipette, carefully layer 10–15 ml of 95% ethanol on top of the suspension of lysed yeast cells by allowing the alcohol to run down the side of the 250-ml beaker. The alcohol should form a distinct layer on the top of the detergent and cell mixture. (The 95% ethanol dehydrates the DNA so it precipitates out of solution.)
8. Insert a glass rod through the 95% ethanol into the cell suspension. Gently stir and roll the rod between your fingers, in one direction, but do not mix the two layers.
9. Observe the glass rod. DNA will stick to the glass rod and appear as slimy white threads (figure 64.1). Note that in this simple isolation procedure, DNA will remain associated with some cellular protein.
10. Scrape the DNA off of the glass rod into a plastic weighing boat and determine its wet weight in μg.

HINTS AND PRECAUTIONS

Check the expiration date on the package of baker's yeast to make sure it has not expired.

Laboratory Report 64

Name: _____

Date: _____

Lab Section: _____

Isolation of Genomic DNA from Saccharomyces cerevisiae

1. Sketch the appearance of the DNA on the glass stirring rod.

2. How much DNA, in µg wet weight, did you isolate from the yeast?

3. In summary, what is the major purpose of this experiment?

Review Questions

1. Where is DNA located in a yeast cell?

2. What is the purpose of the household detergent in this isolation procedure?

3. What are some factors that can affect the amount of DNA isolated?

4. How is DNA organized to fit inside the yeast cell?

5. Describe the chemical structure of DNA.

Isolation and Purification of Genomic DNA from *Escherichia coli*

MATERIALS PER GROUP OF STUDENTS

10–12 hour (log phase) tryptic soy broth culture of *Escherichia coli* (ATCC 11229). Cultures can be maintained in the log phase by immersion in an ice-water bath.

0.5- to 10-μl micropipettes and sterile tips

10- to 100-μl micropipettes and sterile tips

100- to 1,000-μl micropipettes and sterile tips

microcentrifuge tubes

microcentrifuge

1 ml pipettes

Pasteur pipettes

disinfectant (70% ethanol or 3% hydrogen peroxide)

80°C modular dry bath incubator

37°C modular dry bath incubator

heating block modules

65°C water bath

20-mg/ml RNase A solution (DNase free; dissolve 1 mg/ml in 1 × TE [10 mM Tris, pH 7.4 + 1 mM EDTA, pH 8.0], boil 10–30 min, cool to room temperature, store at −20°C)

cell lysate solution (Tris-EDTA-SDS) from Puregene (Gentra Systems, PO Box 13159, Research Triangle Park, NC 27709; 1-800-866-3039)

protein precipitation solution (ammonium acetate) from Puregene

1.5 ml tubes

100% isopropanol

70% ethanol

100% absolute ethanol

vortex mixer

hydration solution (1 × Tris-EDTA hydration solution from Puregene)

Learning Objectives

Each student should be able to

1. Understand how to isolate genomic DNA from the bacterium, *Escherichia coli*
2. Explain why different protocols are needed for DNA isolation from various taxa

SUGGESTED READING IN TEXTBOOK

1. All of chapter 15 on Microbial Genomics.
2. DNA as genetic material, section 11.1.
3. The genetic code, section 11.4.

Pronunciation Guide

Escherichia coli (esh-er-I-ke-a KOH-lee)

Why Is the Above Bacterium Used in This Exercise?

Chemically, DNA is identical in structure from one organism to another. It is the sequence of base pairs that distinguishes DNA from various species. The composition of an organism, whether it is unicellular or multicellular, the exact makeup of the individual cells and tissues, and whether or not a cell wall is present, affects protocols used to isolate DNA. Also, the use for which DNA is required may dictate the specific method employed in its isolation. In exercise 64, a simple procedure was used to quickly isolate and visualize genomic DNA from yeast. Since *Escherichia coli* is one of the most studied procaryotic cells, it will be used in this exercise as a source of DNA. Once the DNA has been obtained and purified, it can

be saved and used in more complex exercises such as quantitation by gel electrophoresis, PCR amplification, digestion with restriction endonucleases, and membrane hybridizations such as Southern blots. To carry out the latter requires a lot of time and expensive equipment that most general microbiology laboratories do not possess. Therefore, this experiment stops with pure DNA from *E. coli* and if instructors wish, depending on time and equipment, they can supplement it with additional genomic protocols.

PRINCIPLES

Isolation and purification of DNA are essential procedures in molecular biology. Extraction and purification of genomic DNA from procaryotic organisms such as *E. coli* have facilitated the study of complex genomes. This has also led to the construction of genomic DNA libraries for many species (see exercise 66).

The genomic DNA isolation and purification technique that will be used in this exercise uses a four-step process (figure 65.1).

1. *E. coli* cells must be lysed to release genome DNA. The most common solution contains the detergent sodium dodecyl sulfate and sodium hydroxide.
2. Ribonuclease is added to the cell bacterial lysate to remove the RNA.
3. The bacterial proteins are removed next by the addition of ammonium or potassium acetate. The acetate precipitates the proteins but leaves the large genomic DNA in solution.
4. The *E. coli* genomic DNA is concentrated and desalted by using isopropanol precipitation followed by an ethanol wash.

Procedure

1. Obtain a 10- to 12-hour tryptic soy broth culture of *E. coli*.
2. Label a microcentrifuge tube. Using a pipette, transfer 1 ml of the *E. coli* culture into the microcentrifuge tube.
3. Pulse the tube at top speed in the microcentrifuge for 5 to 10 seconds to cause the *E. coli* bacteria to form a pellet.
4. Carefully remove most of the supernatant by pipetting it into a container of disinfectant.
5. Add 600 μl of cell lysate solution and gently pipette up and down to resuspend the bacterial pellet. The SDS in the lysis solution is a detergent that disrupts the plasma membrane of the *E. coli* cells.
6. Incubate at 80°C for 5 minutes.

7. Slowly cool the sample in the microcentrifuge tube at room temperature. The sample must reach room temperature before adding RNase since heat will denature the enzyme. Once the sample is at room temperature, add 3 μl of RNase solution to the bacterial cell lysate.
8. Mix by inverting the microcentrifuge tube 25 times and incubate at 37°C for 30 minutes.
9. Cool the sample to room temperature. Add 200 μl of protein precipitation solution and vortex **very gently** for 20 seconds. The ammonium acetate in the solution precipitates the protein leaving the genomic DNA in solution.
10. Microcentrifuge the sample for 3 minutes at 14,000 rpm to pellet the protein.
11. Pour the supernatant into a clean 1.5 ml tube leaving the protein pellet behind.
12. Add 600 μl of 100% isopropanol, cap the tube, and mix **very gently** by inverting the tube at least 50 times. Vigorous mixing will break the DNA into small fragments.
13. Centrifuge in the microcentrifuge at 14,000 rpm for 1 minute to pellet the DNA.
14. Pour off the supernatant and drain the liquid onto an absorbent towel. The pellet will appear off-white to yellow.
15. Add 600 μl of 70% ethanol and invert the tube several times to wash the pelleted DNA. Decant the 70% ethanol and add 600 μl of absolute alcohol. Invert the tube several times to wash the pelleted DNA. The alcohol washes remove any salts from the DNA.
16. Centrifuge at 14,000 rpm for 1 minute.
17. Pour off the supernatant very **slowly and carefully** watching the pellet so that it does not come out of the microcentrifuge tube.
18. Air dry the DNA pellet for at least 15 minutes. Air drying removes any residual alcohol, which will interfere with any subsequent analyses.
19. Add 100 μl of hydration solution to the pellet and place the microcentrifuge tube in a water bath at 65°C for 1 hour or incubate overnight at room temperature.
20. The DNA from *E. coli* is now ready to use. At this point, depending on the available equipment and time, your instructor may have you perform other experiments on the *E. coli* DNA. Examples include:

 • Amplification by the polymerase chain reaction
 • Quantitation by spectrophotometry
 • Quantitation by gel electrophoresis
 • Restriction fragment length polymorphism analysis
 • Southern blotting

Figure 65.1 Summary of Genomic DNA Isolation.

E.coli broth culture

Centrifuge to pellet bacteria

Add cell lysis solution

Add RNase

Incubation at 80°C for 5 minutes

Mix by inversion

Incubate at 37° C for 30 minutes

Add protein precipitation solution and vortex

Centrifuge at 14,000 rpm to pellet protein

Protein debris

Use supernatant

DNA

DNA

Isopropanol precipitation

Mix by inversion 50 times

Centrifuge to pellet DNA

DNA pellet

Wash DNA pellet once with 70% ethanol and once with absolute alcohol

Centrifuge at 14,000 rpm for 1 minute

Air dry pellet for 15 minutes

Resuspend in hydration solution

Incubate for 1 hour at 65°C in a water bath

DNA ready for use

Laboratory Report **65**

Isolation and Purification of Genomic DNA from Escherichia coli

For each of the steps or reagents in the isolation and purification of genomic DNA from *Escherichia coli*, indicate what each accomplishes.

Step/Reagent	Accomplishes
cell lysis solution	_____ _____
incubation	_____ _____
RNase	_____ _____
protein precipitation solution	_____ _____
centrifugation	_____ _____
isopropanol precipitation	_____ _____
alcohol washes	_____ _____
hydration solution	_____ _____

Review Questions

1. Using your knowledge of the cell structure of various procaryotic organisms, hypothesize why specific conditions are used in the cell lysis step.

2. You have just isolated and identified a new bacterium. Design a protocol for isolating its genomic DNA. Explain your rationale.

3. How might you optimize the purity of DNA you extracted from your new organism?

4. Why must you handle DNA gently in this extraction procedure?

5. In summary, what is the major purpose of this experiment?

Identifying Archaea and Bacteria Using the Internet and Computer-Assisted Gene Analysis

MATERIALS PER GROUP OF STUDENTS

 computer with Internet access
 Web browser (such as Netscape or Internet
 Explorer)
 printer

Learning Objectives

Each student should be able to:

1. Understand the importance of bioinformatics in microbiology
2. Understand the significance of the 16S rRNA sequence in the identification of archaea and bacteria
3. Appreciate the importance of identifying archaea and bacteria using the Ribosomal Database Project as well as the availability of other gateway websites containing links to sequence databases and computational centers for sequence analysis

SUGGESTED READING IN TEXTBOOK

1. Chapter 15 on Microbial Genomics.
2. Mapping and Sequencing the Genome, sections 13.7 and 15.3.
3. Bioinformatics, section 15.4.
4. Chapter 19 on Microbial Taxonomy.
5. Chapter 20 on the *Archaea.*
6. Chapters 21–24 on the *Bacteria.*

PRINCIPLES

Bioinformatics is the branch of science devoted to computational information management systems used to collect, store, analyze, and disseminate biological information. It was created with the union of information technology and gene-sequencing projects. Databases for nucleic acids and proteins and the analytical tools needed for management are available to students and scientists on the World Wide Web at no cost.

The classical way for identifying and classifying archaea and bacteria involve morphology and staining, biochemical activities, and rapid multitext systems as presented in parts two, four, and five of this manual. Today, in the new millennium, bioinformatics and the ability to sequence the DNA of organisms has offered new ways of identification. The closer organisms are related to each other, the more similar their **DNA sequences.** These sequences represent the order in which nucleotide bases are found in a piece of DNA. Certain genes common to all organisms can be sequenced and compared. The most useful ones involve the DNA coding for the part of the ribosome called 16S ribosome subunit.

Ribosomes are involved in protein synthesis. Ribosomes are composed of two subunits, each of which consists of protein and a type of RNA called ribosomal RNA (rRNA). Procaryotic ribosomes are made up of a small 30S subunit, consisting of a 16S rRNA particle and 21 polypeptide chains, and a large 50S subunit, which together make up the complete 70S ribosome. Since the 16S rRNA DNA segment is found in all organisms with the same function, ribosomal RNA gene sequences can easily be compared.

Ribosomal RNA gene sequencing is currently being used to study the diversity of procaryotes, as well as all

415

other organisms, and the phylogenetic relationships among them. There are several advantages to using rRNA. First, all cells contain ribosomes and ribosomal RNA. Two closely related organisms will have fewer different bases in their rRNA than two organisms that are more distantly related. A second advantage is that RNA genes have undergone few changes over time and are highly conserved. If the sequences were too different, they could not be compared. Third, when doing rRNA sequencing, cells do not have to be cultured in the laboratory.

This approach to identification of organisms has yielded exciting results. For example, some specific base sequences are always found in particular groups of organisms. These are called **signature sequences.** Signature sequences are DNA sequences about 5–10 bases long found at a specific location in the 16S rRNA that are unique to the *Archaea, Bacteria, Eucarya,* and many major groups within the procaryotes. New relationships between organisms can be determined by comparing base sequences by means of computer programs. The more the sequences diverge, the more the organisms have evolved from one another.

In bioinformatics, these gene sequences often can be used to identify specific organisms. The sequences of more than 16,000 organisms are in public databases. When a new sequence is submitted to a database management organization, in seconds the organization will respond with the most likely identification of the species containing the sequence.

This exercise is designed to enable you to send the DNA sequence of the 16S rRNA of several bacteria and one archaeon to the **Ribosomal Database Project** (*rdp.cme.msu.edu/html/*) at Michigan State University, the Center of Microbial Ecology, to determine the identification of these unknown organisms. This project is sponsored by the National Science Foundation, the U.S. Department of Energy, and Michigan State University. It provides ribosome-related data services to the scientific community, including online data analysis, rRNA derived phylogenetic trees, and aligned and annotated rRNA sequences. Other important gateway websites containing links to sequence databases and computational centers for sequence analysis include the following:

- The Institute for Genomic Research (*http://www. tigr.org*)
- Genomics: A Global Resource (*http://genomics. phrma.org*)

- European Bioinformatics Institute (*http://www.ebi. ac.uk*)
- GeneBank (*http://www.ncbi.nlm.nih.gov*) is a sequence of databases maintained by the National Center of Biotechnology Information, the National Library of Medicine, and the National Institutes of Health.

Procedure

1. Open Netscape or Microsoft Explorer on your computer.
2. Type in the URL *rpd.cme.msu.edu/html/*
3. When the Ribosomal Database Project II comes up, click on "Online Analysis."
4. Find the "Sequence Match" in the purple column. Click on the gray arrow in the run column for the sequence match.
5. In the large clear box that appears at the bottom of the screen, cut and paste the gene sequences (using either Microsoft Word or the sequences found on the McGraw-Hill website, *www.mhhe.com/prescott5/ labsequencedata*) from the four unknown bacteria and one archaeon (figures 66.1–66.5) into the box.
6. Click on "Submit Sequence" a few screens down. Certain parameters will have to be changed in order to complete a successful search.
7. As soon as the analysis has been completed, the genus and species that most closely matches the entered sequence appears. Also appearing is the percentage of the sequences that matched your entered sample nucleotide sequence. Other species and their percentage of similarity are also presented.

HINTS AND PRECAUTIONS

Typing gene sequences is most easily done by one person doing the typing on the keyboard and another reading the letters and spaces. The gene sequences are also available electronically on the McGraw-Hill website, *www.mhhe. com/prescott5/labsequencedata.*

Figure 66.1 Base Sequences for Bacterium A.

1	tctctgatgt tagcggcgga cgggtgagta acacgtggat aacctaccta taagactggg	60
61	ataacttcgg gaaaccggag ctaataccgg ataatatttt gaaccgcatg gttcaaaagt	120
121	gaaagacggt cttgctgtca cttatagatg gatccgcgct gcattagcta gttggtaagg	180
181	taacggctta ccaaggcaac gatgcatagc cgacctgaga gggtgatcgg ccacactgga	240
241	actgagacac ggtccagact cctacgggag gcagcagtag ggaatcttcc gcaatgggcg	300
301	aaagcctgac ggagcaacgc cgcgtgagtg atgaaggtct tcggatcgta aaactctgtt	360
361	attagggaag aacatatgtg taagtaactg tgcacatctt gacggtacct aatcagaaag	420
421	ccacggctaa ctacgtgcca gcagccgcgg taatacgtag gtggcaagcg ttatccggaa	480
481	ttattgggcg taaagcgcgc gtaggcggtt ttttaagtct gatgtgaaag cccacggctc	540
541	aaccgtggag ggtcattgga aactggaaaa cttgagtgca gaagaggaaa gtggaattcc	600
601	atgtgtagcg gttaaatgcg cagagatatg gaggaacacc agtggcgaag gcgactttct	660
661	ggtctgtaac tgacgctgat gtgcgaaagc gtgggaatca acaggatta gataccctgg	720
721	tagtccacgc cgtaaacgat gagtgctaag tgttagggg tttccgcccc ttagtgctgc	780
781	agctaacgca ttaagcactc cgcctgggga gtacgaccgc aaggttgaaa ctcaaaggaa	840
841	ttgacgggga cccgcacaag cggtggagca tgtggtttaa ttcgaagcaa cgcgaagaac	900
901	cttaccaaat cttgacatcc tttgacaact ctagagatag agccttcccc ttcgggggac	960
961	aaagtgacag gtggtgcatg gttgtcgtca gctcgtgtcg tgagatgttg ggttaagtcc	1020
1021	cgcaacgagc gcaacccta agcttagttg ccatcattaa gttgggcact ctaagttgac	1080
1081	tgccggtgac aaaccggagg aaggtgggga tgacgtcaaa tcatcatgcc ccttatgatt	1140
1141	tgggctacac acgtgctaca atggacaata caaagggcag cgaaaccgcg aggtcaagca	1200
1201	aatcccataa agttgttctc agttcggatt gtagtctgca actcgactac atgaagctgg	1260
1261	aatcgctagt aatcgtagat cagcatgcta cggtgaatac gttcccgggt cttgtacaca	1320
1321	ccgcccgtca caccacgaga gtttgtaaca	1351

Figure 66.2 Base Sequences for Bacterium B.

1	taacacgtgg ataacctacc tataagactg ggataacttc gggaaaccgg agctaatacc	60
61	ggataatata ttgaaccgca tggttcaata gtgaaagacg gttttgctgt cacttataga	120
121	tggatccgcg ccgcattagc tagttggtaa ggtaacggct taccaaggca acgatgcgta	180
181	gccgacctga gagggtgatc ggccacactg gaactgagac acggtccaga ctcctacggg	240
241	aggcagcagt agggaatctt ccgcaatggg cgaaagcctg acggagcaac gccgcgtgag	300
301	tgatgaaggt cttcggatcg taaaactctg ttattaggga agaacaaatg tgtaagtaac	360
361	tatgcacgtc ttgacggtac ctaatcagaa agccacggct aactacgtgc	411

Figure 66.3 Base Sequences for Bacterium C.

```
   1    gcctaataca tgcaagtaga acgctgagaa ctggtgcttg caccggttca aggagttgcg     60
  61    aacgggtgag taacgcgtag gtaacctacc tcatagcggg ggataactat tggaaacgat    120
 121    agctaatacc gcataagaga gactaacgca tgttagtaat ttaaaagggg caattgctcc    180
 181    actatgagat ggacctgcgt tgtattagct agttggtgag gtaaaggctc accaaggcga    240
 241    cgatacatag ccgacctgag agggtgatcg gccacactgg gactgagaca cggcccagac    300
 301    tcctacggga ggcagcagta gggaatcttc ggcaatgggg gcaaccctga ccgagcaacg    360
 361    ccgcgtgagt gaagaaggtt ttcggatcgt aaagctctgt tgttagagaa gaatgatggt    420
 421    gggagtggaa aatccaccaa gtgacggtaa ctaaccagaa agggacggct aactacgtgc    480
 481    cagcagccgc ggtaatacgt aggtcccgag cgttgtccgg atttattggg cgtaaagcga    540
 541    gcgcaggcgg tttttaagt ctgaagttaa aggcattggc tcaaccaatg tacgctttgg    600
 601    aaactggaga acttgagtgc agaaggggag agtggaattc catgtgtagc ggtgaaatgc    660
 661    gtagatatat ggaggaacac cggtggcgaa agcggctctc tggtctgtaa ctgacgctga    720
 721    ggctcgaaag cgtggggagc aaagaggatt agataccctg gtagtccacg ccgtaaacga    780
 781    tgagtgctag gtgttaggcc ctttccgggg cttagtgccg gagctaacgc attaagcact    840
 841    ccgcctgggg agtacgaccg caaggttgaa actcaaagga attgacgggg gcccgcacaa    900
 901    gcggtggagc atgtggttta attcgaagca acgcgaagaa ccttaccagg tcttgacatc    960
 961    ccgatgcccg ctctagagat agagttttac ttcggtacat cggtgacagg tggtgcatgg   1020
1021    ttgtcgtcag ctcgtgtcgt gagatgttgg gttaagtccc gcaacgagcg caacccctat   1800
1081    tgttagttgc catcattaag ttgggcactc tag                                 1114
```

Figure 66.4 Base Sequences for Bacterium D.

```
   1    ggtaccactc ggccccgaccg aacgcactcg cgcggatgac cggccgacct ccgcctacgc    60
  61    aatacgctgt ggcgtgtgtc cctggtgtgg gccgccatca cgaagcgctg ctggttcgac   120
 121    ggtgttttat gtaccccacc actcggatga gatgcgaacg acgtgaggtg gctcggtgca   180
 181    cccgacgcca ctgattgacg ccccctcgtc ccgttcggac ggaacccgac tgggttcagt   240
 241    ccgatgccct taagtacaac agggtacttc ggtggaatgc gaacgacaat ggggccgccc   300
 301    ggttacacgg gtggccgacg catgactccg ctgatcggtt cggcgttcgg ccgaactcga   360
 361    ttcgatgccc ttaagtaata acgggtgttc cgatgagatg cgaacgacaa tgaggctatc   420
 421    cggcttcgtc cgggtggctg atgcatctct tcgacgctct ccatggtgtc ggtctcactc   480
 481    tcagtgagtg tgattcgatg cccttaagta ataacgggcg ttacgaggaa ttgcgaacga   540
 541    caatgtggct acctggttct cccaggtggt taacgcgtgt tcctcgccgc cctggtgggc   600
 601    aaacgtcacg ctcgattcga gcgtgattcg atgcccttaa gtaataacgg ggcgttcggg   660
 661    gaaatgcgaa cgtcgtcttg gactgatcgg agtccgatgg gtttatgacc tgtcgaactc   720
 721    tacggtctgg tccgaaggaa tgaggattcc acacctgcgg tccgccgtaa agatggaatc   780
 781    tgatgttagc cttgatggtt tggtgacatc caactggcca cgacgatacg tcgtgtgcta   840
 841    agggacacat tacgtgtccc cgccaaacca agacttgata gtcttggtcg ctgggaacca   900
 901    tcccagcaaa ttccggttga tcctgccgga ggccattgc                           939
```

Figure 66.5 Base Sequences for One Archaeon.

```
1      agagtttgat cctggctcag agcgaacgct ggcggcaggc ttaacacatg caagtcgaac    60
61     gggcgtagca atacgtcagt ggcagacggg tgagtaacgc gtgggaacgt accttttggt   120
121    tcggaacaac acagggaaac ttgtgctaat accggataag cccttacggg gaaagattta   180
181    tcgccgaaag atcggcccgc gtctgattag ctagttggtg aggtaatggc tcaccaaggc   240
241    gacgatcagt agctggtctg agaggatgat cagccacatt gggactgaga cacggcccaa   300
301    actcctacgg gaggcagcag tggggaatat tggacaatgg gcgaaagcct gatccagcca   360
361    tgccgcgtga gtgatgaagg ccctagggtt gtaaagctct tttgtgcggg aagataatga   420
421    cggtaccgca agaataagcc ccggctaact tcgtgccagc agccgcggta atacgaaggg   480
481    ggctagcgtt gctcggaatc actgggcgta aagggtgcgt aggcgggttt ctaagtcaga   540
541    ggtgaaagcc tggagctcaa ctccagaact gcctttgata ctggaagtct tgagtatggc   600
601    agaggtgagt ggaactgcga gtgtagaggt gaaattcgta gatattcgca agaacaccag   660
661    tggcgaaggc ggctcactgg gccattactg acgctgaggc acgaaagcgt ggggagcaaa   720
721    caggattaga taccctggta gtccacgccg taaacgatga atgccagccg ttagtgggtt   780
781    tactcactag tggcgcagct aacgctttaa gcattccgcc tggggagtac ggtcgcaaga   840
841    ttaaaactca aaggaattga cggggggccg cacaagcggt ggagcatgtg gtttaattcg   900
901    acgcaacgcg cagaacctta ccagcccttg acatgtccag gaccggtcgc agagacgtga   960
961    ccttctcttc ggagcctgga gcacaggtgc tgcatggctg tcgtcagctc gtgtcgtgag  1020
1021   atgttgggtt aagtcccgca acgagcgcaa cccccgtcct tagttgctac catttagttg  1080
1081   agcactctaa ggagactgcc ggtgataagc cgcgaggaag gtggggatga cgtcaagtcc  1140
1141   tcatggccct tacgggctgg gctacacacg tgctacaatg gcggtgacaa tgggaagcta  1200
1201   aggggtgacc cttcgcaaat ctcaaaaagc cgtctcagtt cggattgggc tctgcaactc  1260
1261   gagcccatga agttggaatc gctagtaatc gtggatcagc atgccacggt gaatacgttc  1320
1321   ccgggccttg tacacaccgc ccgtcacacc atgggagttg gctttacctg aagacggtgc  1380
1381   gctaaccagc aatggggggca gccggccacg gtagggtcag cgactggggt gaagtcgtaa  1440
1441   caaggtagcc gtaggggaac ctgcggctgg atcacctcct t                       1481
```

Laboratory Report 66

Name: _____

Date: _____

Lab Section: _____

Identifying Archaea and Bacteria Using the Internet and Computer-Assisted Gene Analysis

1. Results for the four bacteria.

 Bacteria A = _____ and _____
 Genus species

 Bacteria B = _____ and _____
 Genus species

 Bacteria C = _____ and _____
 Genus species

 Bacteria D = _____ and _____
 Genus species

2. Results for the archaeon.

 Archaeon = _____ and _____
 Genus species

3. In summary, what is the major purpose of this experiment?

Review Questions

1. What are some advantages of identifying an organism by using the Ribosomal Database Project?

2. What are some disadvantages of identifying an organism by using the Ribosomal Database Project?

3. Why might you obtain different search results from one time to another?

4. What essential terms or search parameters did you find most useful in narrowing your search?

5. Was your search successful? Why or why not?

6. How could a search be performed that accesses only the desired sequence?

7. What is the main purpose of the Ribosomal Database Project?

8. How would you define bioinformatics? Signature sequences?

Appendix A

Dilutions with Sample Problems

Dilutions can be divided into five categories:

1. To dilute a solution to an unspecified final volume in just one step.
2. To dilute a solution to a specified final volume in just one step.
3. To dilute a solution to an unspecified final volume in several steps.
4. To dilute a solution to a specified final volume in several steps.
5. Serial dilutions.

Diluting a Solution to an Unspecified Volume in Just One Step

One must first calculate how many times to dilute (this is called the **dilution factor**) the initial material (**stock solution**) to obtain the final concentration. To accomplish this type of dilution, use the following formula:

$$\frac{\text{Initial Concentration }(IC)}{\text{Final Concentration }(FC)} = \text{Dilution Factor }(DF)$$

For example, if you want to dilute a solution with an initial concentration of solute of 5% down to 1%, using the above formula gives

$$\frac{IC}{FC} = DF \; \frac{5\%}{1\%} = 5$$

Thus, in order to obtain a 1% solution from a 5% solution, the latter must be diluted five times. This can be accomplished by taking one volume (e.g., cc, ml, liter, gallon) of the initial concentration (5%) and adding four volumes (e.g., cc, ml, liter, gallon) of solvent for a total of five volumes. Stated another way, 1 ml of a 5% solution + 4 ml of diluent will give a total of 5 ml, and each ml contains 1% instead of 5%.

Diluting a Solution to a Specified Volume in Just One Step

First, calculate the number of times the initial concentration must be diluted by dividing the final concentration (FC) into the initial concentration (IC).

Second, divide the number of times the initial concentration must be diluted (bottom left paragraph) into the final volume specified to determine the aliquot (or portion) of the initial concentration to be diluted.

Third, dilute the aliquot of the initial concentration calculated in step 2 by the volume specified.

For example, you have a 10% solution and want a 2% solution. However, you need 100 ml of this 2% solution.

$$\frac{IC}{FC} = DF = \frac{10\%}{2\%} = 5$$

Divide the number of times the 10% solution must be diluted (DF) into the final volume specified:

$$\frac{100 \text{ ml}}{5} = 20 \text{ ml}$$

Dilute the portion of 10 to the volume specified:

20 ml of a 10% solution + 80 ml of diluent = 100 ml (each milliliter = 2%)

Another method for performing this type of dilution is to use the following formula:

$$\frac{C_1}{C_2} = \frac{V_2}{V_1} \text{ or } C_1V_1 = C_2V_2$$

C_1 = standard concentration available
C_2 = standard concentration desired
V_2 = final volume of new concentration
V_1 = volume of C_1 required to make the new concentration

For example, if you want to prepare 100 ml of 10% ethyl alcohol from 95% ethyl alcohol, then

$$C_1 = 95\%, C_2 = 10\%,$$
$$V_2 = 100 \text{ ml},$$
$$V_1 = x \text{ and}$$
$$\frac{95}{10} = \frac{100}{x}$$
$$x = 1{,}000/95$$
$$x = 10.5 \text{ ml}$$

Thus, 10.5 ml of 95% ethyl alcohol + 89.5 ml of H_2O = 100 ml of a 10% ethyl alcohol solution.

Diluting a Solution to an Unspecified Volume in Several Steps

Frequently in the microbiology laboratory, large dilutions must be employed. They cannot be done in one step because

they are too large. As a result, they must be done in several steps to conserve not only amounts of diluent to be used but also space. For example, a 0.5 g/ml solution diluted to 1 μg/ml is a 500,000-fold dilution.

$$0.5 \text{ g} = 0.5 \text{ g} \times 10^6 \text{ μg/g}$$
$$= 500,000 \text{ μg}$$

To obtain a solution containing 500,000 μg/ml in one step would require taking 1 ml of 0.5 gm/ml stock solution and adding 499,999 ml of diluent. As you can see, it would be almost impossible to work with such a large fluid volume.

A 500,000 times dilution can be easily performed in two steps by first taking 1 ml of the initial concentration and diluting it to 500 ml and, second, by diluting 1 ml of the first dilution to 1,000 ml.

1 ml of 500,000 μg/ml + 499 ml of diluent = 1,000 μg/ml

1 ml of 1,000 μg/ml + 999 ml of diluent = 1 μg/ml

Thus, by this two-step procedure, we have cut down the volume of diluent used from 499,999 to 1,498 (499 ml + 999 ml).

Dilution Ratios Used in This Manual

According to the *ASM* (*American Society for Microbiology*) *Style Manual*, dilution ratios may be reported with either colons (:) or shills (/), but note there is a difference between them. A shill indicates the ratio of a part to a whole: for example, 1/2 means one of two parts, with a total of two parts. A colon indicates the ratio of one part to two parts, with a total of three parts. Thus, 1/2 equals 1:1, but 1:2 equals 1/3.

Diluting a Solution to a Specified Volume in Several Steps

This type of dilution is identical to all previous dilutions with the exception that the specified final volume must be one factor of the total dilution ratio.

For example, you want a 1/10,000 dilution of whole serum (undiluted) and you need 50 ml.

Divide dilution needed by the volume:

$$\frac{10,000}{50} = 200$$

200 (1/200 dilution) = the first step in the dilution factor; the second is 1/50, obtained as follows:

1 ml of serum + 199 ml of diluent = 1/200 dilution
1 ml of 1/200 dilution + 49 ml of diluent = 1/50
To check: 50 × 200 = 10,000

Serial Dilutions

The usefulness of dilutions becomes most apparent when small volumes of a material are required in serological procedures. For example, if 0.01 ml of serum were required in a certain test, instead of measuring out this small volume with a consequent sacrifice of accuracy, it would be more advantageous to dilute the serum 100 times. One ml of this 1/100 dilution would then contain 0.01 ml of the serum. Each ml of this dilution would be equivalent to the required 0.01 ml of undiluted serum.

Dilutions represent fractional amounts of a material and are generally expressed as the ratio of one volume of material to the final volume of the dilution. Thus, a 1/10 dilution of serum represents 1 volume of serum in 10 volumes of dilution (1 volume of serum + 9 volumes of diluent). Undiluted serum may be expressed as 1/1.

1 ml of serum + 1 ml of saline may be expressed as 1/2. Each milliliter of this dilution is equivalent to 0.5 ml of undiluted serum.

1 ml of serum + 2 ml of saline may be expressed as 1/3. Each milliliter of this dilution is equivalent to 0.33 ml of undiluted serum.

1 ml of serum + 99 ml of saline may be expressed as 1/100. Each milliliter of this dilution is equivalent to 0.01 ml of undiluted serum.

From the above, one can see that dilution expressions are fractions written as ratios where the numerator is unity and the denominator is the dilution value.

Division of the numerator by the denominator will give the amount of material per milliliter of the dilution. For example, in a 1/25 dilution, 1/25 = 0.04. Therefore, each milliliter of this dilution contains 1/25 or 0.04 ml of the original material (e.g., serum).

To convert a ratio into a dilution expression, divide both the numerator and denominator by the value of the numerator. For example, in a mixture consisting of 4 ml of serum and 6 ml of saline,

4 ml of serum + 6 ml of saline
= 10 ml of serum dilution
Ratio of serum dilution = 4/10

Dividing both numerator and denominator by the numerator value (4),

$$\frac{4 \div 4}{10 \div 4} = 1/2.5 = \text{serum dilution}$$

In the preparation of dilutions, any multiple or submultiple of the constituent volumes may be used. For example,

1/30 dilution = 1 ml serum + 29 ml saline
= 0.5 ml serum + 14.5 ml saline
= 2 ml serum + 58.0 ml saline

Serial dilutions indicate that an identical volume of material is being transferred from one vessel to another. The purpose of this procedure is to increase the dilutions of a substance by certain increments. For example, in a twofold dilution, the dilution factor is doubled each time (e.g., 1/2, 1/4, 1/8, etc.). See table on page 425 for further examples.

Tube	Amount of Saline Added to Each Tube	Serum	Protocol	Final Dilution of Serum
1	0	2 ml	Remove 1 ml to tube 2	1/1
2	1 ml	1 ml of 1/1	Mix, remove 1 ml to tube 3	1/2
3	1 ml	1 ml of 1/2	Mix, remove 1 ml to tube 4	1/4
4	1 ml	1 ml of 1/4	Mix, remove 1 ml to tube 5	1/8
5	1 ml	1 ml of 1/8	Mix, remove 1 ml to tube 6	1/16
6	1 ml	1 ml of 1/16	Mix, discard 1 ml	1/32

Sample Problems

Diluting a Solution to an Unspecified Volume in Just One Step

Problem 1

Dilute a solution that has an initial concentration of solute of 10% down to 2%.

Problem 2

Dilute a solution that has an initial concentration of 0.01% to 0.0001%.

Diluting a Solution to a Specified Volume in Just One Step

Problem 3

You have a 0.01% solution. You want a 0.001% solution and you need 25 ml.

Problem 4

Prepare 50 ml of a 3% solution from a 4% solution.

Problem 5

Prepare 100 ml of 10% alcohol from 95% alcohol.

Problem 6

You need 45 ml of 50% alcohol. How can you prepare this from 70% alcohol?

Problem 7

What volume of a 20% dextrose broth solution should be used to prepare 250 ml of 1% dextrose broth?

Problem 8

Prepare a 45 ml of a 2% suspension of RBCs from a 5% suspension.

Problem 9

What volume of a 0.02% solution can be prepared from 25 ml of a 0.1% solution?

Problem 10

What percent concentration of alcohol is prepared when 10 ml of 95% alcohol are diluted with H_2O to make a final volume of 38 ml?

Problem 11

What percent of dextrose is prepared when 3 ml of a 10% dextrose solution are mixed with 12 ml of broth?

Diluting a Solution to an Unspecified Volume in Several Steps

Problem 12

You have a stock solution of protein containing 10 grams per ml. You want a concentration of 2 mg/ml. How would you perform this dilution in just three steps?

Problem 13

You have a stock solution of 10 mg/ml of vitamins and want to obtain a solution of 0.5 μg/ml. How would you perform this dilution in just three steps?

Diluting a Solution to a Specified Volume in Several Steps

Problem 14

You want a 1:128 dilution of serum and you need 4 ml. How would you perform this dilution in several steps?

Problem 15

You want a 1:3,000 dilution of serum and you need 2 ml. How would you perform this dilution in several steps?

Problem 16

How would you prepare 1 ml of a 1:5 dilution of sera?

Problem 17

How would you prepare 8 ml of a 1:20 dilution of sera?

Problem 18

Based on the following dilutions, how many bacteria were present in the original sample? _____

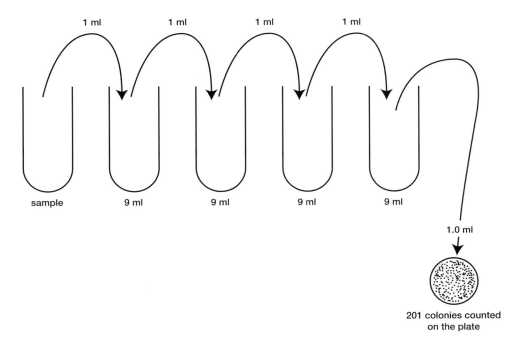

Problem 19

How many bacteria were present in the following sample? _____

Dilution	Amount Plated	Number of Colonies Counted
10^{-4}	1.0 ml	256
10^{-5}	1.0 ml	28
10^{-6}	1.0 ml	7

Problem 20

If 0.1 ml of a urine culture from a 10^{-6} dilution yielded 38 colonies, how many bacteria were there per ml in the original sample? _____

Problem 21

How many bacteria were present in the following sample? _____

Dilution	Volume Plated	Number of Colonies Counted
10^{-7}	0.1 ml	26

Problem 22

How many bacteria were present in the following sample? _____

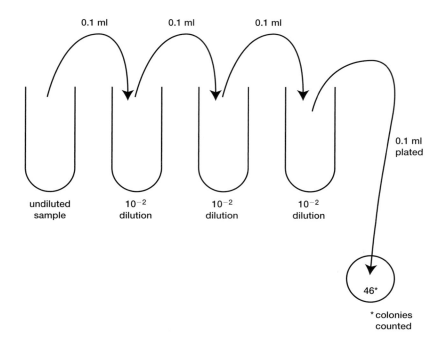

Problem 23

How many bacteria were there in the original sample? _____

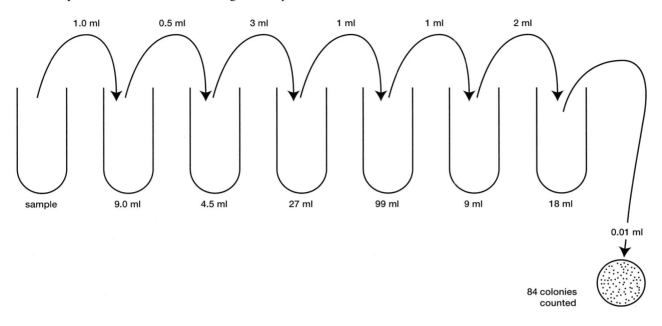

| 1.0 ml | 0.5 ml | 3 ml | 1 ml | 1 ml | 2 ml |

| sample | 9.0 ml | 4.5 ml | 27 ml | 99 ml | 9 ml | 18 ml |

0.01 ml

84 colonies counted

Answers to Sample Problems

Problem 1

Take 1 volume of the initial concentration (10%) and 4 volumes of solvent for a total of 5 volumes: 1 ml of a 10% solution + 4 ml of diluent will give a total of 5 ml and each ml will now contain 2% instead of 10%.

Problem 2

In order to obtain a 0.0001% solution from a 0.01% solution, the latter must be diluted 100 times. Therefore, take 1 volume of the initial concentration (0.01%) and add 99 volumes of solvent for a total of 100 volumes. 1 ml of 0.01% solution + 99 ml of diluent will give a total of 100 ml and each ml now contains 0.0001% instead of 0.01%.

Problem 3

A 0.01% solution is 10 times more concentrated than the 0.001% solution. Divide the number of times the 10 must be diluted into the final volume specified: 25/10 = 2.5. 2.5 ml of a 0.01% solution + 22.5 ml of diluent will give a total of 25 ml and each ml now contains 0.001%.

Problem 4

$C_1 = 4\%$
$C_2 = 3\%$ $\dfrac{4\%}{3\%} = \dfrac{50}{X}$ $4X = 37.5$ ml
$V_2 = 50$ ml
$V_1 = X$

Thus, 37.5 ml of a 4% solution + 12.5 ml of solvent = 50 ml of a 3% solution.

Problem 5

$C_1 = 95\%$
$C_2 = 10\%$ $\dfrac{95\%}{10\%} = \dfrac{100}{X}$ $X = \dfrac{1000}{95}$ $X = 10.5$ ml
$V_2 = 100$ ml
$V_1 = X$

Thus, 10.5 ml of 95% alcohol + 89.5 ml of H_2O = 100 ml of a 10% alcohol solution.

Problem 6

$C_1 = 70\%$
$C_2 = 50\%$ $\dfrac{70\%}{50\%} = \dfrac{45}{X}$ $X = \dfrac{50 \times 45}{70}$ $X = 32.1$ ml
$V_2 = 45$ ml
$V_1 = X$

Thus, 32.1 ml of 70% alcohol + 12.9 ml of H_2O = 45 ml of a 50% alcohol solution.

Problem 7

$C_1 = 20\%$
$C_2 = 1\%$ $\dfrac{20\%}{1\%} = \dfrac{250}{X}$ $20X = 250$ $X = 12.5$ ml
$V_2 = 250$ ml
$V_1 = X$

Thus, 12.5 ml of a 20% solution + 237.5 ml of broth = 250 ml of a 1% dextrose broth solution.

Problem 8

$C_1 = 5\%$
$C_2 = 2\%$ $\dfrac{5\%}{2\%} = \dfrac{45 \text{ ml}}{X}$ $X = \dfrac{90}{5}$ $X = 18$ ml
$V_2 = 45$ ml
$V_1 = X$

Thus, 18 ml of 5% + 27 ml of saline = 45 ml of a 2% red blood cell suspension.

Problem 9

$C_1 = 0.1\%$
$C_2 = 0.02\%$ $\dfrac{0.1\%}{0.02\%} = \dfrac{X}{25 \text{ ml}}$ $0.02X = 2.5$ $X = 125$ ml
$V_2 = X$
$V_1 = 25$ ml

Thus, prepare a total volume of 125 of a 0.02% solution from 25 ml of 0.01% solution by adding 100 ml of diluent to the latter.

Problem 10

$C_1 = 95\%$
$C_2 = X$ $\dfrac{95\%}{X} = \dfrac{38 \text{ ml}}{10 \text{ ml}}$ $38X = 950$ $X = 25\%$
$V_2 = 38$ ml
$V_1 = 10$ ml

Problem 11

$C_1 = 10\%$ $\dfrac{10\%}{X} = \dfrac{15 \text{ ml*}}{3 \text{ ml}}$
$C_2 = X$
$V_2 = 15$ ml *3 + 12 = 15 ml (total volume) $15X = 30$ $X = 2\%$
$V_1 = 3$ ml

Problem 12

In order to make all units equal, you have to convert 10 grams to milligrams.

 stock solution = 10,000 mg/ml

 final concentration = 2 mg/ml

$$\frac{IC}{FC} = \frac{10,000 \text{ mg/ml}}{2 \text{ mg/ml}} = 5,000$$

Thus, the dilution factor is 5,000.

To perform a 1:5000 dilution:

Step #1 1 ml of 10,000 mg/ml stock solution protein + 49 ml of diluent = 1:50 dilution = 200 mg/ml

Step #2 1 ml of 200 mg/ml + 9 ml of diluent = 1:10 dilution = 20 mg/ml

Step #3 1 ml of 20 mg/ml + 9 ml of diluent = 1:10 dilution = 2 mg/ml

To check to make sure the correct dilution was made:

$50 \times 10 \times 10 = 5,000$

Problem 13

Make all units the same.

 10 mg \times 1,000 μg/mg = 10,000 μg

 stock solution = 10,000 μg/ml

 final concentration = 0.5 μg/ml

$$\frac{IC}{FC} = DF = \frac{10,000 \text{ mg/ml}}{0.5 \text{ mg/ml}} = 20,000$$

Thus, the dilution factor is 20,000.

Step #1 1 ml of 10,000 μg/ml + 19 ml of diluent = 1:20 dilution = 500 μg/ml (10,000/20 = 500)

Step #2 1 ml of 500 μg/ml + 99 ml of diluent = 1:100 dilution = 5 μg/ml (500/100 = 5)

Step #3 1 ml of 5 μg/ml + 9 ml of diluent = 1:10 dilution = 0.5 μg/ml (5/10 = 0.5)

To check to make sure the correct dilution was made:

$20 \times 100 \times 10 = 20,000$

To help you with these types of problems, always remember that volumes may change depending on the protocol, but the dilution factor will always remain the same. For example:

1 + 99	
10 + 990	1:100 dilution
0.1 + 9.9	
1 + 4	
10 + 40	1:5 dilution
0.1 + 0.4	
1 + 19	
10 + 190	1:20 dilution
0.1 + 1.9	
1 + 249	
10 + 2,490	1:250 dilution
0.1 + 24.9	
1 + 9	
10 + 90	1:10 dilution
0.5 + 4.5	
1 + 7	
5 + 35	1:8 dilution
0.5 + 3.5	
1 + 14	
5 + 70	1:15 dilution
0.5 + 7.0	

Problem 14

The first step is to establish the initial dilution as follows:

$$\frac{128}{4} = 32$$

1:32 is the first step dilution, and the second is 1:4.
1 ml of serum + 31 ml of diluent = 1:32 (individual dilution)
1 ml of the 1:32 dilution = 3 ml of diluent = 1:4 (individual dilution)
To check to make sure the dilution was correctly made: $32 \times 4 = 128$

Problem 15

We can obtain a 1:3000 dilution in 3 steps by using 1:30 and 1:10 dilutions.

$$\frac{3,000}{2} = 1,500$$

1 ml of serum + 29 ml of diluent = 1:30 (individual dilution)
1 ml of 1:30 dilution + 9 ml of diluent = 1:10 (individual dilution)
1 ml of 1:10 + 9 ml of diluent = 1:10 (individual dilution)
To check to make sure the dilution was correctly made: $30 \times 10 \times 10 = 3,000$

Problem 16

$$\frac{D_1}{D_2} = \frac{V_1}{V_2}$$

$D_1 = 1$
$D_2 = 5$ $\qquad \frac{1}{5} = \frac{X}{1} \qquad 5X = 1 \qquad X = 0.2$ ml
$V_1 = X$
$V_2 = 1$

Thus, 0.2 ml (undiluted sera) + 0.08 ml of saline = 1.0.

Problem 17

$D_1 = 1$
$D_2 = 20$ $\qquad \frac{1}{20} = \frac{X}{8} \qquad 20X = 8 \qquad X = 0.4$ ml
$V_1 = X$
$V_2 = 8$

Thus, 0.4 ml of undiluted sera + 7.6 ml of saline = 1:20.

Problem 18

2.01×10^6

Problem 19

2.8×10^6

Problem 20

3.8×10^8

Problem 21

2.6×10^9

Problem 22

4.6×10^8

Problem 23

8.4×10^{10}

Metric and English Measurement Equivalents

The Metric System

The metric system comprises three basic units of measurement: distance measured in meters, volume measured in liters, and mass measured in grams. In order to designate larger and smaller measures, a system of prefixes based on multiples of ten is used in conjunction with the basic unit of measurement. The most common prefixes are

kilo	$= 10^3$	$= 1,000$	
centi	$= 10^{-2} =$	0.01	$= 1/100$
milli	$= 10^{-3} =$	0.001	$= 1/1,000$
micro	$= 10^{-6} =$	0.000001	$= 1/1,000,000$
nano	$= 10^{-9} =$	0.000000001	$= 1/1,000,000,000$

The English System

The measurements of the English system used in the United States unfortunately are not systematically related. For example, there are 12 inches in a foot and 3 feet in a yard. Quick conversion tables for the metric and English systems are listed below.

Units of Length

Metric to English

1 centimeter (cm) or 10 mm = 0.394 in or 0.0328 ft
1 meter (m) = 100 cm or 1,000 mm = 39.4 in or 3.28 ft or 1.09 yd
1 kilometer (km) = 1,000 m = 3,281 ft or 0.621 mile (mi)

The Number of:	Multiplied by:	Equals:
millimeters	0.04	inches
centimeters	0.4	inches
meters	3.3	feet
meters	1.1	yards
kilometers	0.6	miles

English to Metric

1 in = 2.54 cm
1 ft or 12 in = 30.48 cm
1 yd or 3 ft or 36 in = 91.44 cm or 0.9144 m
1 mi or 5,280 ft or 1,760 yd = 1,609 m or 1.609 km

The Number of:	Multiplied by:	Equals:
inches	2.5	centimeters
feet	30.5	centimeters
yards	0.9	meters
miles	1.6	kilometers

Units of Area

Metric to English

1 square centimeter (cm^2) or 100 mm^2 = 0.155 in^2
1 square meter (m^2) = 1,550 in^2 or 1.196 yd^2
1 hectare (ha) or 10,000 m^2 = 107,600 ft^2 or 2.471 acres (A)
1 square kilometer (km^2) or 1,000,000 m^2 or 100 ha = 247 A or 0.3861 mi^2

The Number of:	Multiplied by:	Equals:
square centimeters	0.16	square inches
square meters	1.2	square yards
square kilometers	0.4	square miles

English to Metric

1 square foot (ft^2) or 144 in^2 = 929 cm^2
1 square yard (yd^2) or 9 ft^2 = 8,361 cm^2 or 0.836 m^2
1 acre or 43,560 ft^2 or 4,840 yd^2 = 4,047 m^2 = 0.405 ha
1 square mile (mi^2) or 27,878 ft^2 or 640 A = 259 ha or 2.59 km^2

The Number of:	Multiplied by:	Equals:
square inches	6.5	square centimeters
square feet	0.09	square meters
square yards	0.8	square meters
square miles	2.6	square kilometers
acres	0.4	hectares

Units of Volume

Metric to English

1 cubic centimeter (cm^3 or cc) or 1,000 mm^3 = 0.061 in^3

1 cubic meter (m^3) or 1,000,000 cm^3 = 61,024 in^3 or 35.31 ft^3 or 1.308 yd^3

The Number of:	Multiplied by:	Equals:
cubic meters	35	cubic feet
cubic meters	1.3	cubic yards

English to Metric

1 cubic ft (ft^3) or 1,728 in^3 = 28,317 cm^3 or 0.02832 m^3

1 cubic yard (yd^3) or 27 ft^3 = 0.7646 m^3

The Number of:	Multiplied by:	Equals:
cubic feet	0.03	cubic meters
cubic yards	0.76	cubic meters

Units of Liquid Capacity

Metric to English

1 milliliter (ml) or 1 cm^3 = 0.06125 in^3 or 0.03 fl oz

1 liter or 1,000 ml = 2.113 pt or 1.06 qt or 0.264 U.S. gal

The Number of:	Multiplied by:	Equals:
milliliters	0.03	fluid ounces
liters	2.10	pints
liters	1.06	quarts

English to Metric

1 fluid ounce or 1.81 in^3 = 29.57 ml

1 pint or 16 fl oz = 473.2 ml

1 qt or 32 fl oz or 2 pt = 946.4 ml

1 gal or 128 fl oz or 4 qt or 8 pt = 3,785 ml or 3.785 liters

The Number of:	Multiplied by:	Equals:
teaspoons	5	milliliters
tablespoons	15	milliliters
fluid ounces[a]	30	milliliters
cups	0.24	liters
pints	0.47	liters
quarts	0.95	liters
gallons	3.8	liters

Units of Mass

Metric to English

1 gram (g) or 1,000 mg = 0.0353 oz

1 kilogram (kg) or 1,000 g = 35.2802 oz or 2.205 lb

1 metric or long ton or 1,000 kg = 2,205 lb or 1.102 short tons

The Number of:	Multiplied by:	Equals:
grams	0.035	ounces
kilograms	2.2	pounds

English to Metric

1 ounce (oz) or 437.5 grains (gr) = 28.35 g

1 pound (1b) or 16 ounces = 453.6 g or 0.454 kg

1 ton (short ton) or 2,000 lb = 907.2 kg or 0.9072 metric ton

The Number of:	Multiplied by:	Equals:
ounces	28	grams
pounds	0.45	kilograms
tons	0.9	metric ton

[a]1 British fluid ounce = 0.961 U.S. fluid ounces or, conversely, 1 U.S. fluid ounce = 1.041 British fluid ounces. The British pint, quart, and gallon = 1.2 U.S. pints, quarts, and gallons, respectively. To convert these U.S. fluid measures, multiply by 0.8327.

Transmission-Absorbance Table for Spectrophotometry

%T	A	%T	A	%T	A	%T	A
1.0	2.000	26.0	0.585	51.0	0.292	76.0	0.119
1.5	1.824	26.5	0.577	51.5	0.288	76.5	0.116
2.0	1.699	27.0	0.569	52.0	0.284	77.0	0.114
2.5	1.602	27.5	0.561	52.5	0.280	77.5	0.111
3.0	1.523	28.0	0.553	53.0	0.276	78.0	0.108
3.5	1.456	28.5	0.545	53.5	0.272	78.5	0.105
4.0	1.398	29.0	0.538	54.0	0.268	79.0	0.102
4.5	1.347	29.5	0.530	54.5	0.264	79.5	0.100
5.0	1.301	30.0	0.523	55.0	0.260	80.0	0.097
5.5	1.260	30.5	0.516	55.5	0.256	80.5	0.094
6.0	1.222	31.0	0.509	56.0	0.252	81.0	0.092
6.5	1.187	31.5	0.502	56.5	0.248	81.5	0.089
7.0	1.155	32.0	0.495	57.0	0.244	82.0	0.086
7.5	1.126	32.5	0.488	57.5	0.240	82.5	0.084
8.0	1.097	33.0	0.482	58.0	0.237	83.0	0.081
8.5	1.071	33.5	0.475	58.5	0.233	83.5	0.078
9.0	1.046	34.0	0.469	59.0	0.229	84.0	0.076
9.5	1.022	34.5	0.462	59.5	0.226	84.5	0.073
10.0	1.000	35.0	0.456	60.0	0.222	85.0	0.071
10.5	0.979	35.5	0.450	60.5	0.218	85.5	0.068
11.0	0.959	36.0	0.444	61.0	0.215	86.0	0.066
11.5	0.939	36.5	0.438	61.5	0.211	86.5	0.063
12.0	0.921	37.0	0.432	62.0	0.208	87.0	0.061
12.5	0.903	37.5	0.426	62.5	0.204	87.5	0.058
13.0	0.886	38.0	0.420	63.0	0.201	88.0	0.056
13.5	0.870	38.5	0.414	63.5	0.197	88.5	0.053
14.0	0.854	39.0	0.409	64.0	0.194	89.0	0.051
14.5	0.838	39.5	0.403	64.5	0.191	89.5	0.048
15.0	0.824	40.0	0.398	65.0	0.187	90.0	0.046
15.5	0.810	40.5	0.392	65.5	0.184	90.5	0.043
16.0	0.796	41.0	0.387	66.0	0.181	91.0	0.041
16.5	0.782	41.5	0.382	66.5	0.177	91.5	0.039
17.0	0.770	42.0	0.377	67.0	0.174	92.0	0.036
17.5	0.757	42.5	0.372	67.5	0.171	92.5	0.034
18.0	0.745	43.0	0.367	68.0	0.168	93.0	0.032
18.5	0.733	43.5	0.362	68.5	0.164	93.5	0.029
19.0	0.721	44.0	0.357	69.0	0.161	94.0	0.027
19.5	0.710	44.5	0.352	69.5	0.158	94.5	0.025
20.0	0.699	45.0	0.347	70.0	0.155	95.0	0.022
20.5	0.688	45.5	0.342	70.5	0.152	95.5	0.020
21.0	0.678	46.0	0.337	71.0	0.149	96.0	0.018
21.5	0.668	46.5	0.332	71.5	0.146	96.5	0.016
22.0	0.658	47.0	0.328	72.0	0.143	97.0	0.013
22.5	0.648	47.5	0.323	72.5	0.140	97.5	0.011
23.0	0.638	48.0	0.319	73.0	0.137	98.0	0.009
23.5	0.629	48.5	0.314	73.5	0.134	98.5	0.007
24.0	0.620	49.0	0.310	74.0	0.131	99.0	0.004
24.5	0.611	49.5	0.305	74.5	0.128	99.5	0.002
25.0	0.602	50.0	0.301	75.0	0.125	100.0	0.000
25.5	0.594	50.5	0.297	75.5	0.122		

Appendix D

Logarithms

A logarithm is the exponent of 10, indicating the power to which 10 must be raised to produce a given number. Since 1 is 10^0 and 10 is 10^1, it is evident that the numbers between 1 and 10 must be greater than 10^0. Likewise, numbers between 10 and 100 must be greater than 10^1 but less than 10^2. These numbers will then have fractional exponents expressed as mixed fractions. If they are in fractional forms, they present difficulties in addition or subtraction, so it is best to express them as a decimal; for example, $10^{0.3010}$.

A number written in the form b^n is said to be in exponential form where b is the **base** and n is a **logarithm.** For example, in the following equation,

$$N = b^n$$

the number N is equal to the base b to the exponent n. Let us say that b is equal to 2 and n is equal to 4. Written in exponential form, we would have 2^4. Two to the fourth power equals 16.

In logarithmic form, we would write the log of N to the base b is n ($\log_b N = n$). So if we take $2^4 = 16$ and place it in logarithmic form, we would have

$$\log_2 16 = 4$$

In logarithm tables the logarithms are located in the body of the table, and the numbers from 1.0 to 9.9 are given in the left-hand column and the top row. For example, to locate the logarithm of 4.7 in a logarithm table, read down the left-hand column to 47 and across the column to 0 to find 0.6721.

Finally, the following relationships should be remembered when working with logarithms:

$$\log 1 = \log 10^0 = 0$$
$$\log 10 = \log 10^1 = 1$$
$$\log 100 = \log 10^2 = 2$$
$$\log 1{,}000 = \log 10^3 = 3$$
$$\log 10{,}000 = \log 10^4 = 4$$
$$\log 0.1 = \log 10^{-1} = -1$$
$$\log 0.01 = \log 10^{-2} = -2$$
$$\log 0.001 = \log 10^{-3} = -3$$
$$\log 0.0001 = \log 10^{-4} = -4$$

Logarithms are particularly useful in graphical relations that extend over a wide range of values since they have the property of giving equal relative weight to all parts of the scale. This is valuable in "spreading out" the values that would otherwise be concentrated at the lower end of the scale; for example, in graphing the growth of microbial populations in a culture versus time. Logarithms are also used in pH calculations.

pH and pH Indicators

pH is a measure of hydrogen ion (H^+) activity. In dilute solutions, the H^+ activity is essentially equal to the concentration. In such instances, $pH = -\log [H^+]$. The pH scale ranges from 0 ($[H^+] = 1.0^0$ M) to 14 ($[H^+] = 10^{-14}$ M).

A pH meter should be used for accurate pH determinations, observing the following precautions:

1. Adjust the temperature of the buffer used for pH meter standardization to the same temperature as the sample. Buffer pH changes with temperature; for example, the pH of standard phosphate buffer is 6.98 at 0°C, 6.88 at 20°C, and 6.85 at 37°C.
2. It is important to stir solutions while measuring their pH. If the sample is to be stirred with a magnetic mixer, stir the calibrating buffer in the same way.

3. Be sure that the electrodes used with tris buffers are recommended for such use by the manufacturer. This is necessary because some pH electrodes do not give accurate readings with tris (hydroxymethyl) aminomethane buffers.

In instances where precision is not required, such as in the preparation of routine media, the pH may be checked by the use of pH indicator solutions. By the proper selection, the pH can be estimated within ± 0.2 pH units. Some common pH indicators and their useful pH ranges are listed in the following table.

pH Indicator	pH Range	Full Acidic Color	Full Basic Color
Brilliant green	0.0–2.6	Yellow	Green
Bromcresol green	3.8–5.4	Yellow	Blue-green
Bromcresol purple	5.2–6.8	Yellow	Purple
Bromophenol blue	3.0–4.6	Yellow	Blue
Bromothymol blue	6.0–7.6	Yellow	Blue
Congo red	3.0–5.0	Blue-violet	Red
Cresol red	2.3–8.8	Orange	Red
Cresolphthalein	8.2–9.8	Colorless	Red
2,4-dinitrophenol	2.8–4.0	Colorless	Red
Ethyl violet	0.0–2.4	Yellow	Blue
Litmus	4.5–8.3	Red	Blue
Malachite green	0.2–1.8	Yellow	Blue-green
Methyl green	0.2–1.8	Yellow	Blue
Methyl red	4.4–6.4	Red	Yellow
Neutral red	6.8–8.0	Red	Amber
Phenolphthalein	8.2–10.0	Colorless	Pink
Phenol red	6.8–8.4	Yellow	Red
Resazurin	3.8–6.4	Orange	Violet
Thymol blue	8.0–9.6	Yellow	Blue

Note: All of the above indicators can be made by (1) dissolving 0.04 g of indicator in 500 ml of 95% ethanol, (2) adding 500 ml of distilled water, and (3) filtering through Whatman No. 1 filter paper. Indicators should be stored in a dark, tightly closed bottle.

Scientific Notation

Microbiologists often have to deal with either very large or very small numbers, such as 5,550,000,000 or 0.00000082. The mere manipulation of these numbers is cumbersome. As a result, it is more convenient to express such numbers in **scientific notation (standard exponential notation).** Scientific notation is a set of rules involving a shorthand method for writing these numbers and performing simple manipulations with them. Scientific notation uses the fact that every number can be expressed as the product of two numbers—one of which is a power of the number 10.

Numbers greater than 1 can be expressed as follows:

$$1 = 10^0 \qquad 100,000 = 10^5$$
$$10 = 10^1 \qquad 1,000,000 = 10^6$$
$$100 = 10^2 \qquad 10,000,000 = 10^7$$
$$1,000 = 10^3 \qquad 100,000,000 = 10^8$$
$$10,000 = 10^4 \qquad 1,000,000,000 = 10^9$$

In the above notations, the **exponent** to which the 10 is raised is equal to the number of **zeros** following the 1.

Numbers less than 1 can be expressed as follows:

$$0.1 = 10^{-1} \qquad 0.000001 = 10^{-6}$$
$$0.01 = 10^{-2} \qquad 0.0000001 = 10^{-7}$$
$$0.001 = 10^{-3} \qquad 0.00000001 = 10^{-8}$$
$$0.0001 = 10^{-4} \qquad 0.000000001 = 10^{-9}$$
$$0.00001 = 10^{-5}$$

In the above notations, the number of the negative exponent to which 10 is raised is equal to the number of digits to the right of the decimal point.

Numbers that are not an exact power of 10 can also be dealt with in scientific notation. For example, a number such as 1234, which is greater than 1, can be expressed in the following ways:

$$123.4 \times 10 \qquad 0.1234 \times 10,000$$
$$12.34 \times 100 \qquad 0.01234 \times 100,000$$
$$1.234 \times 1,000$$

The same numbers can be expressed in scientific notation as follows:

$$123.4 \times 10^1 \qquad 0.1234 \times 10^4$$
$$12.34 \times 10^2 \qquad 0.01234 \times 10^5$$
$$1.234 \times 10^3$$

The same reasoning is followed for numbers less than 1. Consider the number 0.1234; it can be expressed in the following ways:

$$1.234 \times 0.1 \qquad 1,234 \times 0.0001$$
$$12.34 \times 0.01 \qquad 12,340 \times 0.00001$$
$$123.4 \times 0.001$$

The same numbers can be expressed in scientific notation as follows:

$$1.234 \times 10^{-1} \qquad 1,234 \times 10^{-4}$$
$$12.34 \times 10^{-2} \qquad 12,340 \times 10^{-5}$$
$$123.4 \times 10^{-3}$$

Multiplication can also be done in scientific notation. Consider the following multiplication:

$$50 \times 250$$

First, rewriting each number in scientific notation:

$$(5.0 \times 10^1) \times (2.5 \times 10^2)$$

To multiply, multiply the first two numbers:

$$5.0 \times 2.5 = 12.5$$

To multiply the second part, add the exponents:

$$10^1 + 10^2 = 10^3$$

The answer is written as 12.5×10^3. It can also be written as 1.25×10^4. These same two steps are done in every case of multiplication, even with numbers less than 1. For example, to multiply 0.5×0.25:

$$(5 \times 10^{-1}) \times (2.5 \times 10^{-1})$$
$$= 12.5 \times 10^{22}$$
$$= 1.25 \times 10^{-1} = 0.125$$

When multiplying numbers greater than 1 by numbers less than 1, express the numbers in convenient form, multiply the first part, add the exponents of the second part, and then express the answer in scientific notation. For example, multiply $0.125 \times 5,000$:

$$(1.25 \times 10^{-1}) = (5 \times 10^3) = 6.25 \times 10^2$$

When adding a negative number to a positive number, always subtract the negative number from the positive number.

Dividing in scientific notation is similar to multiplying. Consider dividing $2,500/500$.

First, rewriting in scientific notation gives

$$\frac{2.5 \times 10^3}{5 \times 10^2}$$

Second, divide the first two numbers as follows:

$$\frac{2.5}{5} = 0.5$$

Third, subtract the bottom exponent from the top exponent:

$$10^3 - 10^2 = 10^1$$

The answer in scientific notation is expressed as 0.5×10^1. Always remember that when you subtract one negative number from another negative number, you add the numbers and express the answer as a negative number. When subtracting a negative number from a positive number, it is the same as adding a positive number to a positive number. To subtract a positive number from a negative number, add the positive number to the negative number and express the answer as a negative number.

Microbiologists use scientific notation continuously. For example, in this laboratory manual, it is used to describe the number of bacteria in a population and to express concentrations of chemicals in solution, of disinfectants, and of antibiotics.

Appendix G

Identification Charts*

Chart I Characterization of Oxidative-Fermentative Gram-Negative Rods

Legend
+ = Positive
− = Negative
V = Variable (11% to 89% positive)
Data based on literature only

		OF Anaerobic Dextrose	Arginine Dihydrolase	Nitrogen Gas	H₂S	Indole	OF Xylose	Urea	Citrate	OF Dextrose	Oxidase
Achromobacter	Species Bio. 1	−	−	+	−	−	V	−	V	V	+
	Bio. 2	−	V	V	−	−	V	V	+	V	+
	Xylosoxidans	−	−	V	−	−	V	−	−	+	+
Acinetobacter	Anitratus	V	−	−	−	−	+	+	V	V	−
	Lwoffii	−	−	−	−	−	−	−	V	V	+
Aeromonas hydrophila		+	V	−	−	V	−	+	−	V	+
Alcaligenes faecalis		−	−	V	−	−	−	−	V	V	+
Bordetella bronchiseptica		−	−	−	−	−	−	−	+	+	+
Flavobacterium species		V	−	−	−	V	−	V	−	−	+
Group	2F Flavobacter-	−	−	−	−	+	−	−	−	−	+
	2J IUM-Like	−	−	−	−	+	−	−	+	−	+
	2K-1 Pseudomonas-	−	−	−	−	−	−	−	−	−	V
	2K-2 Like	+	−	−	−	−	+	V	+	+	+
	4E Alcaligenes-Like	−	−	−	−	−	−	−	V	V	+
	5A-1	−	−	+	−	−	+	+	V	+	+
	5A-2 Pseudomonas	−	−	V	−	−	V	+	V	V	+
	5E-1 Like	+	+	−	−	−	+	+	V	+	−
	5E-2	V	−	−	−	−	+	+	V	V	+
	M-4 Moraxella-	−	−	−	−	−	−	−	−	+	+
	M-4f Like	−	−	−	−	−	−	−	+	+	+
Moraxella	Species	−	−	−	−	−	−	−	−	−	+
	Phenylpyruvica	−	−	−	−	−	−	−	+	−	+
Pasteurella	Haemolytica	V	−	−	−	−	−	−	−	−	+
	Multocida	−	−	−	−	+	−	−	−	−	+
	Ureae	V	−	−	−	−	−	−	V	−	+
Plesiomonas shigelloides		+	+	−	−	+	−	+	−	−	+
Pseudomonas	Aeruginosa	−	+	V	−	−	V	V	V	V	+
	Acidovorans	−	−	−	−	−	−	−	−	V	+
	Alcaligenes	−	−	−	−	−	−	−	V	V	+
	Cepacia	V	−	−	−	−	V	V	V	+	V
	Diminuta	−	−	−	−	−	−	−	−	V	+
	Fluorescens	−	V	−	−	−	V	V	V	+	+
	Mallei	−	+	−	−	−	−	+	−	−	−
	Maltophilia	−	−	−	−	−	−	V	V	−	V
	Pseudoalcaligenes	−	−	−	−	−	−	−	−	V	+
	Pseudomallei	−	+	V	−	−	−	+	V	+	+
	Putida	V	+	−	−	−	+	+	V	+	+
	Putrefaciens	−	−	−	+	−	−	−	V	−	+
	Stutzeri	−	V	V	−	−	V	V	−	+	+
	Testosteroni	−	−	−	−	−	−	−	−	V	+
	Vesicularis	V	−	−	−	−	V	V	−	−	+
Vibrio	Alginolyticus	+	−	−	−	V	−	+	−	−	+
	Cholerae	+	−	−	−	+	V	+	−	V	+
	Parahaemolyticus	+	−	−	−	+	−	V	−	V	+
Chromobacterium violaceum		+	+	−	−	−	−	+	−	V	V

*These charts are based on rapid test systems. At times these test results may differ from results obtained with so-called conventional tests.

Chart II Characterization of Enterobacteriaceae—The Enterotube II System

Tribe	Group	Organism	Reactions	Glucose	Gas Production	Lysine	Ornithine	H₂S	Indole	Adonitol	Lactose	Arabinose	Sorbitol	Voges-Proskauer	Dulcitol	Phenylalanine Deaminase	Urea	Citrate
	Escherichieae	Escherichia		+	+J	d	d	−K	+	−	+J	+	±	−	d	−	−	−
				100.0	92.0	80.6	57.8	4.0	96.3	5.2	91.6	91.3	80.3	0.0	49.3	0.1	0.1	0.2
		Shigella		+	−A	−	∓B	−	∓	−	−B	±	∓	−	d	−	−	−
				100.0	2.1	0.0	20.0	0.0	37.8	0.0	0.3	67.8	29.1	0.0	5.4	0.0	0.0	0.0
	Edwardsielleae	Edwardsiella		+	+	+	+	+	+	−	−	∓	−	−	−	−	−	−
				100.0	99.4	100.0	99.0	99.6	99.0	0.0	0.0	10.7	0.2	0.0	0.0	0.0	0.0	0.0
Salmonelleae		Salmonella		+	+C	+H	+I	+E	−	−	−	±	+	−	dD	−	−	dF
				100.0	91.9	94.6	92.7	91.6	1.1	0.0	0.8	89.2	94.1	0.0	86.5	0.0	0.0	80.1
		Arizona		+	+	+	+	+	+	−	d	+	+	−	−	−	−	+
				100.0	99.7	99.4	100.0	98.7	2.0	0.0	69.8	99.1	97.1	0.0	0.0	0.0	0.0	96.8
	Citrobacter	freundii		+	+	−	d	±	−	−	d	+	+	−	d	−	dw	+
				100.0	91.4	0.0	17.2	81.6	6.7	0.0	39.3	100.0	98.2	0.0	59.8	0.0	89.4	90.4
		amalonaticus		+	+	−	+	−	+	−	±	+	+	−	∓	−	±	+
				100.0	97.0	0.0	97.0	0.0	99.0	0.0	70.0	99.0	97.0	0.0	11.0	0.0	81.0	94.0
		diversus		+	+	−	+	−	+	+	d	+	+	−	±	−	dw	+
				100.0	97.3	0.0	99.8	0.0	100.0	100.0	40.3	98.0	98.2	0.0	52.2	0.0	85.8	99.7
Proteeae	Proteus	vulgaris		+	±G	−	−	+	+	−	−	−	−	−	−	+	+	d
				100.0	86.0	0.0	0.0	95.0	91.4	0.0	0.0	0.0	0.0	0.0	0.0	100.0	95.0	10.5
		mirabilis		+	+G	−	+	+	−	−	−	−	−	∓	−	+	±	±
				100.0	96.0	0.0	99.0	94.5	3.2	0.0	2.0	0.0	0.0	16.0	0.0	99.6	89.3	58.7
	Morganella	morganii		+	±G	−	+	−	+	−	−	−	−	−	−	+	+	−L
				100.0	86.0	0.0	97.0	0.0	99.5	0.0	0.0	0.0	0.0	0.0	0.0	95.0	97.1	0.0
	Providencia	alcalifaciens		+	dG	−	−	−	+	+	−	−	−	−	−	+	−	+
				100.0	85.2	0.0	1.2	0.0	99.4	94.3	0.3	0.7	0.6	0.0	0.0	97.4	0.0	97.9
		stuartii		+	−	−	−	−	+	∓	−	−	−	−	−	+	∓	+
				100.0	0.0	0.0	0.0	0.0	98.6	12.4	3.6	4.0	3.4	0.0	0.0	94.5	20.0	93.7
		rettgeri		+	∓G	−	−	−	+	+	d	−	−	−	−	+	+	+
				100.0	12.2	0.0	0.0	0.0	95.9	99.0	10.0	0.0	1.0	0.0	0.0	98.0	100.0	96.0
Klebsielleae	Enterobacter	cloacae		+	+	−	+	−	−	∓	±	+	+	+	d	−	±	+
				100.0	99.3	0.0	93.7	0.0	0.0	28.0	94.0	99.4	100.0	100.0	15.2	0.0	74.6	98.9
		sakazakii		+	+	−	+	−	∓	−	+	+	−	+	−	−	−	+
				100.0	97.0	0.0	97.0	0.0	16.0	0.0	100.0	100.0	0.0	97.0	6.0	0.0	0.0	94.0
		gergoviae		+	+	±	+	−	−	−	∓	+	−	+	−	−	+	+
				100.0	93.0	64.0	100.0	0.0	0.0	0.0	42.0	100.0	0.0	100.0	0.0	0.0	100.0	96.0
		aerogenes		+	+	+	+	−	−	+	+	+	+	+	−	−	−	+
				100.0	95.9	97.5	95.9	0.0	0.8	97.5	92.5	100.0	98.3	100.0	4.1	0.0	0.0	92.6
		agglomerans		+	∓	−	−	−	∓	−	d	+	d	∓	d	∓	d	d
				100.0	24.1	0.0	0.0	0.0	19.7	7.5	52.9	97.5	26.3	64.8	12.9	27.6	34.1	84.2
	Hafnia	alvei		+	+	+	+	−	−	−	d	+	−	±	−	−	d	d
				100.0	98.9	99.6	98.6	0.0	0.0	0.0	2.8	99.3	0.0	65.0	2.4	0.0	3.0	5.6
	Serratia	marcescens		+	±G	+	+	−	−w	∓	−	−	+	+	−	−	dw	+
				100.0	52.6	99.6	99.6	0.0	0.1	56.0	1.3	0.0	99.1	98.7	0.0	0.0	39.7	97.6
		liquefaciens		+	d	±	+	−	−w	−	d	+	+	∓	−	−	dw	+
				100.0	72.5	64.2	100.0	0.0	1.8	8.3	15.6	97.3	97.3	49.5	0.0	0.9	3.7	93.6
		rubidaea		+	dG	+	−	−	−w	+	+	+	−	+	−	−	dw	±
				100.0	35.0	61.0	0.0	0.0	2.0	88.0	100.0	100.0	8.0	92.0	0.0	0.0	4.0	88.0
	Klebsiella	pneumoniae		+	+	+	−	−	−	±	+	+	+	+	∓	−	+	+
				100.0	96.0	97.2	0.0	0.0	0.0	89.0	98.7	99.9	99.4	93.7	33.0	0.0	95.4	96.8
		oxytoca		+	+	+	−	−	+	±	∓	+	+	∓	−	−	∓	∓
				100.0	96.0	97.2	0.0	0.0	100.0	89.0	98.7	100.0	98.0	93.7	33.0	0.0	95.4	96.8
		ozaenae		+	d	∓	−	−	−	+	d	+	±	−	−	−	d	d
				100.0	55.0	35.8	1.0	0.0	0.0	91.8	26.2	100.0	78.0	0.0	0.0	0.0	14.8	28.1
		rhinoschieromatis		+	−	−	−	−	−	+	d	+	+	−	−	−	−	−
				100.0	0.0	0.0	0.0	0.0	0.0	98.0	6.0	100.0	98.0	0.0	0.0	0.0	0.0	0.0
Yersiniae	**Yersinia**	enterocolitica		+	−	−	+	−	∓	−	−	+	+	−	−	−	+	−
				100.0	0.0	0.0	90.7	0.0	26.7	0.0	0.0	98.7	98.7	0.1	0.0	0.0	90.7	0.0
		pseudotuberculosis		+	−	−	−	−	−	−	−	±	−	−	−	−	+	−
				100.0	0.0	0.0	0.0	0.0	0.0	0.0	0.0	55.0	0.0	0.0	0.0	0.0	100.0	0.0

Source: Copyright © Becton Dickinson Microbiology Systems. Reprinted by permission.

Note: E. *S. enteritidis* bioserotype Paratyphi A and some rare biotypes may be H₂S negative.

F. *S. typhi*, *S. enteritidis* bioserotype Paratyphi A and some rare biotypes are citrate-negative and *S. cholerae-suis* is usually delayed positive.

G. The amount of gas produced by *Serratia*, *Proteus*, and *Providencia alcalifaciens* is slight; therefore, gas production may not be evident in the Enterotube II.

H. *S. enteritidis* bioserotype Paratyphi A is negative for lysine decarboxylase.

I. *S. typhi* and *S. gallinarum* are ornithine decarboxylase-negative.

J. The Alkalescens-Dispar (A–D) group is included as a biotype of *E. coli*. Members of the A–D group are generally anaerogenic and nonmotile, and do not ferment lactose.

K. An occasional strain may produce hydrogen sulfide.

L. An occasional strain may appear to utilize citrate.

Chart III Characterization of Gram-Negative Rods—The API 20E System

Organism	ONPG	ADH	LDC	ODC	CIT	H₂S	URE	TDA	IND	VP	GEL	GLU	MAN	INO	SOR	RHA	SAC	MEL	AMY	ARA	OXI
E. coli	98.2	1.0	90.2	67.3	0	1.0	0	0	85.0	0	0	100	98.4	0.1	95.5	84.5	41.1	88.4	0.1	95.0	0
Shigella dysenteriae	27.8	0	0	0	0	0	0	0	33.0	0	0	100	0.1	0	0	22.2	0	61.1	0	16.7	0
Sh. flexneri	5.3	0	0	0	0	0	0	0	15.0	0	0	100	94.7	0	78.9	0	0	21.1	0	36.8	0
Sh. boydii	5.0	0	0	0	0	0	0	0	20.0	0	0	100	60.0	0	53.3	1.0	0	33.3	0	66.7	0
Sh. sonnei	96.7	0	0	80.0	0	0	0	0	0	0	0	100	99.0	0	39.9	80.0	0	50.0	0	96.7	0
Edwardsiella tarda	0	0	99.0	99.0	0	55.0	0	0	100	0	0	100	0	0	0	0	0	50.0	0	1.1	0
Salmonella enteritidis	1.9	1.0	89.2	95.4	15.4	76.9	0	0	3.1	0	0	100	98.7	4.6	95.2	95.4	4.6	96.9	0	94.5	0
Sal. typhi	0	0	90.0	0	0	0.1	0	0	0	0	0	100	99.0	0	99.0	1.8	0	100	0	27.0	0
Sal. paratyphi A	0	0	0	100	0	0.2	0	0	0	0	0	100	99.0	0	99.0	99.0	0	40.0	0	80.0	0
Arizona-S. arizonae	94.7	1.0	95.0	98.5	15.0	85.0	0	0	0	0	0	100	99.0	0	87.0	96.1	0	89.5	0	95.0	0
Citrobacter freundii	97.0	10.0	0	60.0	10.0	81.0	0	0	6.0	0	0	100	98.0	1.0	96.0	87.0	59.0	77.0	30.0	98.0	0
C. diversus-Levinea	97.0	10.0	0	90.0	10.0	0	0	0	91.0	0	0	100	97.0	14.5	88.0	99.0	51.0	47.0	34.0	99.0	0
C. amalonaticus	97.0	10.0	0	95.0	10.0	0	0	0	99.0	0	0	100	97.0	0.1	93.0	99.0	29.4	53.0	80.0	93.8	0
Klebsiella pneumoniae	99.0	0	80.0	0	13.9	0	10.0	0	0	72.0	0	100	98.0	30.0	95.0	91.0	99.0	99.0	98.0	99.0	0
K. oxytoca	98.0	0	83.0	0	13.0	0	10.0	0	100	60.0	1.0	100	99.0	29.0	92.0	98.0	99.0	99.0	98.0	99.0	0
K. ozaenae	85.0	0	38.0	0	1.0	0	0	0	0	0	0	100	69.0	1.0	76.0	69.0	15.0	92.0	99.0	84.0	0
K. rhinoscleromatis	0	0	0	0	0	0	0	0	0	0	0	100	99.0	1.0	86.0	53.0	33.0	66.0	99.0	95.0	0
Enterobacter aerogenes	99.0	0	98.0	98.0	8.9	0	0	0	0	56.0	0	100	99.0	28.0	90.0	90.0	85.0	97.0	96.0	98.0	0
Ent. cloacae	97.0	51.9	0	65.0	9.0	0	0	0	0	80.0	0	100	99.0	1.0	92.0	90.0	98.0	92.0	65.0	95.0	0
Ent. agglomerans	90.0	0	0	0	5.4	0	0	0	50.0	20.0	0	100	99.0	1.0	80.0	60.0	60.0	70.0	70.0	95.0	0
Ent. gergoviae	99.0	0	61.0	99.0	8.2	0	75.0	0	0	75.0	0	100	99.0	1.0	8.3	99.0	99.0	99.0	99.0	99.0	0
Ent. sakazakii	97.0	51.6	0	59.0	8.6	0	0	0	4.0	85.0	0	100	99.0	4.0	8.5	90.0	95.0	90.0	76.0	95.0	0
Serratia liquefaciens	85.0	0	85.0	95.0	8.9	0	1.0	0	0	50.0	60.0	100	99.0	1.0	99.0	30.0	85.0	80.7	80.0	92.9	0
Ser. marcescens	83.0	0	88.0	94.0	8.0	0	1.0	0	0	58.0	72.0	100	96.0	1.0	97.0	2.0	98.0	37.0	72.0	18.0	0

Row groups (left margin): *Escherichiae*, *Salmonellae*, *Klebsiellae*

Chart III *continued*

Organism	ONPG	ADH	LDC	ODC	CIT	H₂S	URE	TDA	IND	VP	GEL	GLU	MAN	INO	SOR	RHA	SAC	MEL	AMY	ARA	OXI
Ser. rubidaea	96.0	0	60.5	0.1	8.2	0	0	0	0	70.0	75.6	100	99.0	10.0	75.0	13.4	99.0	82.6	96.0	85.8	0
Ser. odorifera 1	99.0	0	95.0	100	9.5	0	0	0	90.0	63.0	62.0	100	99.0	10.0	99.0	85.0	100	99.0	90.0	99.0	0
Ser. odorifera 2	99.0	0	92.0	0	9.1	0	0	0	90.0	80.0	78.0	100	99.0	10.0	99.0	95.0	0	99.0	85.4	99.0	0
Hafnia alvei	60.0	0	99.0	99.0	1.0	0	0	0	0	25.0	0	99.0	99.0	0.1	35.0	75.0	0	50.0	30.0	95.0	0
Proteus vulgaris	0.5	0	0	0	4.1	75.3	91.0	95.0	75.3	0	75.3	100	0	0	0	0	83.0	1.0	20.0	4.0	0
Prot. mirabilis	1.0	0	0	90.0	5.8	66.0	97.0	90.0	1.0	0	93.0	100	0	0	0	1.0	9.6	10.0	1.0	27.0	0
Providencia alcalifaciens	0	0	0	0	9.8	0	0	95.0	94.0	0	0	100	0	0	0	0	0	0	0	25.0	0
Prov. stuartii	1.0	0	0	0	8.5	0	0	95.0	86.0	0	0	100	0	8.0	0	0.8	3.7	34.0	0	30.0	0
Prov. stuartii URE 1	1.0	0	0	0	6.9	0	99.0	99.0	95.0	0	0	100	15.0	5.0	0	0.5	65.0	20.0	0	20.0	0
Prov. rettgeri	1.0	0	0	0	7.1	0	80.0	95.0	90.0	0	0	100	85.0	1.0	30.0	40.0	5.0	0	40.0	10.0	0
Morganella morganii	1.0	0	0	87.0	0.2	0	78.0	92.0	92.0	0	0	98.0	0	0	0	0	0	0	0	1.0	0
Yersinia enterocolitica	81.0	0	0	36.0	0	0	59.0	0	54.0	0.4	0	100	99.0	1.0	95.0	9.0	78.0	40.4	31.0	76.6	0
Y. pseudotuberculosis	80.0	0	0	0	0	0	88.0	0	0	0	0	100	94.0	0	76.0	58.0	0	5.0	0	52.0	0
Y. pestis	93.0	0	0	0	0	0	0	0	0	1.0	0	93.0	87.0	0	56.0	0	0	0.6	25.0	87.0	0
API Group 1	99.0	0	58.8	99.0	9.2	0	0	0	99.0	0	0	100	99.0	0	75.4	82.4	82.4	94.1	97.0	94.1	0
API Group 2	99.0	2.0	7.3	0	0	0	0	0	0	0	0	100	99.0	0	2.3	30.8	5.6	90.0	38.5	92.3	0
Pseudomonas maltophilia	62.0	0	5.0	0	7.6	0	0	0	0	0	50.0	0.5	0	0	0	0	0	0	0	22.0	4.8
Ps. cepacia	61.0	0	5.0	5.0	7.5	0	0	0	0	1.0	46.0	33.0	1.0	0	1.0	0	7.0	0	1.0	1.0	90.7
Ps. paucimobilis	40.0	0	0	0	1.0	0	0	0	0	0	0	0.5	0	0	0	0	0.5	0	0	0.5	50.0
A. calco. var. anitratus	0	0	0	0	2.8	0	0	0	0	1.0	0.1	85.0	0	0	0	0	0.1	77.0	0	60.0	0
A. calco. var. lwoffii	0	0	0	0	0	0	0	0	0	0.1	0	0	0	0	0	0	0	0	0		0
CDC Group VE-1	90.0	1.0	0	0	7.7	0	0	0	0	1.0	1.3	33.0	0	1.0	0	1.0	0.1	1.0	1.0	16.0	0
CDC Group VE-2	0	0	0	0	7.9	0	0	0	0	1.0	0.1	4.5	0	1.0	0	0	0	1.0	0	5.0	0

Group labels (left margin): *Klebsielleae*, *Proteeae*, *Yersiniae*, *Other Gram-negatives*

Source: Copyright BioMérieux Vitek, Inc., Hazelwood, MO. Reprinted by permission.
Note: Figures indicate the percentage of positive reactions after 18 to 24 hours of incubation at 35° to 37°C.

443

Appendix H

Reagents, Solutions, Stains, and Tests

Reagents and stains appear in this appendix as the authors have presented the material in the individual laboratory exercises and are listed in alphabetical order. When necessary, methodology is given with the reagents, stains, or tests. The detailed procedures, however, are presented in the exercise in which their use is discussed.

Acid-Alcohol (for Ziehl-Neelsen stain)
Concentrated hydrochloric acid 3 ml
95% ethyl alcohol . 97 ml

Alcohol, 90%, 500 ml (from 95%)
95% alcohol . 473 ml
Distilled water . 27 ml

Alcohol, 80%, 500 ml (from 95%)
95% alcohol . 421 ml
Distilled water . 79 ml

Alcohol, 75%, 500 ml (from 95%)
95% alcohol . 395 ml
Distilled water . 105 ml

Alcohol, 70%, 500 ml (from 95%)
95% alcohol . 368 ml
Distilled water . 132 ml

Barritt's Reagent (for Voges-Proskauer test)
Solution A: 6 g of α-naphthol in 100 ml of 95% ethyl alcohol.
Dissolve the α-naphthol in the ethanol with constant stirring.
Solution B: 40 g of potassium hydroxide in 100 ml of water. Store in the refrigerator.

Bile Solubility Test (10% bile)
Sodium deoxycholate 1 g
Sterile distilled water 9 ml

To test for bile solubility, prepare two tubes, each containing a sample of fresh culture (a light suspension of the bacterium in buffered broth, pH 7.4). To one tube add a few drops of a 10% solution of sodium deoxycholate. The same volume of sterile physiological saline is added to the second tube. If the bacterial cells are bile soluble, the tube containing the bile salt will lose its turbidity in 5 to 15 minutes and show an increase in viscosity.

Cleaning Solution for Glassware Strong:
Potassium dichromate20 g
Distilled water .200 ml
Dissolve dichromate in water; when cool, add very slowly:
Concentrated sulfuric acid 9 parts

2% aqueous potassium dichromate 1 part

Copper Sulfate Solution (20%)
Copper sulfate ($CuSO_4 \cdot 5H_2O$) 20 g
Distilled water . 80 ml

Crystal Violet Capsule Stain (1%)
Crystal violet (85% dye content) 1 g
Distilled water . 100 ml

Decolorizers (for Gram stain)

1. Intermediate agent, 95% ethyl alcohol.
2. Fastest agent, acetone.
3. Slowest agent, acetone-isopropyl alcohol (isopropyl alcohol, 300 ml; acetone, 100 ml). For the experienced microbiologist, any one of the three decolorizing agents will yield good results.

Diphenylamine Reagent (for the nitrate test)
Working in a fume hood, dissolve 0.7 g of diphenylamine in a mixture of 60 ml of concentrated sulfuric acid and 28.8 ml of distilled water. Allow to cool. Slowly add 11.3 ml of concentrated hydrochloric acid. After the solution has stood for 12 to 24 hours, some of the base will separate. This indicates that the reagent is saturated.

Eosin Blue
Eosin blue stain . 1 g
Distilled water . 99 ml

Ferric Chloride Reagent
$FeCl_3 \cdot 6H_2O$. 12 g
2% aqueous HCl . 100 ml
The 2% aqueous HCl is made by adding 5.4 ml of concentrated HCl (37%) to 94.6 ml of distilled H_2O.

Gram's Iodine (Lugol's)
According to the *ASM Manual for Clinical Microbiology,* dissolve 2 g of potassium iodide in 300 ml of distilled water and then add 1 g of iodine crystals. Rinse the solution into an amber bottle with the remainder of the distilled water. Discard when the color begins to fade.

Gram Stain
(A) **Crystal violet**
(Hucker modification)
1. Crystal violet 85% dye 2 g
 95% ethyl alcohol 20 ml
 Mix and dissolve.
2. Ammonium oxalate 0.8 g
 Distilled water . 80.0 ml

Add solution A to solution B. Let stand for a day, then filter. If the crystal violet is too concentrated, solution A may be diluted as much as 10 times.

(B) Gram's Iodine Solution (mordant)

Iodine crystals . 1 g
Potassium iodide 2 g
Distilled water . 300 ml

Store in an amber bottle; discard when the color begins to fade.

(C) Safranin (counterstain) **Solution**

Safranin . 2.5 g
95% ethyl alcohol 100.0 ml

For a working solution, dilute stock solution 1/10 (10 ml of stock safranin to 90 ml of distilled water).

India Ink (for capsule stain)

Mix the specimen with a small drop of India ink on a clean slide. If the India ink is too dark, dilute it to 50% with distilled water.

Kinyoun Acid-Fast Stain

(A) Kinyoun Carbolfuchsin

Basic fuchsin . 4 g
95% alcohol . 20 g
Phenol crystals 8 g
Distilled water 100 ml

(B) Acid-alcohol

Concentrated hydrochloric acid 3 ml
95% ethyl alcohol 97 ml

Methylene Blue Counterstain

Methylene blue 0.3 g
Distilled water 100 ml

Kovacs' Reagent (for the indole test)

N-amyl or isoamyl alcohol 150 ml
Concentrated hydrochloric acid 50 ml
p-dimethylaminobenzaldehyde 10 g

Working in a fume hood, dissolve the aldehyde in alcohol and then slowly add the acid. The dry aldehyde should be light in color. Alcohols that result in indole reagents that become deep brown should not be used. Store in a dark bottle with a glass stopper in a refrigerator when not in use.

Malachite Green Solution (for endospore stain)

Malachite green oxalate 5 g
Distilled water . 100 ml

Methylene Blue (Löffler's alkaline)

Solution A: Dissolve 0.3 g of methylene blue (90% dye content) in 30 ml of 95% ethyl alcohol.

Solution B: Dissolve 0.01 g of potassium hydroxide in 100 ml of distilled water.

Mix solutions A and B. Filter with Whatman No. 1 filter paper before use.

Methylene Blue Stain (simple staining)

Methylene blue . 0.3 g
Distilled water . 100.0 ml

Methyl Red Reagent (for detection of acid)

Methyl red . 0.1 g
95% ethyl alcohol 300.0 ml

Dissolve the dye in alcohol and add sufficient distilled water to make 500 ml. Positive tests are red-orange, and negative tests are yellow.

Naphthol, Alpha (for the Enterotube II System)

5% α-naphthol in 95% ethyl alcohol.

Nessler's Reagent (for the ammonia test)

Working in a fume hood, dissolve 50 g of potassium iodide in 35 ml of cold (ammonia-free) distilled water. Add mercuric chloride drop by drop until a slight precipitate forms. Add 400 ml of a 50% solution of potassium hydroxide. Dilute to 1 liter, allow to settle, and decant the supernatant for use. Store in a tightly closed dark bottle.

Alternate procedure:

Solution A:

Mercuric chloride 1 g
Distilled water . 6 ml

Dissolve completely.

Solution B:

Potassium iodide 2.5 g
Distilled water . 6.0 ml

Solution C:

Potassium hydroxide 6 g
Distilled water . 6 ml

Dissolve solution C completely and add to the mixture of solutions A and B. Add 13 ml of distilled water. Mix well and filter through Whatman No. 1 filter paper before use. Store in a dark, stoppered bottle.

Nigrosin Solution (Dorner's, for negative staining)

Water-soluble nigrosin 10.0 g
Distilled water . 100.0 ml
Formalin (40% formaldehyde) 0.5 ml

Gently boil the nigrosin and water approximately 30 minutes. Add 0.5 ml of 40% formaldehyde as a preservative. Filter twice through Whatman No. 1 filter paper and store in a dark bottle in the refrigerator.

Nitrate Test Reagent (see under diphenylamine)

Nitrite Test Reagents (Caution—solution B may be carcinogenic. Use safety precautions such as the avoidance of aerosols, mouth pipetting, and contact with skin.)

(A) Solution A: Dissolve 8 g of sulfanilic acid in 1 liter of 5 N acetic acid (1 part glacial acetic acid to 2.5 parts distilled water).

(B) Solution B: Dissolve 6 ml of N, N,-dimethyl-1-naphthylamine in 1 L of 5 N acetic acid.

DO NOT MIX SOLUTIONS.

Oxidase Test Reagent

Mix 1 g of dimethyl-p-phenylenediamine hydrochloride in 100 ml of distilled water. This reagent should be made fresh daily and stored in a dark bottle in the refrigerator.

O-nitrophenyl-β-D-Galactoside (ONPG)

0.1 M sodium phosphate buffer 50.0 ml
ONPG (8×10^{-4} M) 12.5 mg

Phosphate Buffers

Stock buffers:

Alkaline buffer, 0.067 M Na_2HPO_4 solution. Dissolve 9.5 g of Na_2HPO_4 in 1 liter of distilled water.

Acid buffer, 0.067 M NaH_2PO_4 solution. Dissolve 9.2 g of $NaH_2PO_4 \cdot H_2O$ in 1 liter of distilled water.

Buffered water (pH 7.0 to 7.2)

Acid buffer (NaH_2PO_4)	39 ml
Alkaline buffer (Na_2HPO_4)	61 ml
Distilled water .	900 ml

BE SURE GLASSWARE IS CLEAN. Buffered water, if sealed, is stable for several weeks.

Physiological Saline

Dissolve 8.5 g of sodium chloride in 1 liter of distilled water (0.85%) or 9 g in 1 liter of distilled water (0.9%).

Physiological Saline (Buffered)

Sodium chloride (0.85%; 8.5 g in 1 liter of distilled water) is buffered to pH 7.2 with 0.067 M potassium phosphate mixture.

Phosphate-Buffered Saline

10× stock solution, 1 liter
80 g NaCl
2 g KCl
11 g $Na_2HPO_4 \cdot 7H_2O$
2 g KH_2PO_4

Stock Solutions

1 M $CaCl_2$
147 g
$CaCl_2 \cdot 2H_2O$
H_2O to 1 liter

1 M HCl
Mix in the following order:
913.8 ml H_2O
86.2 ml concentrated HCl

1 M KCL
74.6 g KCl
H_2O to 1 liter

1 M $MgCl_2$
20.3 g
$MgCl_2 \cdot 6H_2O$
H_2O to 100 ml

5 M NaCl
292 g NaCl
H_2O to 1 liter

10 M NaOH
Dissolve 400 g
NaOH in 450 ml of
distilled water.
Add water to 1 liter.

Triton X-100 Stock Solution (10%)

Triton X-100 .	10 ml
Distilled water .	90 ml

Mix and store in a tightly stoppered bottle at room temperature; the solution will keep indefinitely.

Trommsdorf's Reagent (for the nitrite test)

Working in a fume hood with a beaker on a hot plate, slowly add, with constant stirring, 100 ml of a 20% aqueous zinc chloride solution to a mixture of 4 g of starch in water. Continue heating until the starch is completely dissolved and the solution is clear. Dilute with water and add 2 g of potassium iodide. Dilute to 1 liter with distilled water, filter once through Whatman No. 1 filter paper, and store in a capped, dark bottle.

Vaspar

Melt 1 pound of Vaseline and 1 pound of paraffin. Store in small student-use bottles.

Voges-Proskauer Reagent (see Barritt's reagent)

West Stain (flagella)

Solution A:

Mordant: 50 ml of saturated aqueous aluminum potassium sulfate + 100 ml of 10% tannic acid solution + 10 ml of 5% ferric chloride. This solution should be stored in an aluminum foil-covered bottle at 5°C until used.

Solution B:

Stain: 7.5 g of silver nitrate ($AgNO_3$) in 150 ml of distilled water. While working in a fume hood, add concentrated NH_4OH dropwise to 140 ml of the silver nitrate solution while it is being stirred on a magnetic mixer. A brown precipitate will form at the start of NH_4OH addition. Enough NH_4OH should be added so that the brown precipitate just dissolves. Finally, add 5% silver nitrate dropwise until a faint cloudiness persists. This solution should be stored at 5°C in an aluminum foil-covered bottle until used.

Ziehl-Neelsen Acid-Fast Stain

(A) **Solution A:** Dissolve 0.3 g of basic fuchsin (90% dye content) in 10 ml of 95% ethyl alcohol.

Solution B: Dissolve 5 g of phenol in 95 ml of distilled water.

Mix solutions A and B. Note: Add either 1 drop of Tergitol No. 4 per 30 ml of carbolfuchsin or 2 drops of Triton X-100 per 100 ml of stain for use in the heatless method. Tergitol No. 4 and Triton X act as detergents, emulsifiers, and wetting agents.

(B) **Acid-alcohol, 3%**

Concentrated hydrochloric acid	3 ml
95% alcohol .	97 ml

Culture Media

Sterilization of all tubed media is accomplished at 15 lb pressure (121°C) for 15 minutes unless otherwise specified. Longer sterilization times will be required for large volumes of media. Most of the media are available commercially in powdered form, with specific instructions for their preparation and sterilization.

Sources of Microbiological Media

Difco Laboratories
Division of Becton Dickinson Company
1 Becton Drive
Franklin Lakes, NJ 07417
1-202-847-6800
FAX 1-410-584-7121
www.bd.com/microbiology

ICN Biomedicals Inc.
1263 South Chillicothe
Aura, Ohio 44202
1-800-854-0530
FAX 1-800-334-6999
www.icnbiomed.com

Thomas Scientific
PO Box 99
Swedesboro, NJ 08085
1-800-345-2100
FAX 1-800-345-5232
www.thomassci.com

EM Industries
480 Democrat Road
Gibbstown, NJ 08027
1-800-222-0342
FAX 1-856-423-4389
www.emdchemicals.com

In addition to making media from commercially prepared supplies, companies such as Oxoid Unipath, 800 Proctor Avenue, Ogdensburg, NY 13669-2205; Scott Laboratories, West Warwick, RI 02893 and Carson, California, 90746; Fisher Scientific, 711 Forbes Avenue, Pittsburgh, PA 15219; The Scientific Products Division of Baxter Healthcare Corporation, 1430 Waukegan Road, McGraw Park, IL 60085; Wards Natural Science Establishment, 5100 West Henrietta Road, P.O. Box 92912, Rochester, NY; and Carolina Biological Supply, 2700 York Road, Burlington, NC 27215 can supply most of the media used in this manual already prepared in tubes, bottles, and plates. Some offer special services for diagnostic media.

Actidione (Cycloheximide) Agar (pH 5.5)

Glucose	50.0 g
Agar	15.0 g
Pancreatic digest of casein	5.0 g
Yeast extract	4.0 g
Potassium dihydrogen phosphate	0.5 g
Potassium chloride	0.42 g
Calcium chloride	0.12 g
Magnesium sulfate	0.12 g
Bromcresol green	22.0 mg
Actidione (cycloheximide)	10.0 mg
Ferric chloride	2.5 mg
Distilled water	1,000.0 ml

Agar, Noble

Noble agar is carefully washed agar that is purified and essentially free from impurities. It is used in electrophoretic procedures, nutritional studies, and wherever an agar of increased purity is needed.

Ammonium Sulfate API Broth (pH 7.5)

Bacto yeast extract	1.0 g
Ascorbic acid	0.1 g
Sodium lactate	5.2 g
Magnesium sulfate	0.2 g
Dipotassium phosphate	0.01 g
Ferrous ammonium sulfate	0.1 g
Sodium chloride	10.0 g
Distilled water	1,000.0 ml

Azotobacter Nitrogen-Free Broth (pH 7.2)

Dipotassium phosphate	1.0 g
Magnesium sulfate	0.2 g
Sodium chloride	0.2 g
Ferrous sulfate	5.0 mg
Distilled water	1,000.0 ml

Bile Esculin Agar (pH 6.8)

Oxgall	20.0 g
Agar	15.0 g
Pancreatic digest of gelatin	5.0 g
Beef extract	3.0 g
Esculin	1.0 g
Ferric citrate	0.5 g
Distilled water	1,000.0 ml

Blood Agar (pH 7.3)

Infusion from beef heart	500.0 g
Tryptose	10.0 g
Sodium chloride	5.0 g
Agar	15.0 g
Distilled water	1,000.0 ml

Note: Dissolve the above ingredients and autoclave. Cool the sterile blood agar base to 45° to 50°C and aseptically add 50 ml of sterile, defibrinated blood. Mix thoroughly and then dispense into plates while a liquid. Blood agar base for use in making blood agar also can be purchased. A combination of hemoglobin and a commercial nutrient supplement can be used in place of defibrinated blood.

Bottom Agar (pH 7.0)

Use 12-ml sterile nutrient agar pours to prepare plates.

Brain-Heart Infusion Agar (pH 7.4)

Calf brains, infusion from	200.0 g
Beef hearts, infusion from	250.0 g
Proteose peptone	10.0 g

Dextrose . 2.0 g
Sodium chloride 5.0 g
Disodium phosphate 2.5 g
Agar . 15.0 g
Distilled water 1,000.0 ml

Brewer's Anaerobic Agar (pH 7.2)
Bacto tryptone 5.0 g
Proteose peptone 10.0 g
Bacto yeast extract 5.0 g
Bacto dextrose 10.0 g
Sodium chloride 5.0 g
Agar . 20.0 g
Sodium thioglycollate 2.0 g
Sodium formaldehyde sulfoxylate 1.0 g
Resazurin 0.002 g
Distilled water 1,000.0 ml

Brilliant Green Bile Lactose (2%) Broth
Peptone . 10.0 g
Oxgall . 20.0 g
Lactose . 10.0 g
Brilliant green 0.0133 g
Distilled water 1,000.0 ml

Chocolate Agar (pH 7.0)
Proteose peptone 20.0 g
Dextrose 0.5 g
Sodium chloride 5.0 g
Disodium phosphate 5.0 g
Agar . 15.0 g
Distilled water 1,000.0 ml
Note: Aseptically add 5% sterile, defibrinated sheep blood to the sterile and molten agar. Heat at 80°C for 15 minutes or until a chocolate color develops.

Cystine Tryptic Agar (pH 7.3)
Tryptose 20.0 g
L-cystine 0.5 g
Sodium chloride 5.0 g
Sodium sulfite 0.5 g
Agar . 2.5 g
Phenol red 0.017 g
Distilled water 1,000.0 ml
After autoclaving and cooling to 50°C, add appropriate Bacto Differentiation Disk Carbohydrate (e.g., dextrose, fructose, maltose, sucrose). Allow to cool unslanted in an upright position.

Deoxyribonuclease (DNase Test) Agar (pH 7.3)
Deoxyribonucleic acid 2.0 g
Phytone peptone 5.0 g
Sodium chloride 5.0 g
Trypticase 15.0 g
Agar . 15.0 g
Distilled water 1,000.0 ml

Endo Agar (pH 7.5)
Peptone . 10.0 g
Lactose . 10.0 g
Dipotassium phosphate 3.5 g
Sodium sulfite 2.5 g
Basic fuchsin 0.4 g
Agar . 15.0 g
Distilled water 1,000.0 ml

Enriched Nitrate Broth
See Nitrate Broth.

Eosin-Methylene Blue (EMB) Agar (pH 7.2)

Peptone . 10.0 g
Lactose . 5.0 g
Sucrose . 5.0 g
Dipotassium phosphate 2.0 g
Agar . 13.5 g
Eosin Y . 0.4 g
Methylene blue 0.06 g
Distilled water 1,000.0 ml

Eugon Agar (pH 7.0)
Tryptose 15.0 g
Soytone . 5.0 g
Dextrose 5.0 g
L-cystine 0.2 g
Sodium chloride 4.0 g
Sodium sulfite 0.2 g
Agar . 15.0 g
Distilled water 1,000.0 ml

Eugon Broth (pH 7.0)
Tryptose 15.0 g
Soytone . 5.0 g
Dextrose 5.0 g
L-cystine 0.2 g
Sodium chloride 4.0 g
Sodium sulfite 0.2 g
Distilled water 1,000.0 ml

Gel Diffusion Agar
Noble agar 10.0 g
Distilled water 1,000.0 ml
(A 1/1,000 dilution of merthiolate can be added as a preservative. Dispense in appropriate dishes.)

Glucose–Minimal Salts
Prepare solutions A–D.
A. Minimal salts
$(NH_4)_2SO_4$ 20.0 g
K_2HPO_4 140.0 g
KH_2PO_4 60.0 g
Sodium citrate · $2H_2O$ 10.0 g
$MgSO_4 \cdot 7H_2O$ 2.0 g
Distilled water 1,000.0 ml
Autoclave under standard conditions.
B. 50% glucose
Glucose 50.0 g
Distilled water 100.0 ml
Autoclave under standard conditions.
C. Amino acids
Prepare solutions of individual amino acids to give 2 mg/ml and filter sterilize.
D. Agar solution
Agar . 15.0 g
Distilled water 880.0 ml
To prepare final glucose–minimal salts mix aseptically:
Agar solution (50°C)—
solution D 880.0 ml
Minimal salts—solution A 100.0 ml
50% glucose—solution B 10.0 ml
Amino acids—solution C 10.0 ml

KF Streptococcus Agar (pH 7.2)
Proteose Peptone, No. 3 Difco 10.0 g
Yeast extract 10.0 g
Sodium chloride 5.0 g
Sodium glycerophosphate 10.0 g
Maltose . 20.0 g

Lactose 1.0 g
Sodium azide 0.4 g
Bromcresol purple 0.015 g
Agar 20.0 g
Distilled water1,000.0 ml

Kligler Iron Agar (pH 7.4)
Beef extract 3.0 g
Yeast extract 3.0 g
Peptone 15.0 g
Proteose peptone 5.0 g
Lactose 10.0 g
Dextrose 1.0 g
Ferrous sulfate 0.2 g
Sodium chloride 5.0 g
Sodium thiosulfate 0.3 g
Agar 12.0 g
Phenol red 0.024 g
Distilled water1,000.0 ml

Lactose Fermentation Broth (1× and 2×, pH 6.9)
Beef extract 3.0 g
Peptone 5.0 g
Lactose 5.0 g
Distilled water1,000.0 ml
Note: For the 2×, use twice the ingredients.

Lauryl Tryptose Broth (pH 6.8)
Tryptose 20.0 g
Lactose 5.0 g
Potassium phosphate, dibasic 2.75 g
Potassium phosphate, monobasic 2.75 g
Sodium chloride 5.0 g
Sodium lauryl sulfate 0.1 g
Distilled water1,000.0 ml

Levine EMB Agar (pH 7.1)
Peptone 10.0 g
Lactose 10.0 g
Dipotassium phosphate 2.0 g
Agar 15.0 g
Eosin Y 0.4 g
Methylene blue 0.065 g
Distilled water1,000.0 ml

Litmus Milk
Skim milk powder 100.0 g
Litmus 0.75 g
Distilled water1,000.0 ml
Note: Autoclave at 12 lb pressure for 15 minutes.

Löwenstein–Jensen Medium
Asparagine 3.6 g
Monopotassium phosphate 2.4 g
Magnesium sulfate 0.24 g
Magnesium citrate 0.6 g
Potato flour 30.0 g
Malachite green 0.4 g
Distilled water 600.0 ml

Lysine Iron Agar (pH 6.7)
Peptone 5.0 g
Yeast extract 3.0 g
Dextrose 1.0 g
L-lysine hydrochloride 10.0 g
Ferric ammonium citrate 0.5 g
Sodium thiosulfate 0.04 g
Bromcresol purple 0.02 g
Agar 15.0 g

Distilled water1,000.0 ml

Mannitol Salt Agar (pH 7.4)
Beef extract 1.0 g
Peptone 10.0 g
Sodium chloride 75.0 g
D-mannitol 10.0 g
Agar 15.0 g
Phenol red 0.025 g
Distilled water1,000.0 ml

M-Endo Broth (pH 7.5)
Yeast extract 6.0 g
Thiotone peptone 20.0 g
Lactose 25.0 g
Dipotassium phosphate 7.0 g
Sodium sulfite 2.5 g
Basic fuchsin 1.0 g
Distilled water1,000.0 ml
Note: Heat until boiling but do not autoclave.

M-FC Broth (pH 7.4)
Biosate peptone or tryptose 10.0 g
Polypeptone peptone or proteose
 peptone 5.0 g
Yeast extract 3.0 g
Sodium chloride 5.0 g
Lactose 12.5 g
Bile salts 1.5 g
Aniline blue 0.1 g
Distilled water1,000.0 ml
Note: Add 10 ml of rosolic acid (1% in 0.2 N sodium hydroxide). Heat to boiling with gentle agitation. Do not autoclave.

MM 1 Medium (pH 7.4)
Spizizen's salts supplemented with the following:
Vitamin-free casein hydrolysate 0.2 g
L-tryptophan 0.05 g
Dextrose 5.0 g
Distilled water1,000.0 ml

MM 2 Medium (pH 7.4)
Spizizen's salts supplemented with the following:
Vitamin-free casein hydrolysate 0.2 g
L-tryptophan 0.005 g
Dextrose 5.0 g
Magnesium sulfate to a final
 concentration of 5.0 mM
Distilled water1,000.0 ml

MacConkey's Agar (pH 7.1)
Bacto peptone 17.0 g
Proteose peptone 3.0 g
Lactose 10.0 g
Bile salts mixture 1.5 g
Sodium chloride 5.0 g
Agar 13.5 g
Neutral red 0.03 g
Crystal violet 0.001 g
Distilled water1,000.0 ml

Moeller's Decarboxylase Broth with Ornithine (pH 7.2)
Peptone 5.0 g
Beef extract 5.0 g
Dextrose 0.5 g
Bromcresol purple 0.01 g
Cresol red 0.005 g

Pyridoxal . 0.005 g
L-ornithine . 10.0 g
Distilled water1,000.0 ml
In the place of 10 g (1%) of L-ornithine, 10 g (1%) of L-lysine or L-arginine can be used.

Motility Test Media (pH 7.2)
Tryptose . 10.0 g
Sodium chloride 5.0 g
Agar . 5.0 g
Distilled water1,000.0 ml

MR-VP Broth (pH 6.9)
Peptone . 7.0 g
Dextrose . 5.0 g
Potassium phosphate 5.0 g
Distilled water1,000.0 ml

Mueller-Hinton Agar (pH 7.4)
Beef, infusion 300.0 g
Casamino acids 17.5 g
Starch . 1.5 g
Agar . 17.0 g
Distilled water1,000.0 ml

Nitrate Agar Slants (pH 6.8)
Peptone . 5.0 g
Beef extract . 3.0 g
Potassium nitrate 1.0 g
Agar . 12.0 g
Distilled water1,000.0 ml

Nitrate Broth (pH 7.2)
Peptone . 5.0 g
Beef extract . 3.0 g
Potassium nitrate 1.0 g
Distilled water1,000.0 ml

Nitrate-Free Broth (pH 7.2)
Peptone . 5.0 g
Beef extract . 3.0 g
Distilled water1,000.0 ml

Nutrient Agar (pH 7.0)
Peptone . 5.0 g
Beef extract . 3.0 g
Agar . 15.0 g
Distilled water1,000.0 ml
Note: Autoclave at 121 lb pressure for 15 minutes.

Nutrient Broth (pH 7.0)
Peptone . 5.0 g
Beef extract . 3.0 g
Distilled water1,000.0 ml

Nutrient Gelatin (pH 6.8)
Peptone . 5.0 g
Beef extract . 3.0 g
Gelatin . 120.0 g
Distilled water1,000.0 ml

Peptone Broth (pH 7.2)
Peptone . 10.0 g
Sodium chloride 5.0 g
Distilled water1,000.0 ml

Phenol Red Dextrose Broth (pH 7.4)
Trypticase (proteose peptone) 10.0 g
Beef extract . 1.0 g
Dextrose . 5.0 g
Sodium chloride 5.0 g
Phenol red . 0.025 g
Distilled water1,000.0 ml

Note: Autoclave at 15 lb pressure for 15 minutes. Be careful not to autoclave longer.

Phenol Red Lactose Broth (pH 7.4)
Trypticase (proteose peptone) 10.0 g
Beef extract . 1.0 g
Lactose . 10.0 g
Sodium chloride 5.0 g
Phenol red . 0.025 g
Distilled water1,000.0 ml
Note: Autoclave at 15 lb pressure for 15 minutes. Be careful not to autoclave longer.

Phenol Red Sucrose (Saccharose) Broth (pH 7.4)
Trypticase (proteose peptone) 10.0 g
Beef extract . 1.0 g
Sucrose (saccharose) 10.0 g
Sodium chloride 5.0 g
Phenol red . 0.025 g
Distilled water1,000.0 ml
Note: Autoclave at 15 lb pressure for 15 minutes. Be careful not to autoclave longer.

Phenylalanine Deaminase (Phenylalanine) Agar (pH 7.3)
Yeast extract 3.0 g
Dipotassium phosphate 1.0 g
Sodium chloride 5.0 g
DL-phenylalanine 2.0 g
Agar . 12.0 g
Distilled water1,000.0 ml

Plate Count Agar (Standard Methods Agar, Tryptone Glucose Yeast Agar; pH 7.0)
Tryptone . 5.0 g
Yeast extract 2.5 g
Dextrose (glucose) 1.0 g
Agar . 15.0 g
Distilled water1,000.0 ml

Presence-Absence Broth (P-A Broth)
Pancreatic digest of casein 10.0 g
Lactose . 7.5 g
Pancreatic digest of gelatin 5.0 g
Beef extract . 3.0 g
Sodium chloride 2.5 g
Dipotassium phosphate (K_2HPO_4) 1.375 g
Potassium dihydrogen phosphate (KH_2PO_4) 1.375 g
Sodium lauryl sulfate 0.05 g
Bromcresol purple 8.5 mg
Distilled water1,000.0 ml
Add components to distilled/deionized water and bring volume to 333.0 ml. Mix thoroughly.
Distribute into screw-capped 250.0-ml milk dilution bottles in 50.0-ml volumes. Autoclave for 15 min at 15 psi pressure—121°C.

Potato Dextrose Agar (pH 5.6)
Potatoes, infusion from 200.0 g
Dextrose . 20.0 g
Agar . 15.0 g
Distilled water1,000.0 g

Sabouraud (Dextrose) Agar (pH 5.6)
Peptone . 10.0 g
Dextrose . 40.0 g
Agar . 15.0 g
Distilled water1,000.0 ml

Salt Medium–Halobacterium

Sodium chloride	250.0 g
Magnesium sulfate · $7H_2O$	10.0 g
Potassium chloride	5.0 g
Copper chloride	0.2 g
Yeast extract	10.0 g
Tryptone	2.5 g
Agar	20.0 g
Distilled water	1,000.0 ml

SIM Agar (pH 7.3)

Peptone	30.0 g
Beef extract	3.0 g
Ferrous ammonium sulfate	0.2 g
Sodium thiosulfate	0.025 g
Agar	3.0 g
Distilled water	1,000.0 ml

Simmons Citrate Agar (pH 6.9)

Ammonium dihydrogen phosphate	1.0 g
Dipotassium phosphate	1.0 g
Sodium chloride	5.0 g
Sodium citrate	2.0 g
Magnesium sulfate	0.2 g
Agar	15.0 g
Bromothymol blue	0.08 g
Distilled water	1,000.0 ml

Spirit Blue Agar with 3% Lipase (pH 6.8)

Tryptone	10.0 g
Yeast extract	5.0 g
Agar	20.0 g
Spirit blue	0.15 g
Distilled water	1,000.0 ml

After the above has been autoclaved and allowed to cool to 50° to 55°C, add 30 ml of Bacto lipase reagent (Difco) slowly while agitating the medium in the flask to obtain an even distribution.

Spizizen's Salts

Sodium sulfate	2.0 g
Dipotassium phosphate	14.0 g
Monopotassium phosphate	6.0 g
Sodium citrate	1.0 g
Magnesium sulfate	0.2 g
Distilled water	1,000.0 ml

SS Agar (pH 7.0)

Beef extract	5.0 g
Peptone	5.0 g
Lactose	10.0 g
Bile salts no. 3	8.5 g
Sodium citrate	8.5 g
Sodium thiosulfate	8.5 g
Ferric citrate	1.0 g
Agar	13.5 g
Brilliant green	0.33 mg
Neutral red	0.025 g
Distilled water	1,000.0 ml

Standard Methods Agar

See Plate Count Agar.

Starch Agar (pH 7.5)

Beef extract	3.0 g
Soluble starch	10.0 g
Agar	12.0 g
Distilled water	1,000.0 ml

T-Broth (pH 7.3)

Nutrient broth	8.0 g
Peptone	5.0 g
Sodium chloride	5.0 g
Glucose	1.0 g
Distilled water	1,000.0 ml

To make 2× concentrated T-broth, double all of the above ingredients except the distilled water. Adjust the pH to 7.3 with 0.1 M NaOH.

Thayer-Martin Modified Medium (pH 7.0)

Bacto GC medium base	36.0 g
Hemoglobin	10.0 g
Bacto supplement B or VX	10.0 ml
Bacto antimicrobic vial CNVT	10.0 ml
Distilled water	1,000.0 ml

Thioglycollate Broth (pH 7.1)

Peptone	15.0 g
Yeast extract	5.0 g
Dextrose	5.0 g
L-cystine	0.75 g
Thioglycollic acid	0.5 g
Agar	0.75 g
Sodium chloride	2.5 g
Resazurin	0.001 g
Distilled water	1,000.0 ml

Todd-Hewitt Broth (pH 7.8)

Beef heart, infusion	500.0 g
Neopeptone	20.0 g
Dextrose	2.0 g
Sodium chloride	2.0 g
Disodium phosphate	0.4 g
Sodium carbonate	2.5 g
Distilled water	1,000.0 ml

Top Agar (pH 7.0)

Nutrient broth plus 0.75% agar.
Prepare 4.5 ml pours.

Triple Sugar Iron Agar (pH 7.4)

Beef extract	3.0 g
Yeast extract	3.0 g
Peptone	15.0 g
Peptose-peptone	5.0 g
Lactose	10.0 g
Saccharose	10.0 g
Dextrose	1.0 g
Ferrous sulfate	0.2 g
Sodium chloride	5.0 g
Sodium thiosulfate	0.3 g
Phenol red	0.024 g
Agar	12.0 g
Distilled water	1,000.0 ml

Tryptic Nitrate Broth (pH 7.2)

Tryptose	20.0 g
Dextrose	1.0 g
Disodium phosphate	2.0 g
Potassium nitrate	1.0 g
Distilled water	1,000.0 ml

Trypticase (Tryptic) Soy Agar (pH 7.3)

Trypticase (tryptone)	15.0 g
Phytone (soytone)	5.0 g
Sodium chloride	5.0 g
Agar	15.0 g
Distilled water	1,000.0 ml

Trypticase Soy Agar with Lecithin and Polysorbate 80 (pH 7.3)

Tryptone	15.0 g
Soy peptone	5.0 g
Sodium chloride	5.0 g
Lecithin	0.7 g
Sorbitan monooleate complex	5.0 g
Agar	15.0 g
Distilled water	1,000.0 ml

Trypticase (Tryptic) Soy Broth (pH 7.3)

Tryptone	17.0 g
Soytone	3.0 g
Dextrose	2.5 g
Sodium chloride	5.0 g
Dipotassium phosphate	2.5 g
Distilled water	1,000.0 ml

Tryptone Agar

Tryptone	10.0 g
Calcium chloride (reagent)	0.01 M
Sodium chloride	5.0 g
Agar	11.0 g
Distilled water	1,000.0 ml

Tryptone Broth

Tryptone	10.0 g
Calcium chloride (reagent)	0.01 M
Sodium chloride	5.0 g
Distilled water	1,000.0 ml

Tryptone Glucose Yeast Agar

See Plate Count Agar.

Tween 80 (Polysorbate 80)

Tween 80 (polysorbate 80) is a surface-active agent that lowers the interfacial tension around bacteria suspended in the medium. This permits more rapid entry of desired compounds into the bacterial cell. As a result, there may be more rapid growth or other activity between the bacteria and the reactive compounds in the medium.

Urea Broth (pH 6.9)

Yeast extract	0.1 g
Monopotassium phosphate	0.091 g
Disodium phosphate	0.095 g
Urea	20.0 g

Phenol red	0.01 g
Distilled water (sterile)	1,000.0 ml

Violet Red Bile Agar (pH 7.4)

Yeast extract	3.0 g
Peptone	7.0 g
Bile salts no. 3	1.5 g
Lactose	10.0 g
Sodium chloride	5.0 g
Agar	15.0 g
Neutral red	0.03 g
Crystal violet	0.002 g
Distilled water	1,000.0 ml

Vogel-Johnson Agar (pH 7.2)

Tryptone	10.0 g
Yeast extract	5.0 g
Mannitol	10.0 g
Dipotassium phosphate	5.0 g
Lithium chloride	5.0 g
Glycine	10.0 g
Agar	15.0 g
Phenol red	0.025 g
Distilled water	1,000.0 ml

Yeast Extract

Yeast extract is the water-soluble portion of autolyzed yeast. The autolysis is carefully controlled to preserve the naturally occurring B-complex vitamins. In concentrations of 0.3 to 0.5%, as it is generally employed, it forms sparklingly clear solutions with a pH of approximately 6.6.

YM Agar

Yeast extract	3.0 g
Malt extract	3.0 g
Peptone	5.0 g
Dextrose	10.0 g
Agar	20.0 g
Distilled water	1,000.0 ml

YM Broth

Yeast extract	3.0 g
Malt extract	3.0 g
Peptone	5.0 g
Dextrose	10.0 g
Distilled water	1,000.0 ml

Sources and Maintenance of Microbiological Stock Cultures

Sources of Microbiological Cultures

In addition to the major biological supply houses such as Carolina, Fisher, and Wards, the following organizations also supply microbiological cultures:

American Type Culture Collection
12301 Parklawn Drive
Rockville, MD 20852-1776
USA/Canada 1-800-638-6597
Outside USA/Canada 1-301-881-2600
FAX 1-301-231-5826
www.atcc.org

REMEL
PO Box 14428
12076 Santa Fe Drive
Lenexa, KA 66215-3594
1-800-255-6730
FAX 1-800-447-3643
www.remel.com

Difco Laboratories
Division of Becton Dickinson Company
1 Becton Drive
Franklin Lakes, NJ 07417
1-202-847-6800
FAX 1-410-584-7121
www.bd.com/microbiology

ICN Biomedicals Inc.
1263 South Chillicothe
Aura, OH 44202
1-800-854-0530
FAX 1-800-334-6999
www.icnbiomed.com

Thomas Scientific
PO Box 99
Swedesboro, NJ 08085
1-800-345-2100
FAX 1-800-345-5232
www.thomassci.com

VWR Scientific Products
Educational Division
1310 Goshen Parkway

West Chester, PA 19380-5985
1-800-932-5000
FAX 1-610-436-1761
www.vwrsp.com

Becton Dickinson
Microbiology Systems
PO Box 243
Cockeysville, MD 21030-0248
1-201-818-8900
FAX 410-584-7121
www.bd.com

Maintenance of Microbiological Stock Cultures

In microbiology, a stock culture is a standard strain that conforms to typical morphological, biochemical, physiological, and serological characteristics of the species it represents. Over time, the culture will possess sufficient stability to display or retain these characteristics.

Stock cultures can be maintained at little expense (compared to buying new ones each time a strain is needed) and require small amounts of time with respect to upkeep. The two following general procedures are commonly used:

I. **Freeze-drying** (lyophilization) or **quick-freeze** methods requiring specific equipment
 A. **Freeze-drying** (lyophilization): In this procedure, microorganisms in a liquid medium are quick-frozen in dry ice with a solvent and dried under high vacuum from the frozen state. Many microorganisms can be preserved by freeze-drying almost indefinitely.
 B. **Quick-freeze method:** This method is recommended for both anaerobic and aerobic microorganisms. Most will survive for 6 months or longer.
 1. Grow the microbiological culture on either an agar plate or slant for 24 hours.
 2. Label small, sterile, screw-cap vials with the name of the microorganism and date.

453

3. Add 1.0 ml of sterile defibrinated sheep's blood to the vials.
4. Using aseptic technique, remove a loopful of culture and suspend in the blood within the vial.
5. Prepare a dry-ice bath by using crushed dry ice and acetone.
6. Place the vials in the dry-ice bath for about 15 seconds. This rapidly freezes the culture in the vial.
7. Store at −40°C or colder.
8. To reconstitute the culture, place the frozen vial in a 37°C water bath for only 1 to 2 minutes. Using aseptic technique, remove a loopful and subculture onto an appropriate growth medium.

II. **Special media maintenance** methods involve holding at room temperature, incubator temperature, or refrigeration

 A. **CTA (cysteine-trypticase agar)** without carbohydrates is available commercially. This medium will support growth for a long period of time at room temperature.
 1. Aseptically inoculate a loopful of culture into a tube of tryptic soy broth. Incubate at 35°C for 18 to 24 hours.
 2. Using a sterile 1-ml pipette, inoculate a few drops of the tryptic soy broth into the CTA tube.
 3. Usually, this stock culture can be maintained up to 6 months at room temperature.
 4. Every 2 to 3 months check for viability by subculturing onto appropriate growth media.
 5. If the subculture growth is scanty but displays typical characteristics of the microorganism, make a transfer from the new culture to a new CTA tube.
 6. If the subculture is not in good condition, the stock culture should be discarded and a new one made.

 B. **More fastidious microorganisms:** Pneumococcus, α- and β-hemolytic streptococci, enterococcus, and others can be maintained on blood agar or tryptose agar slants in screw-cap tubes streaked for heavy growth. Keep the tubes in a refrigerator and transfer the cultures every 2 weeks to a fresh slant.

 C. **Cooked meat medium:** Commercially available cooked meat medium is an excellent medium for both aerobes and anaerobes. Many gram-negative bacteria (e.g., *Salmonella, Shigella, Proteus*) can remain viable for years. Coryne bacteria and staphylococci can remain viable up to 6 months and must then be subcultured.

The microbiological stock culture is inoculated into the meat layer of a tube of cooked meat medium. To maintain anaerobes, first boil the medium to drive off the dissolved oxygen. After inoculation and incubation for 24 to 48 hours, add a layer of 1 ml of sterile paraffin oil. Check the culture for viability every 2 to 3 months.

 D. **TSA slants:** TSA slants with screw caps can be inoculated with a specific microorganism and incubated for 24 to 48 hours. It is then covered with a small amount of sterile sheep or horse serum and frozen at −50°C. Most microorganisms can be maintained this way for 6 or more months.

Overall considerations. Regardless of the procedure used to maintain specific microbiological cultures, the following special considerations should always be followed:
 1. Do not use media that contain fermentable carbohydrates if you plan to keep the cultures for long periods.
 2. Never use selective media.
 3. Never allow a culture to dry out. Always use tightly closed screw-cap tubes for storage.
 4. Do not refrigerate temperature-sensitive microorganisms (e.g., *N. gonorrhoeae, N. meningitidis*) although they survive well by quick-freezing.

Rehydration of freeze-dried cultures. Recipients of ATCC (or from any of the previously listed sources) microbiological cultures should use those media and incubation temperatures specified in the catalog entry for each strain when first subculturing to ensure optimal conditions for recovery.

It is suggested that each culture be thoroughly checked upon receipt. If a culture is found to be unsatisfactory in any respect, the supplier should be notified so the strain in question can be investigated.

The technique (p. 456) in the accompanying illustrations should be followed to recover bacteria, fungi, or algae from freeze-dried cultures.

III. **Use of PROTECT Bacterial Preservers** to store bacteria for future use.

The PROTECT organism preservation system—marketed by Key Scientific Products, 149C Texas Avenue, Round Rock, Texas 78664—is a method for preserving bacterial cultures without freeze drying or frequent transfers. The system consists of a sterile vial containing porous beads in a cryopreservative fluid. Each bead acts as a carrier for bacteria during storage. If the vials are kept at −70°C, cultures may remain viable for several years. In a −20°C freezer that does not self-defrost, hardy bacteria such as *E. coli* may survive for two years or more. *Neisseria* and other fastidious bacteria will last about a year at −20°C. Survival times are shorter if a self-defrosting refrigerator must be used. When working with fastidious bacteria, several vials should be stored frozen

and a new one used each time one needs to start fresh cultures from the stocks.

PROTECT bacterial preservers are used as described below.

A. **Preparation of the cultures:**

1. Aseptically transfer bacteria with a loop from an 18- to 24-hour slant or plate culture to a labelled PROTECT vial. Enough paste should be added to make the cryopreservative fluid moderately turbid. One loopful will probably be sufficient. Completely emulsify the paste in the fluid so that there are no visible clumps of bacteria.

2. Cap the tube and invert it six times to mix. DO NOT VORTEX THE VIAL.

3. After mixing by inversion, allow the vial to stand upright for 30 to 60 seconds. During this interval, the bacteria bind to the beads.

4. Carefully and aseptically remove all excess cryopreservative fluid with a sterile Pasteur pipette. The beads should be as free of fluid as possible. Close the vial finger tight.

5. It is essential to freeze the vials immediately after preparation. Freeze the vial as rapidly as possible in a slanted position. If a $-70°C$ freezer is available, it should be used. The vials will cool more rapidly in a regular $-20°C$ freezer if they are placed in a polypropylene or Styrofoam vial container that has been frozen overnight. Where at all possible, use a freezer that is NOT self-defrosting.

B. **Recovery and use of culture beads:**

1. When a culture is needed, take a vial from the freezer and carefully open it.

2. Remove a single bead using a sterile needle, sterile forceps, or the special PROTECT hook. If the remainder of the vial's contents is to be saved for future use, it should immediately be returned to the freezer before it thaws. (A special aluminum cryoblock may be used to keep vials cold when they are outside the freezer.)

3. After removing the bead from the vial, drop it into the appropriate broth, such as tryptic soy broth, or rub it over solid culture medium. Incubate at the appropriate temperature.

Double-vial preparations

- Tip
- Insulator
- Cotton plug
- Outer vial (soft glass)
- Inner vial
- Freeze-dried pellet
- Cotton
- Desiccant with indicator

Opening the vial

1. Heat the tip of the outer vial in a flame.

2. Squirt a few drops of water on the hot tip to crack glass.

3. Strike with file or pencil to remove tip.

4. Remove insulation and inner vial.

5. With forceps, gently raise cotton plug.

Use a sterile Pasteur pipette to rehydrate the freeze-dried culture

Inner vial

Single-vial preparations

- Borosilicate glass
- Freeze dried cells

Opening the vial

1. These preparations may be enclosed in a thin skin of cellulose. This skin must be removed (either with a sharp blade or by soaking in water for a few minutes). Score the ampule once briskly with a sharp file about one inch from the tip.

2. Disinfect the ampule with alcohol-dampened gauze.

3. Wrap gauze around the ampule, and break at the scored area. Care should be taken not to have the gauze too wet, or alcohol could be sucked into the culture when the vacuum is broken. Rehydrate material at once.

Single vial

Redrawn with permission from *American Type Culture Collection Catalogue of Bacteria and Bacteriophages,* Eighteenth Edition, 1992. American Type Culture Collection, Rockville, MD 20852.

Nitrate reductase, 205–6
Nitrate reduction test, 205–7, 214–15, 236
 micrococci, 341
 neisseriae, 362–63
 reagents, 445
 staphylococci, 340–41
 streptococci, 341
Nitrite reduction test, 212–16
 neisseriae, 362–63
Nitrite test reagent A, 205–7
Nitrite test reagent B, 205–7
Noble agar, 447
Nocardia asteroides, acid-fast staining, 54
Nocardia brasiliensis, acid-fast staining, 54
Nocardiosis, 54
Non-acid-fast cells, 54
Nonmotile bacteria, 15–16
Nosocomial infection, 206, 267
Novobiocin, 264
Novobiocin resistance, staphylococci, 341,
 344
Nuclease, streptococcal, 354
Nucleus, 373
Nutrient agar, 79, 251–52, 303–6
 recipe, 450–51
Nutrient broth, 399–402
 recipe, 450–51
Nutrient gelatin deep, 167–68, 233–36, 339
 recipe, 450–51

O

Objective lens, 3, 19
 high-dry (40$_\times$), 3
 low-power (10$_\times$), 3–4
 oil immersion (100$_\times$), 3
Obligate aerobe, 112
Obligate anaerobe, 112
Ocular lens. *See* Eyepiece
Ocular micrometer, 5
 calibration, 5–6
Oil immersion (100×) objective, 3
Oleandomycin, 264
ONPG test, 128–32, 214–15
 KONFIRM test for fecal coliforms,
 297–98
 reagents, 445–46
Optimum growth temperature, 243
Optochin test, pneumococci, 348–49
Organelles, 373
Organic mercury compounds, as antiseptics,
 256
Ornithine decarboxylase, 193–95
Ornithine decarboxylase test, 193–96,
 214–15
Osmotic pressure, 252
 effect on growth of microorganisms,
 251–52
Osteomyelitis, *S. aureus,* 339
Otitis media, *S. pneumoniae,* 347
Oxidase test, 181–83, 212–16, 236
 micrococci, 341
 neisseriae, 362–63
 reagents, 445
 staphylococci, 341
 streptococci, 341
Oxidase test disk, 181–83, 233–36
OxyDish, 111, 114, 116
Oxygen
 damaging effect on anaerobes, 112
 relationship to bacterial growth, 115
 as terminal electron acceptor, 181–83
OxyPlate, 111, 114–15
Oxyrase For Agar, 111, 114, 116

Oxyrase For Broth, 111, 114, 116
Oxyrase Plate, 367–69

P

Paramecium caudatum, phase-contrast
 microscopy, 24
Paratyphoid fever, 295
Pasteur, Louis, 287
Pasteurella, identification, 440
Pasteurization, 287
 milk, 288, 319
P-A coliform test. *See* Presence-absence
 coliform test
Pathogen, potential, 325
Penicillin, discovery, 127
Penicillin G, 264
Penicillium, 379
 soil, 309
Penicillium notatum, morphology and life
 cycle, 379–82
Peptone, 248
Peptone broth, 450–51
Peptostreptococcus, 112
Periplasmic flagella, 19–20
Peritonitis, *S. pneumoniae,* 347
Peritrichous flagella, 15, 71–73
Peroxidase, 172
Petri plate, 79
pH, 247, 437
 effect on growth of microorganisms,
 247–48
 growth optimum, 247
 pH indicators, 437
 pH meter, 247–48
 pH paper, 247–48
Phage
 E. coli, 288
 isolation from sewage, 303–6
 plaques, 304–6
Phage titer, 303–6
Phase-contrast microscope, 23–25
 bright-phase-contrast microscope, 24
 dark-phase-contrast microscope, 24
Phenol, as disinfectant, 256–57
Phenol coefficient, 255–57
Phenolphthalein, 437
Phenol red, 130, 136, 188, 194, 437
Phenol red dextrose broth, 233–36
 recipe, 450–51
Phenol red lactose broth, 233–36
 recipe, 450–51
Phenol red sucrose broth, 233–36
 recipe, 450–51
Phenylalanine deaminase, 199–200
Phenylalanine deaminase agar, 199–201
 recipe, 450–51
Phenylalanine deaminase test, 199–201,
 220–23
Phosphate buffer, 393–95
 recipe, 446
Phosphate-buffered saline, 446
Phospholipids, 145
Phycomycetes, 379–82
Physiological saline, 446
Pipette, 85–86
 blow-out (serological), 86
 sterilization, 86
 to-deliver (Mohr measuring), 86
Pipette bulb, 86
Pipette can, 86
Pipetting, 88
Pipetting device, electronically powered, 86
Plague, 156, 212

Plaque, virus, 304
Plasmid, 387, 399
 antibiotic-resistant, 399–402
 conjugative, 399–402
Plasmolysis, 252
Plate count, 119–22
 on food product, 315–16
Plate count agar, 119–22, 163–64, 315–16
 recipe, 450–51
Pleomorphic bacteria, 41
Plesiomonas shigelloides, identification,
 440
Pneumococci, 347–49
 bile solubility test, 348–49
 capsule, 65–66, 348
 culture on blood agar, 347–49
 hemolysis pattern, 348–49
 identification, 347–49
 morphology, 348
 mouse susceptibility test, 347–49
 normal microbiota of human body, 336
 optochin test, 348–49
 quellung reaction, 348–49
Pneumonia
 K. pneumoniae, 156, 205–6
 P. vulgaris, 72
 S. aureus, 339–40
 S. pneumoniae, 65–66, 347
Polar flagella, 15, 71–73
Poliomyelitis virus, 295
Polymyxin B, 264
Polyporus, 379, 381
Pond water, 23–25
Potable water, 287
Potassium permanganate, as antiseptic, 256
Potato dextrose agar, 379–80
 recipe, 450–51
Potential pathogen, 325
Pour-plate technique, 107–8
Presence-absence (P-A) broth, 289–92
 recipe, 450–51
Presence-absence (P-A) coliform test,
 289–92
Primary stain, 46
Prodigiosin, 85
Profile index number, 213
Prontosil red, 325
Propidium iodine, 30
Propionibacterium, fermentation pathways,
 129
Propionic acid fermentation, 129
Proskauer, Bernhard, 155
PROTECT Bacterial Preservers, 454–55
Proteins
 casein hydrolysis test, 163–64
 catalase test, 171–72
 coagulase test, 175–77
 DNase test, 175–77
 gelatin hydrolysis test, 167–68
 hydrogen sulfide production and
 motility test, 149–51
 IMViC tests, 155–60
 lysine and ornithine decarboxylase
 tests, 193–96
 nitrate reduction test, 205–7
 oxidase test, 181–83
 phenylalanine deaminase test, 199–201
 urease test, 187–89
Proteolytic enzymes, 163–64
Proteus
 fermentation pathways, 129
 identification, 235, 441, 443
 maintenance of culture, 87–88
 ornithine decarboxylase test, 194
 phenylalanine deaminase test, 199